LOW-LEVEL ENVIRONMENTAL RADIOACTIVITY

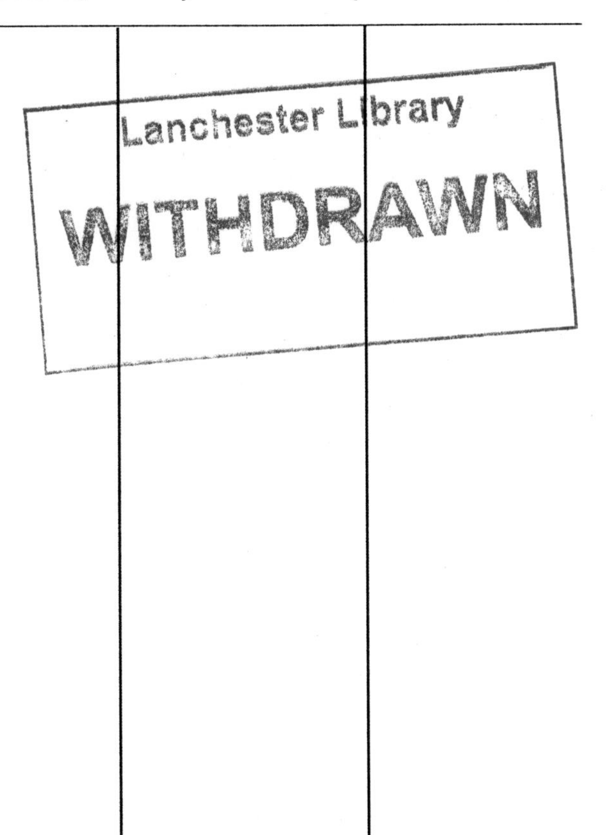

HOW TO ORDER THIS BOOK

BY PHONE: 800-233-9936 or 717-291-5609, 8AM–5PM Eastern Time

BY FAX: 717-295-4538

BY MAIL: Order Department
Technomic Publishing Company, Inc.
851 New Holland Avenue, Box 3535
Lancaster, PA 17604, U.S.A.

BY CREDIT CARD: American Express, VISA, MasterCard

LOW-LEVEL ENVIRONMENTAL RADIOACTIVITY

Sources and Evaluation

Richard Tykva, D.Sc.
Head, Department of Radioisotopes
Institute of Organic Chemistry and Biochemistry
Academy of Sciences of the Czech Republic
Prague, Czech Republic

Josef Sabol, D.Sc.
Associate Professor, Department of Dosimetry
and Applications of Ionizing Radiation
Faculty of Nuclear Sciences and Physical Engineering
Czech Technical University
Prague, Czech Republic

LANCASTER · BASEL

Low-Level Environmental Radioactivity
a **TECHNOMIC**® publication

Published in the Western Hemisphere by
Technomic Publishing Company, Inc.
851 New Holland Avenue, Box 3535
Lancaster, Pennsylvania 17604 U.S.A.

Distributed in the Rest of the World by
Technomic Publishing AG
Missionsstrasse 44
CH-4055 Basel, Switzerland

Printed in the United States of America
10 9 8 7 6 5 4 3 2 1

Main entry under title:
Low-Level Environmental Radioactivity: Sources and Evaluation

A Technomic Publishing Company book
Bibliography: p.
Includes index p. 323

Library of Congress Catalog Card No. 95-60049
ISBN No. 1-56676-189-1

TABLE OF CONTENTS

INTRODUCTION

The term *radioactivity* was used for the first time by Marie Curie a few years after the discovery of this phenomenon by Henri Becquerel in 1896. Actually, the names of both scientists are closely identified with radioactivity—the conventional unit of activity was until recently Ci (curie), now replaced by Bq (becquerel) consistent with the SI system of units.

Radioactive nuclides—*radionuclides*—are present virtually everywhere; any material contains certain amounts of trace radioactive elements. Some natural substances such as soil, rocks, or water sometimes contain rather high concentrations of natural radionuclides, which are of terrestrial or extraterrestrial origin. However, the concentrations of these elements in different samples are quite variable. In addition to the natural radioactivity in our environment, many man-made radionuclides are also present in small amounts. Beyond their presence in the environment, many radionuclides are often used as a versatile methodical tool in many fields such as medicine or life sciences.

Low-level radioactivity is related to those radioactive sources of ionizing radiation that are characterized by low activities. Sometimes activity here does not represent total amount of radionuclides but rather their concentration. In other cases, the total activity may be quite high, but we can measure only a relatively small portion of the material. "Low" may have, for different situations and circumstances, not only considerably different meanings but also different absolute values as far as the activity or activity concentration is concerned. For example, one can refer to low activity in the case of radiocarbon dating, where the concentration of ^{14}C is actually lower than its natural concentration, and also in the case of radon monitoring where, especially in mines or in some enclosed spaces, its concentration may be several thousand times higher than the outdoor "natural" concentrations.

Emphasis is now being placed on the analysis of naturally occurring radionuclides in the environment or on the release of radionuclides from their different man-made sources because liquid and aerial discharge level controls have become more rigorous. In addition, the applicability of low-level methodology increases the extent of different radionuclide applications considerably.

Since individual radionuclides differ in their decay scheme and particles emitted as well in their energies, there is no universal method for the accurate measurement of all radioactive sources. Moreover, there is usually a mixture of radionuclides in a sample and this always causes some difficulties in a selective evaluation of a given radionuclide. Due to the random nature of radioactive disintegrations, the appropriate interpretation of the experimental results would be, in most cases, impossible without elaborate statistical treatment and evaluation of the data obtained. Thanks to the availability of computer-based instrumentation, the measuring data can be in most cases processed and evaluated on-line, which makes it possible to control and optimize the experiment in order to extract the maximum amount of information carried by the detector response.

The purpose of this book is to provide an introduction to low-level radioactivity assessment and to clarify the nature of its sources, as well as the principal methods used in its measurement. Our evaluation is concentrated on the present-day aspects of low-level methodology. The book may be useful for all who need highly sensitive analysis of natural or artificial radioactivity both within and outside the nuclear field.

The attempt of this book is to summarize the sources of environmental radioactivity and their possible radiological impact in terms of resulting doses to the population, and to present a sound review of the measuring methods and techniques for the evaluation of low-level radioactivities encountered in both the environment and in a number of applications where radioactive sources are used as a means of obtaining important information.

CHAPTER 1

Radionuclides and Radiation Emitted

1.1 RADIONUCLIDES AND RADIOACTIVITY

Initially, the different elements were recognized by some of their specific properties and characteristics, e.g., state, color, density, etc. Later the elements became characterized by their physical and chemical parameters and properties. The elements are subdivided into *nuclides,* where a nuclide can be defined as any species of atom having in its nucleus certain numbers of protons and neutrons. The sum of the numbers of protons and neutrons represents the mass number of the nuclide.

As a matter of fact, the nuclides can be either stable or unstable. The unstable nuclides—*radionuclides*—decay to stable or unstable products with lower atomic mass, the difference in mass being emitted in the form of energetic radiation according to Einstein's well-known mass-energy equation. Radioactive material or substances may consist of one or more radionuclides.

Radioactivity is the spontaneous emission of subatomic particles and high-frequency electromagnetic radiation by radioactive elements. It was one of the first and most important phenomena that led to our present understanding of nuclear structure. The process was first announced in February 1896 by the French physicist Antoine Henri Becquerel (1852–1908) shortly after the discovery of X-rays by the German professor Konrad Wilhelm Röntgen (1845–1923) in 1895. In fact, radioactivity was accidentally discovered by the exposure-producing effect on a photographic plate by a uranium-containing mineral—pitchblende—when wrapped in a black paper and kept in the dark.

Soon after these fundamental discoveries, Pierre Curie (1859–1906) together with his Polish born wife Maria Curie-Sklodowska (1867–1934) extracted from Jáchymov's (now in the Czech Republic) pitchblende two new elemental radioactive sources—radium and polonium. The name of the latter radioactive element was chosen in honor of Poland, the country of Mme. Curie's birth. P. and M. Curie detected the presence of radioactivity by a charged gold-leaf electroscope, then one of few available tools to study radiation.

Ernest Rutherford (1871–1937) found in 1898 that there are at least two components in radiation emitted by these elements. In 1899 Rutherford distinguished alpha and beta particles. The following year Paul Villard (1860–1934) discovered and described gamma rays emitted by radium.

In 1919 E. Rutherford carried out the first artificial transmutation of one element to another. In his experiments he was able to produce stable oxygen ^{17}O and ^{1}H by bombarding stable nitrogen ^{14}N with alpha particles.

At the beginning of the 1930s other important discoveries contributed considerably to our knowledge of the microworld. The neutron was discovered in 1932 by James Chadwick (1891–1974), the positron in the same year by Carl David Anderson (1905–1991), and deuterium (^{2}H) in 1933 by Harold C. Urey (1893–1981). In addition, the first charged-particle accelerators were built at that time in several laboratories.

A further significant step in the study of radioactivity was the discovery of *artificial radioactivity* by Frédéric Joliot (1900–1958) and Irène Curie (1897–1956) in 1934. F. Joliot was an assistant and I. Curie a daughter of Marie Curie; both were her close collaborators. They bombarded boron and aluminum with alpha particles from polonium and obtained radioactive nitrogen (^{13}N) and sulphur (^{30}P), respectively. These two reactions may be represented in condensed notation simply as $^{10}B(\alpha,n)^{13}N$ and $^{27}Al(\alpha,n)^{30}P$. Both ^{13}N, half-life 9.96 min, and ^{30}P, half-life 2.50 min, decay by positron emission.

Very soon after the discovery of artificial radioactivity, scientists began to bombard practically every element of the periodic system with accelerated protons, deuterons and alpha particles—using early Cockcroft-Walton accelerators, Van de Graaff accelerators and cyclotrons. In this way there were able to produce and identify hundreds of new radionuclides. The use of the electron linear accelerator provides the additional possibility of producing new radionuclides.

By now we know something like 2600 nuclides: 260 stable nuclides, 25 very long-lived naturally occurring radionuclides, and more than 2300 man-made radionuclides. A high percentage of all man-made radionuclides, however, have very short half-lives (in the range of min, s, ms, μs and even shorter). The recent advances in particle accelerators, nuclear instrumentation, and experimental techniques have led to an increased ability to synthesize new nuclides.

It may be interesting to note that at the time when artificial radioactivity was discovered, only four chemical elements were missing in the periodic system (Draganic et al., 1990) based on the idea of a Russian scientist Dmitri Ivanovich Mendeleev (1834–1907). Besides francium, which was the last element to be found in nature, the remaining three elements (technetium, astatine, and promethium) are of artificial (synthetic) origin. After the completion of Mendeleev's periodic system, scientists successfully produced and identified some new elements located in the "terra incognita" beyond uranium. The first of them, neptunium, was discovered in 1940, followed by plutonium in 1941. Today we know of fourteen more transuranium elements: americium (atomic or Z number 95), curium (96), berkelium (97), californium (98), einsteinium (99), fermium (100), mendelevium (101), nobelium (102), lawrencium (103), and further elements with Z from 104 up to 109. The proposed names for the element 104 are rutherfordium (backed by Americans) and kurchatovium (suggested by the Russians), while for element 105 two other names are being considered: hahnium and nielsbohrium. The rest of these new elements have no names so far; moreover, the claims of different research centers and laboratories regarding their discoveries have not been officially accepted.

1.1.1 Properties and Quantities Characterizing Radionuclides

The main quantity used for quantification of radioactive sources is *activity,* which is defined as a quotient of the number of radioactive transformations in a radionuclide $\mathrm{d}N$, and the time interval $\mathrm{d}t$ in which these transformations (sometimes called disintegrations) occurred, i.e.,

$$A = \frac{\mathrm{d}N}{\mathrm{d}t} \tag{1.1}$$

In fact, a nucleus can undergo one of the following basic nuclear transformations: α (alpha) decay, β^- (beta minus) decay, β^+ (beta plus) decay, orbital electron capture, internal conversion, isomeric transmission, spontaneous fission, neutron emission and other. All these processes are characterized by statistical fluctuations governed usually by a simple *Poisson distribution.* Although the actual number of repeated observations of nuclear transformations differ, the average number of such observations shows less fluctuation. In fact, as the number of these observations increases, e.g., the number of readings or measurements, the average value approaches the expected value of the quantity. In our case this quantity is *activity,* which actually corresponds to the *mean value* of the number of spontaneous nuclear transformations dN divided by the time interval dt, during which they occurred.

The activity of an amount of a radionuclide is numerically equal to the decay rate of this radionuclide and is given by the following fundamental law of radioactivity:

$$\frac{dN}{dt} = -\lambda N \tag{1.2}$$

where N is the number of radioactive (unstable) nuclei and λ is the *decay constant.* After integration a very useful equation can be obtained

$$N(t) = N(0)e^{-\lambda t} \tag{1.3}$$

where $N(0)$ is the number of radioactive nuclei present at some reference time ($t_0 = 0$) and N is their number present at some time t later, and e is the base of the natural logarithms. This exponential decay relationship is of fundamental concern in working with radioactive materials. Taking into account Equations (1.1) and (1.2), the last equation can be rewritten in the form

$$A(t) = A(0)e^{-\lambda t} \tag{1.4}$$

where $A(0)$ and $A(t)$ now represent activity in time $t = 0$ and t, respectively.

In practice, radionuclides are characterized by their half-life rather than decay constant. The *half-life* is defined as the time period required for the activity to decrease to half of its initial value. It is actually the time interval over which the chance of survival of a particular radioactive atom is exactly one-half. From Equation (1.4) we obtain

$$\frac{1}{2} = e^{-\lambda T_{1/2}} \tag{1.5}$$

where $T_{1/2}$ is the half-life, for which we get

$$T_{1/2} = \frac{\ln 2}{\lambda} = \frac{0.693}{\lambda} \tag{1.6}$$

Sometimes, also a *mean life* may be used; it can be introduced on the basis of the probability $p(t)\mathrm{d}t$ that a radioactive nucleus survives up to the time t and decays in the interval between t and $t + \mathrm{d}t$, i.e.,

$$p(t)\mathrm{d}t = e^{-\lambda t}\mathrm{d}t \tag{1.7}$$

from which we can calculate the mean life T_{ml}

$$T_{ml} = \frac{\int_0^\infty tp(t)\mathrm{d}t}{\int_0^\infty p(t)\mathrm{d}t} = \frac{1}{\lambda} \tag{1.8}$$

The mean life T_{ml} and the half-life (also called *half-period*) are bound together as

$$T_{ml} = \frac{T_{1/2}}{0.693} \tag{1.9}$$

The relations between the mean-life half-life T_{ml}, the half-life $T_{1/2}$ and the decay constant as well as the initial number of radioactive atoms N_0 is graphically illustrated in Figure 1.1.

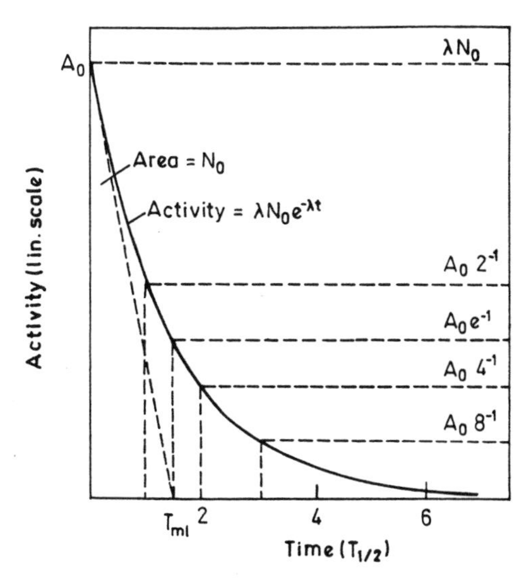

FIGURE 1.1. Graphical illustration of relationships in radioactive decay.

It is quite common that in the decay of a radioactive atom the resulting atom (daughter or decay product) is not stable and can again undergo a nuclear transformation. The process may continue in a series until it comes to an end stable product. In general, a radioactive nuclide A decays into a nuclide B, which is also radioactive. Then the nuclide B decays into a radioactive nuclide C, and so on. For example, ^{232}Th decays into a series of ten successive radioactive nuclides.

Let us consider the initial part of a radioactive series consisting of three nuclides A, B, C, and D (the first three are radioactive, the last one is stable) with decay constants λ_A, λ_B, and λ_C, i.e.,

$$A \xrightarrow{\lambda_A} B \xrightarrow{\lambda_B} C \xrightarrow{\lambda_C} D \tag{1.10}$$

If there are at the beginning, at time $t = 0$, N_{A0} atoms of type A, the numbers N_A, N_B, and N_C of atoms of types A, B, and C which will be present at a later time t, are given by the equations

$$N_A = N_{A0}e^{-\lambda_A t} \tag{1.11}$$

$$N_B = N_{A0}\frac{\lambda_A}{\lambda_B - \lambda_A}(e^{-\lambda_A t} - e^{-\lambda_B t}) \tag{1.12}$$

$$N_C = N_{A0}\left(\frac{\lambda_A}{\lambda_C - \lambda_A}\frac{\lambda_B}{\lambda_B - \lambda_A}e^{-\lambda_A t} + \frac{\lambda_A}{\lambda_A - \lambda_B}\frac{\lambda_B}{\lambda_C - \lambda_B}e^{-\lambda_B t}\right.$$

$$\left. + \frac{\lambda_A}{\lambda_A - \lambda_C}\frac{\lambda_B}{\lambda_B - \lambda_C}e^{-\lambda_C t}\right) \tag{1.13}$$

whereas the activities of individual radionuclides A, B, and C are given as

$$A_A = \lambda_A N_A, A_B = \lambda_B N_B, A_C = \lambda_C N_C \tag{1.14}$$

As long as a radioactive parent is initially a pure source whose activity at $t = 0$ is $A_{A0} = N_{A0}\lambda_A$, the ratio of the activities of the parent A and the daughter B will be changing with time following the equation

$$\frac{A_B(t)}{A_A(t)} = \frac{\lambda_B}{\lambda_B - \lambda_A}(1 - e^{-(\lambda_B - \lambda_A)t}) \tag{1.15}$$

If the half-life of the radionuclide B is greater than that of A, i.e., $\lambda_A > \lambda_B$, then this ratio will continuously increase with time. On the other

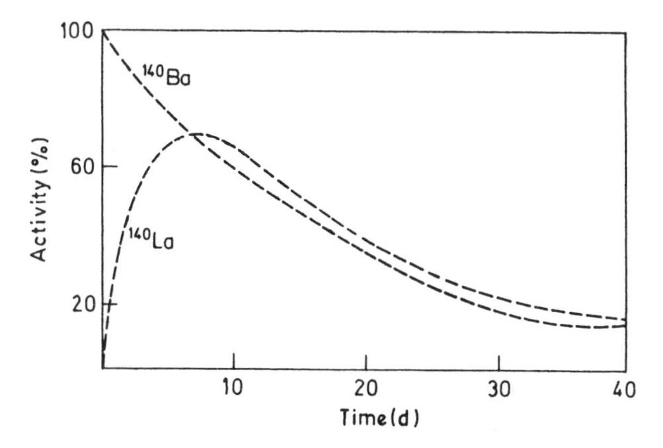

FIGURE 1.2. Growth and decay of ^{140}La ($T_{1/2}$ = 40.272 h) from freshly separated ^{140}Ba ($T_{1/2}$ = 12.74 d) and the decay of ^{140}Ba as a function of time. The maximum of ^{140}La activity is at T_m = 6.57 d.

hand, if the half-life of the parent A is greater than the half-life of its daughter B, i.e., $\lambda_B > \lambda_A$, as t increases, the ratio $A_B(t)/A_A(t)$ also increases, but as t becomes large, this ratio becomes a constant greater than one. Consequently, for sufficiently large values of t the daughter will decay with the half-life of the parent, but its activity will be greater than the parent activity by the factor $\lambda_B/(\lambda_B - \lambda_A)$. Such a situation is referred to as a *transient equilibrium.* An example of this kind of equilibrium is shown in Figure 1.2 (Mann et al., 1980) for ^{140}Ba, $T_{1/2}$ = 12.74 d, which decays to ^{140}La, $T_{1/2}$ = 40.272 h. It can be seen that the activity of ^{140}La is zero initially and then grows to a maximum at some intermediate time T_m, for which one can derive the generally valid relationship

$$T_m = \frac{\ln (\lambda_B/\lambda_A)}{\lambda_B - \lambda_A} \qquad (1.16)$$

In the illustrated case the activity of ^{140}La rises to the maximum T_m = 135 h, then decreases and after a few hundred hours reaches the transient equilibrium with its parent.

The other type of equilibrium occurs when the parent is very much longer-lived than the daughter so that $\lambda_B \gg \lambda_A$. In this case, taking into account Equations (1.11) and (1.12), for the ratio of the activities A_B/A_A one obtains

$$\frac{A_B(t)}{A_A(t)} = 1 - e^{-\lambda_B t} \qquad (1.17)$$

The situation represented by the above equation, where after a sufficiently long time the daughter activity will reach the parent activity, is called *secular equilibrium.* An example of such secular equilibrium is the decay of ^{226}Ra ($T_{1/2}$ = 1600 y) to ^{222}Rn ($T_{1/2}$ = 3.82 d).

In addition to the previous two cases characterized with a certain degree of equilibrium, there is a third case of interest where $\lambda_A > \lambda_B$ (a short-lived parent and a long-lived daughter) and no equilibrium occurs.

In the SI (System International of Units) the unit of activity is the *becquerel* with symbol Bq and 1 Bq corresponds to one nuclear transformation per second, or 1 Bq = 1 s^{-1}. The unit becquerel is actually a special name for the reciprocal second (second to the power of minus one). Some seventy years ago, when radium was the most important radioactive source, the amount of radioactivity was given in mass, usually in terms of mg or g of radium. The former unit of activity *curie* (Ci) was related to the disintegration rate of 1 gram of ^{226}Ra, for which early experiments gave values nearer to 3.7×10^{10} per second.

Before, when the unit Ci was in use, the small amount of radionuclides was expressed by means of sub-multiples formed by the normalized SI prefixes such as pico- (10^{-12}), nano- (10^{-9}), micro- (10^{-6}), or milli- (10^{-3}). This was because 1 Ci was too big a unit not only in the case of low-level radioactivities, but also for most other applications and related fields. With the unit Bq the situation is quite the opposite: even for low-level counting, it is sometimes considered to be too small, so that often its multiples are used.

A radioactive source can be sometimes considered as a *point source* whose dimensions are small enough that they may be neglected. Quite often, however, the sources have different sizes, shapes, and forms, so other quantities derived from *activity* may be more useful.

This is especially true in the case of environmental radioactivity assessment and measurement, where it is often desirable to relate the activity of a given radionuclide to a unit volume or mass of its sample, or to express it in terms of activity per unit of area.

The first two such quantities are usually referred to as *specific activity* or *activity concentration* and they are defined as

$$a_m = \frac{A}{m} \tag{1.18}$$

$$a_V = \frac{A}{V} \tag{1.19}$$

where A is activity of the material having mass m and volume V. The basic units of mass activity and volume activity are $Bq \cdot kg^{-1}$ and $Bq \cdot m^{-3}$ (sometimes also $Bq \cdot L^{-1}$), respectively.

The activity concentration per unit volume is used mainly for radioactive gases where other parameters, such as temperature and pressure, must also be stated. It may sometimes be useful to give the activity of a gas in terms of its value per mole as a unit of substance.

The specific activity of a pure (carrier-free, unmixed with any other nuclear species) radioactive source can be calculated using the following formula

$$a_m = \frac{\lambda N}{(N \cdot M)/N_A} \tag{1.20}$$

where N is the number of radioactive atoms and M is the molar mass of the source, and N_A is the Avogadro number.

Radionuclides seldom occur in a pure form; they are usually diluted in a much higher concentration of stable atoms of the same element. In addition, other stable atoms may be included from different elements that may be chemically combined with those of the source.

When a radionuclide is spread on a surface, e.g., in the case of a surface contamination, the appropriate quantity would be *area activity* given by the quotient of the activity A and the relevant area or surface S, i.e.,

$$a_S = \frac{A}{S} \tag{1.21}$$

with the unit $Bq \cdot m^{-2}$.

During the last decade or so, much attention has been paid to the radon problem since the radiation dose from its inhaled decay products is the dominant component of natural radiation exposures of the general population. For radon decay products, the collective quantity of most use is the *equilibrium-equivalent decay-product concentration*$-a_{V_{eq}}$ (Rn), which is also called the *equilibrium-equivalent radon concentration,* but it is used as a measure of the decay-product concentration. The concentration of short-lived decay products in air is ordinarily not given in terms of individual decay-product concentrations, but rather by the quantity that is normalized to the amount of alpha decay energy that will ultimately result from the mixture of the decay products that are present (Nero, 1988). The equilibrium-equivalent radon concentrations for ^{222}Rn and ^{220}Rn are given in terms of their individual decay-product concentrations as

$$a_{V_{eq}}(^{222}\mathrm{Rn}) = 0.106\, a_V(^{218}\mathrm{Po}) + 0.513\, a_V(^{214}\mathrm{Pb}) + 0.381\, a_V(^{214}\mathrm{Bi}) \tag{1.22}$$

and

$$a_{V_{eq}}(^{220}\mathrm{Rn}) = 0.913\, a_V(^{212}\mathrm{Pb}) + 0.087\, a_V(^{212}\mathrm{Bi}) \qquad (1.23)$$

At secular equilibrium (the case where the half-life of a decay product is much shorter than that of the parent; this situation will be illustrated further in the text) between radon and its decay products, the equilibrium-equivalent radon concentration is equal to the radon concentration. In both the above equations these concentrations are given in $\mathrm{Bq \cdot m^{-3}}$.

The extent to which the concentrations of short-lived decay products are under real conditions lower than the values of their concentrations corresponding to secular equilibrium with radon as a parent radionuclide can be characterized by the *equilibrium factor* (F) defined as the ratio between the equilibrium-equivalent radon concentration and the actual radon concentration at the place of interest, i.e.,

$$F = \frac{a_{V_{eq}}(\mathrm{Rn})}{a_V(\mathrm{Rn})} \qquad (1.24)$$

For the assessment of radiological hazards due to exposure to radon (and especially to its decay products) it is very important to have information about the *potential alpha-energy concentration* (PAEC). This quantity represents the energy that would eventually be released in a specified volume of undisturbed air by the short-lived decay products of radon through the emission of alpha particles. The PAEC is expressed in units of $\mathrm{J \cdot m^{-3}}$ or sometimes in a special unit, widely used in radon monitoring, the so called *working level* (WL). The working level is defined as any combination of short-lived radon products in one liter of air that will result in the ultimate emission of potential alpha particle energy equal to 1.3×10^5 MeV. At secular equilibrium, 1 WL corresponds to the ^{222}Rn concentration 3.7 $\mathrm{kBq \cdot m^{-3}}$ (originally this was related to the radon concentration equal to 100 $\mathrm{pCi \cdot L^{-1}}$).

The concentrations of individual radon decay products can be expressed in WL units as

$$WL = \frac{13.69\, n_V(^{218}\mathrm{Po}) + 7.69\, \{n_V(^{214}\mathrm{Pb}) + n_V(^{214}\mathrm{Bi})\}}{1.3 \cdot 10^5} \qquad (1.25)$$

where $n_V(^{218}\mathrm{Po})$, $n_V(^{214}\mathrm{Pb})$, and $n_V(^{214}\mathrm{Bi})$ are the concentrations of the relevant decay products in numbers of their atoms per liter.

In some applications we may be interested in the number of particles emitted out of a radiation source per unit of time. This approach is useful especially in the case of neutron sources where the *emission rate* is often used. The emission rate of a neutron source is defined as the number of neutrons emitted by the source per second. Obviously, the unit is 1 s^{-1}, i.e., one per second or *reciprocal second.*

When using quantities and units, one always has to realize that any statement without a reference to the type of source or radionuclide would be incomplete. Presumably, there is a big difference between 10 kBq of ^{3}H and the same activity of other radionuclides with much higher toxicity (as well as many other different properties), say ^{226}Ra.

1.1.2 Original Radionuclides in the Environment

1.1.2.1 INTRODUCTION

At present more than 2500 radionuclides of various elements are known. Approximately eighty of these are found in nature, the rest are produced artificially as the direct products of many types of different nuclear reactions, or are produced indirectly as the radioactive progenies of these products.

The list of naturally occurring radionuclides, together with their abundance and half-life, is presented in Table 1.1. Most of these radionuclides occur in extremely tiny concentrations. Some of them were believed to have been created together with the earth more than 3×10^9 years ago. They may be considered as naturally occurring parent radionuclides, having such long half-lives that their detectable residual activity is still observable today. As a matter of fact, we can detect the presence of a radioactive substance for only about ten half-lives. This means that those original radionuclides with half-lives of about 0.3×10^9 years or less should not be found in the earth at present. Some of the parent radionuclides found in nature are given in Table 1.2.

Some of the common elements contain long-lived, natural radionuclides. The most prominent is terrestrial potassium containing 0.0117% of the radioactive isotope ^{40}K. This radionuclide is the principal radiation source present in the human body which on average contains about 4.4 kBq of ^{40}K.

In addition to natural parent radionuclides and their decay products, usually referred to as *terrestrial* or *primordial radionuclides,* there are some radionuclides of extraterrestrial origin, known as *cosmogenic radionuclides.* Such radionuclides as ^{3}H, ^{7}Be, and ^{14}C are examples of radionuclides that are continuously produced by the interactions of cosmic rays with some atoms present in the atmosphere.

Naturally occurring radionuclides are by far the largest contributor to radiation doses received by human beings. Exposures to natural sources of radiation are a subject of interest in their own right, as well as a suitable

TABLE 1.1. Natural Radionuclides and Their Main Parameters.

Radio-nuclide	Abundance (%)	Half-Life (Years)	Radio-nuclide	Abundance (%)	Half-Life (Years)
^{14}C	trace	5.73×10^{3}	^{149}Sm	13.9	$\approx 4 \times 10^{14}$
^{40}K	0.012	1.26×10^{9}	^{152}Gd	0.20	1.1×10^{14}
^{48}Ca	0.187	$>2 \times 10^{16}$	^{159}Tb	100	5×10^{16}
^{50}V	0.25	6×10^{15}	^{165}Ho	100	$>6 \times 10^{16}$
^{64}Zn	48.6	$>8 \times 10^{15}$	^{169}Tm	100	$>5 \times 10^{16}$
^{70}Zn	0.62	$>1 \times 10^{15}$	^{175}Lu	97.41	$>1 \times 10^{17}$
^{76}Ge	7.8	$>5 \times 10^{21}$	^{176}Lu	2.60	3.0×10^{10}
^{82}Se	9.19	1.4×10^{20}	^{174}Hf	0.18	2.0×10^{15}
^{87}Rb	27.85	5×10^{11}	^{180}Ta	0.012	$>1 \times 10^{13}$
^{96}Zr	2.80	$>3.6 \times 10^{17}$	^{180}W	0.14	$>1.1 \times 10^{15}$
^{92}Mo	14.8	4×10^{18}	^{182}W	26.3	$>2 \times 10^{17}$
^{100}Mo	9.6	$\geq 3 \times 10^{17}$	^{183}W	14.3	$>1.1 \times 10^{17}$
^{113}Cd	12.26	9.3×10^{15}	^{186}W	28.6	$>6 \times 10^{15}$
^{116}Cd	7.58	$>1 \times 10^{17}$	^{187}Re	62.6	7×10^{10}
^{115}In	95.72	5.1×10^{14}	^{192}Os	41	$>1 \times 10^{14}$
^{124}Sn	5.64	$>2.4 \times 10^{17}$	^{190}Pt	0.013	7×10^{11}
^{123}Sb	42.75	1×10^{16}	^{192}Pt	0.78	1×10^{15}
^{123}Te	0.89	1.2×10^{13}	^{198}Pt	7.21	1×10^{15}
^{130}Te	34.48	2.51×10^{21}	^{196}Hg	0.146	$>1 \times 10^{14}$
^{138}La	0.09	1.1×10^{11}	^{209}Bi	100	$>2 \times 10^{18}$
^{126}Ce	0.193	3×10^{11}	^{222}Rn	From ^{226}Ra	3.82 d
^{142}Ce	11.07	$>5 \times 10^{16}$	^{226}Ra	From ^{230}Th	1622
^{141}Pr	100	$>2 \times 10^{16}$	^{230}Th	From ^{234}U	8×10^{4}
^{144}Nd	23.85	$\approx 5 \times 10^{15}$	^{232}Th	100	1.41×10^{10}
^{150}Nd	5.62	$>5 \times 10^{18}$	^{234}U	0.005	2.47×10^{5}
^{147}Sm	15.1	1.06×10^{11}	^{235}U	0.72	7.1×10^{8}
^{148}Sm	11.3	1.2×10^{13}	^{238}U	99.275	4.51×10^{9}

TABLE 1.2. Original Natural Parent Radionuclides and Some of Their Properties (EC—Electron Capture).

Radionuclide			Decay Type (Transitions)	Disintegration Energy (MeV)
Symbol	Z	A		
K	19	40	β^-, EC	1.3 (β^-), 1.5 (EC)
Se	34	82	β^-, β^+	3.0
Rb	37	87	β^-	0.3
Cd	48	113	β^-	0.3
In	49	115	β^-	0.5
La	57	138	β^-, EC	1.0 (β^-), 1.75 (EC)
Nd	60	144	α	1.9
Sm	62	147	α	2.3
Sm	62	148	α	1.99
Gd	64	152	α	2.2
Lu	71	176	β^-, gamma	0.6
Hf	72	174	α	2.5
Re	75	187	β^-	0.003
Pt	78	190	α	3.24
Th	90	232	α	4.08
U	92	235	α	4.68
U	92	238	α	4.27

background or reference level against which we can compare and evaluate the significance of exposures from artficial sources of radiation.

The contributions of various natural sources to the average effective dose equivalent are shown in Table 1.3 (UNSCEAR, 1988).

The quantity *effective dose equivalent,* used in the Table 1.3 for quantification of radiation exposures, is one of the most important radiation protection quantities. It is based on the *absorbed dose* and the *dose equivalent* which have been defined as follows. The *absorbed dose* of radiation is the energy imparted per unit mass of the irradiated material. The basic unit of this quantity is *gray* (Gy), which is a special name for SI unit $J \cdot kg^{-1}$. On the other side the *dose equivalent H* represents the product of the absorbed dose *D* in tissue and so-called *quality factor Q,* which depends on the ionization abilities of the radiation in question, i.e.,

$$H = D \cdot Q \tag{1.26}$$

The *effective dose equivalent* H_E is introduced by the definition

$$H_E = \sum_T w_T \cdot H_T \tag{1.27}$$

TABLE 1.3. Estimated per Capita Annual Effective Dose Equivalent from Natural Sources of Radiation in Areas with Normal Background.

Source of Irradiation	Effective Dose Equivalent (μSv/y) External	Internal	Total
Cosmic rays			
Ionizing component	300		300
Neutron component	55		55
Cosmogenic radionuclides		15	15
Primordial radionuclides			
^{40}K	150	180	330
^{87}Rb		6	6
^{238}U series:	100	1240	1340
$^{238}U \rightarrow {}^{234}U$		5	5
^{230}Th		7	7
^{226}Ra		7	7
$^{222}Rn \rightarrow {}^{214}Po$		1100	1100
$^{210}Pb \rightarrow {}^{210}Po$		120	120
^{232}Th series:	160	180	340
^{232}Th		3	3
$^{228}Ra \rightarrow {}^{224}Ra$		13	13
$^{220}Rn \rightarrow {}^{208}Tl$		160	160
Total (rounded)	800	1600	2400

Adapted from UNSCEAR (1988).

In other words, it is the sum over the selected tissues of the product of the dose equivalent H_T in a tissue T and the weighting factor w_T representing its proportion of the total stochastic effects resulting from irradiation of this tissue to the total risk when the whole body is exposed uniformly.

From Table 1.3 one can see that inhalation of short-lived decay products of radon is the most important contribution to the annual effective dose equivalent from natural sources of radiation. The second largest contribution comes from external irradiation, which is responsible for almost 800 μSv, divided approximately equally between cosmic radiation and terrestrial sources. Less significant are doses due to the ingestion of ^{40}K (180 μSv), inhalation of decay products of thoron (160 μSv) and internal contamination by ^{210}Pb-^{210}Po (120 μSv). The other natural radionuclides contribute very little to the total annual effective dose equivalent.

1.1.2.2 TERRESTRIAL RADIONUCLIDES

Terrestrial radionuclides have existed in the earth since its formation several billion years ago and have not substantially decayed. The most abundant and important of these so-called *primordial radionuclides* are ^{238}U, ^{235}U, ^{232}Th and their decay products, and ^{40}K as well as ^{87}Rb.

The heavy elements uranium and thorium are parents of three radioactive series (families, decay chains), headed by naturally occurring unstable nuclei ^{238}U, ^{235}U, and ^{232}Th, respectively. These three classical sets, fully delineated by 1935, are usually called the *uranium series, actinium series,* and *thorium series.* The fourth series, the *neptunium series,* commences with ^{237}Np, which has a relatively short half-life (about 2×10^6 years). Its members can be produced artificially by nuclear reactions, but they do not occur naturally because their half-lives are short compared with the age of the earth.

The three naturally occurring transformation series and the radioactive properties of their members are summarized in Table 1.4. The *uranium series* begins with ^{238}U and ends with the stable nuclide ^{206}Pb. The *thorium series* is headed with ^{232}Th and ends with the stable isotope of lead, ^{208}Pb. The last of these series is the *actinium series,* named for its first-discovered member, ^{227}Ac.

As we can see, the products of radioactive decay are often also radioactive daughter products. Because mass number changes are only possible by alpha transformation, all heavy nuclides can be grouped into the families with the atomic mass numbers $A = 4n + 2$ (Uranium series), $4n$ (Thorium series), and $4n + 3$ (Actinium series).

Many different graphical illustrations of radioactive series have been used. One of the possible presentations is shown in Figure 1.3 (Uranium series) and Figure 1.4 (Thorium series); the other approach is shown in Figure 1.5 (Actinium series) and Figure 1.6 (Neptunium series). Figures 1.3 and 1.4 are based on ICRP (1983), while Figures 1.5 and 1.6 have been taken from Mann et al. (1980).

The average concentration of primordial radionuclides in the earth's crust is quite low, but it may vary from place to place significantly. The concentrations of ^{40}K, ^{232}Th, and ^{238}U in some rocks and soil are given in Table 1.5 (Liden and Holm, 1985). These radionuclides as well as their decay products are expected to be present in various environmental samples (biologic, hydrologic, geologic, etc.) through dust deposition, wash-out, weathering, sedimentation, and biological and other transfer mechanisms and processes (Holm et al., 1981).

The impact of natural radionuclides in soil on the exposure of persons can be illustrated by some data from the USA. This is documented in Table 1.6 where the values of absorbed dose rates in air due to the contamination of the soil by some primordial radionuclides are shown (Myrick et al., 1983).

The concentrations of radionuclides in oceans and seawaters is quite low, partly because of sedimentation processes and partly because of dilution in

TABLE 1.4. Naturally Occurring Radioactive Series and Some of Their Parameters.

Series	Member	Atomic Number	Half-Life	Decay Type
Uranium	^{238}U	92	4.5×10^{9} y	α
	^{234}Th	90	24 d	β^-
	^{234m}Pa	91	1.2 m	IT, β^-
	^{234}Pa	91	6.7 h	β^-
	^{234}U	92	2.5×10^{5} y	α
	^{230}Th	90	8×10^{4} y	α
	^{226}Ra	88	1600 y	α
	^{222}Rn	86	3.8 d	α
	^{218}Po	84	3.0 m	α
	^{214}Pb	82	27 m	β^-
	^{214}Bi	83	20 m	β^-, α
	^{214}Po	84	160 μs	α
	^{210}Tl	81	1.3 m	β^-
	^{210}Pb	82	22 y	β^-
	^{210}Bi	83	5.0 d	β^-
	^{210}Po	84	138 d	α
	^{206}Pb	82	Stable	Stable
Thorium	^{232}Th	90	1.4×10^{10} y	α
	^{228}Ra	88	5.8 y	β^-
	^{228}Ac	89	6.1 h	β^-
	^{228}Th	90	1.9 y	α
	^{224}Ra	88	3.7 d	α
	^{220}Rn	86	56 s	α
	^{216}Po	84	0.15 s	α
	^{212}Pb	82	10.6 h	β^-
	^{212}Bi	83	1.0 h	β^-, α
	^{212}Po	84	0.3 μs	α
	^{208}Tl	81	3.1 m	β^-
	^{208}Pb	82	Stable	Stable
Actinium	^{235}U	92	7.1×10^{8} y	α
	^{231}Th	90	26 h	β^-
	^{231}Pa	91	3.3×10^{4} y	α
	^{227}Ac	89	22 y	β^-, α
	^{227}Th	90	19 d	α
	^{223}Fr	87	22 m	β^-, α
	^{223}Ra	88	11 d	α
	^{219}At	85	0.9 m	α, β^-
	^{219}Rn	86	4.0 s	α
	^{215}Po	84	1.8 ms	α
	^{211}Pb	82	36 m	α
	^{211}Bi	83	2.2 m	α, β^-
	^{211}Po	84	0.5 s	α
	^{207}Tl	81	4.8 m	β
	^{207}Pb	82	Stable	Stable

Note: y = year; d = days; h = hours; m = minutes; s = seconds.

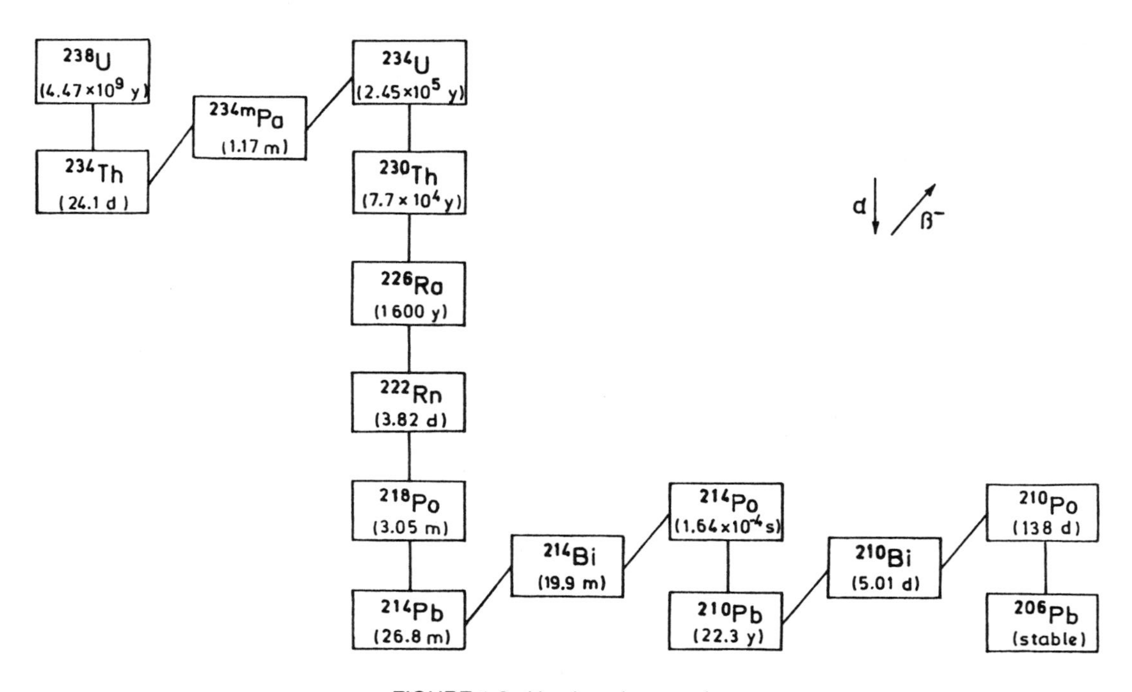

FIGURE 1.3. Uranium decay series.

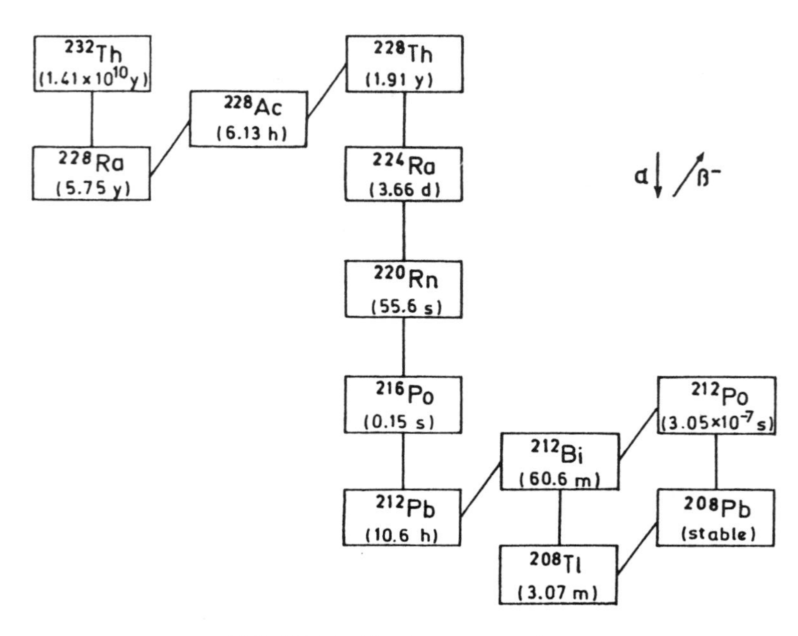

FIGURE 1.4. Thorium decay series.

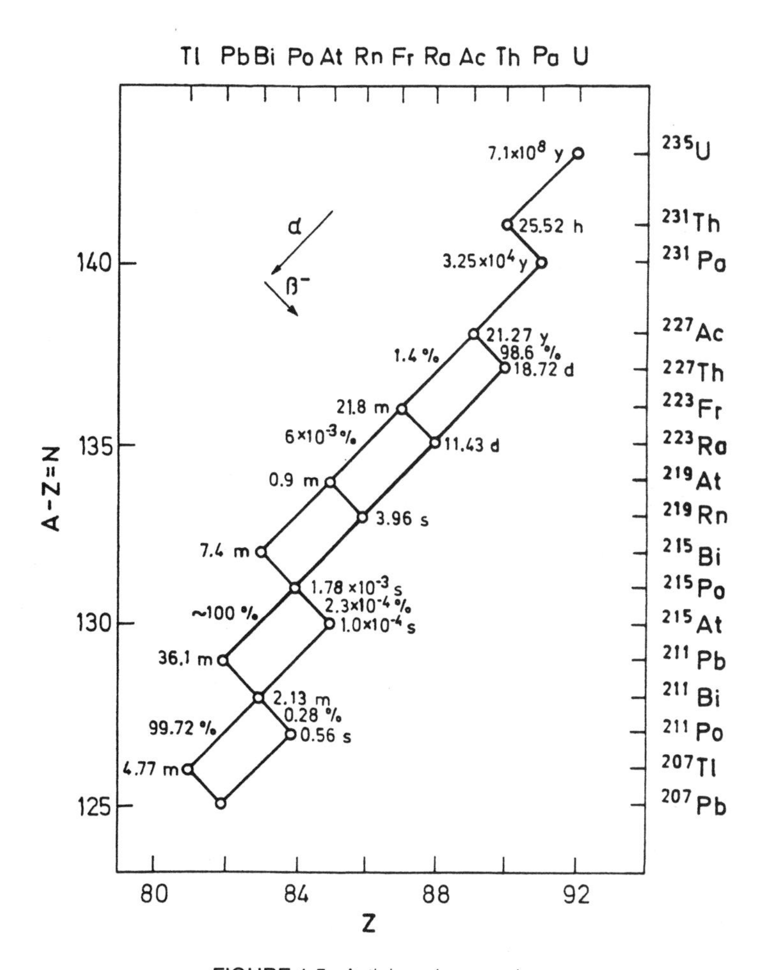

FIGURE 1.5. Actinium decay series.

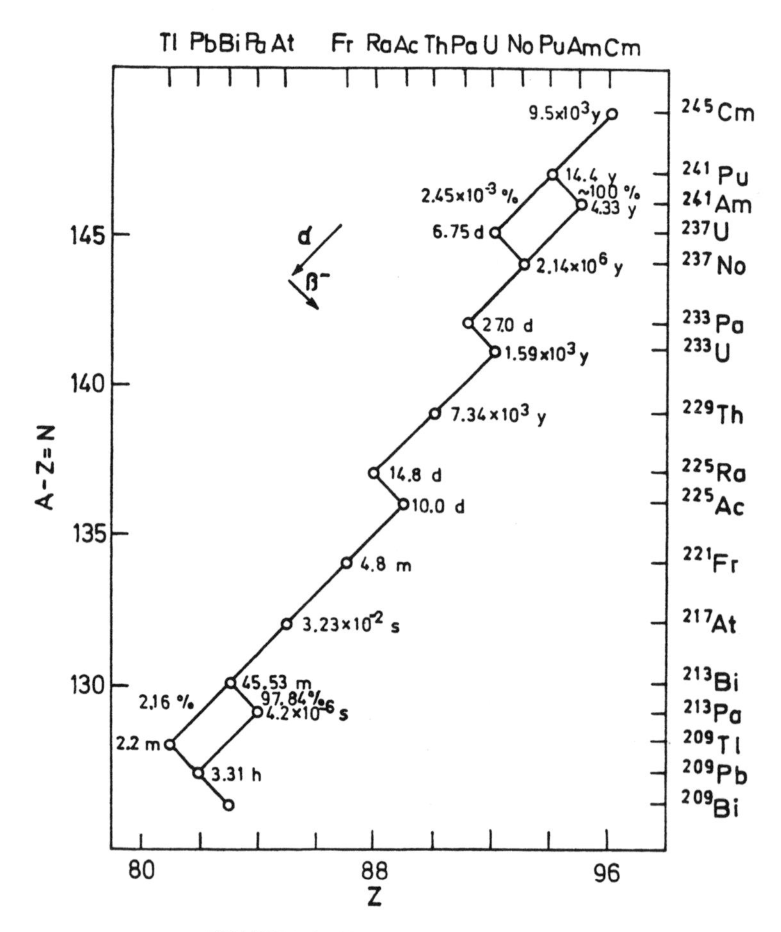

FIGURE 1.6. Neptunium decay series.

TABLE 1.5. Mass and Activity Concentrations of ^{40}K, ^{232}Th and ^{238}U in Some Rocks and Soil.

	^{40}K		^{232}Th		^{238}U	
Type of Rock	mg/kg	Bq/kg	mg/kg	Bq/kg	mg/kg	Bq/kg
Igneous						
Granite	5	1200	17	70	3	35
Basalt	0.9	230	4	15	1	12
Sedimentary						
Limestone	0.3	70	2	8	2	25
Sandstone	1.3	300	3	11	1.5	18
Shale	3.0	700	12	45	4	50
Soil						
World average	1.4	370	6	25	2	25

From Liden and Holm (1985).

TABLE 1.6. Activity Concentrations of Natural Radionuclides in Soil and Associated Absorbed Dose Rate in Air.

	Concentration (Bq/kg)		Dose Rate (nGy/h)	
Radionuclide	Average	Range	Average	Range
^{40}K	370	100–700	15	4–29
^{232}Th series	35	4–130	22	2–81
^{238}U series	35	4–140		
^{226}Ra sub-series	40	8–160	18	4–74

the huge water volumes. For example, the concentration of potassium and uranium are 0.35 g/L (11 Bq/L) and 3 μg/L (36 mBq/L), respectively. In groundwater ^{222}Rn frequently appears at quite high concentrations (10–200 Bq/L), particularly in deep well water. Its precursor ^{226}Ra, however, is found in much lower concentrations (Liden and Holm, 1985). The transfer of terrestrial radionuclides to the hydrosphere depends strongly on various chemical and physical properties of the rock-soil-water-atmosphere system. Strong local deviations from the general pattern of transport and deposition of various radionuclides are observed.

In addition to their significant contributions as external emitters, ^{40}K as well as the members of the uranium and thorium series represent the main sources of internal irradiation in the human body. The intake of these natural radionuclides arises from inhalation and ingestion. In order to assess the exposure due to the intake of various radionuclides, the International Commission on Radiological Protection (ICRP) introduced such quantities as the *committed dose equivalent* and *committed effective dose equivalent,* which will soon be replaced by the recently adopted quantity *committed effective dose.*

The *committed dose equivalent* (ICRP, 1977) to a given organ or tissue T from a single intake of radioactive material into the body was introduced by the expression

$$H_T(50) = \int_{t_0}^{t_0+50} \dot{H}_T(t)\mathrm{d}t \tag{1.28}$$

where $\dot{H}_T(t)$ is the relevant dose equivalent rate at time t and t_0 is the time of intake. The period of integration was then taken as fifty years.

The *committed effective dose equivalent* $H_E(50)$ was defined as

$$H_E(50) = \sum_T H_T(50) \tag{1.29}$$

In the latest ICRP Recommendations (ICRP, 1991) new radiation protection quantities have been suggested, namely the *equivalent dose* and the *effective dose,* and especially the *committed effective dose,* which is now the main quantity for the assessment of irradiation of the human body from incorporated radionuclides.

The *equivalent dose* is derived from the absorbed dose averaged over a tissue or organ T by the expression

$$H_T = \sum_R w_R \cdot D_{T,R} \tag{1.30}$$

where w_R is the radiation weighting factor, depending on the type and energy of radiation, and $D_{T,R}$ is the average absorbed dose from radiation R in tissue T.

It has been found that the relationship between the probability of stochastic effects and equivalent dose depends on the organ or tissue irradiated. This is why a further quantity, based on the equivalent dose, had to be introduced to indicate the combination of different doses to several different organs or tissues in such a way that it would be likely to correlate well with the total stochastic effects. Following this philosophy, the *effective dose* E is defined as the sum of the weighted equivalent doses in all the relevant tissues and organs of the body, i.e.,

$$E = \sum_T w_T \cdot H_T \tag{1.31}$$

where H_T is the equivalent dose in tissue or organ T and w_T is the weighting factor for this tissue or organ.

Terrestrial and other radionuclides can irradiate the human body both externally and internally. While the externally applied radiation results in the simultaneous deposition of energy in irradiated tissues or organs, the irradiation due to the incorporated radionuclides is spread out in time. The time distribution of energy deposition will vary with the physicochemical form of the radionuclide and its subsequent biokinetics. In order to take into account this distribution, the ICRP recommended the use of the *committed equivalent dose* $H_T(\tau)$ (for a single intake of activity at time t_0) defined as

$$H_T(\tau) = \int_{t_0}^{t_0+\tau} \dot{H}_T(t)\mathrm{d}t \tag{1.32}$$

where $\dot{H}_T(t)$ is the relevant equivalent dose rate in an organ or tissue T at time t and τ is the time period over which the integration is performed. This period is now usually taken as fifty years for adults and seventy years for children and infants.

If the committed tissue or organ equivalent doses resulting from an intake are multiplied by the appropriate weightig factors w_T, and then summed, the result will be the *committed effective dose* $E(\tau)$

$$E(\tau) = \sum_T w_T \cdot H_T(\tau) \tag{1.33}$$

When referring to an equivalent or effective dose accumulated in a given period of time, it is implicit that any committed doses from intakes occurring in that same period are included (ICRP, 1991).

Now, after the introduction of relevant radiation protection quantities, we can present in Table 1.7 (NRPB, 1991) the *dose coefficients* for some terrestrial radionuclides in terms of the *committed effective dose* per unit activity of a given radionuclide. The activity concentrations of some natural radionuclides in common types of foods and air are given in Table 1.8.

Very useful data appears in Table 1.9 where the annual intakes by ingestion and inhalation for adults together with the associated doses are given. The effective doses to which infants are exposed from annual intake by ingestion increases by about 40% compared to roughly equal values for children and adults. It can be shown from the tabulated data that the most important radionuclide is ^{210}Po. As to intake by inhalation, the values decrease with decreasing age at inhalation by about 5% for children compared to adults and by 50% for infants. Also in the case of inhalation the largest contribution is from ^{210}Po.

The doses to adults from reference annual intakes of the long-lived series radionuclides can be compared to the annual doses re-estimated from the UNSCEAR 1988 Report (1988) with new ICRP tissue weighting factors (ICRP, 1991). For radionuclides of the uranium and thorium families, the doses are 80 μSv due to annual intake and 92 μSv per year from average body content. For potassium ^{40}K the same doses are about 165 μSv and 180 μSv, respectively. The total dose is 245 μSv by intake and 272 μSv by body content.

TABLE 1.7. Committed Effective Dose per Unit Activity Intake of Natural Radionuclides (Dose Integrated to Age Seventy Years for Children and Infants and over Fifty Years for Adults per Unit Intake).

	Dose Coefficient (μSv·Bq^{-1})					
	Ingestion			Inhalation		
Radionuclide	Adults	Children	Infants	Adults	Children	Infants
^{238}U	0.036	0.059	0.15	31	50	130
^{234}U	0.039	0.065	0.17	35	56	150
^{230}Th	0.076	0.10	0.20	51	74	170
^{226}Ra	0.22	0.35	0.91	2.1	4.7	14
^{210}Pb	0.86	1.1	2.5	2.2	2.8	6.3
^{210}Po	0.21	0.48	1.5	1.0	2.4	7.2
^{232}Th	0.37	0.43	0.64	210	260	440
^{228}Ra	0.27	0.40	1.1	1.1	2.3	6.7
^{228}Th	0.067	0.13	0.38	66	120	370
^{235}U	0.038	0.063	0.16	33	52	130
^{231}Pa	1.4	1.8	2.6	160	220	370
^{227}Ac	2.2	2.7	5.4	280	410	1000

TABLE 1.8. Reference Activity of Some Terrestrial Radionuclides in Food and Air (Data for ^{228}Th Are the Same as for ^{232}Th).

	Activity Concentration ($mBq \cdot kg^{-1}$)						
Intake	$^{238}U \rightarrow ^{234}U$	^{230}Th	^{226}Ra	^{210}Pb	^{210}Po	^{228}Ra	^{232}Th
Milk products	1	0.5	5	40	60	5	0.3
Meat products	2	2	15	80	60	10	1
Grain products	20	10	80	100	240	60	3
Leafy vegetables	20	20	50	30	30	40	15
Roots and fruits	3	0.5	30	25	30	20	0.5
Fish products	30	–	100	200	5000	–	–
Water supplies	1	0.1	0.5	10	5	0.5	00.05
	Activity Concentration ($\mu Bq \cdot kg^{-1}$)						
Intake	$^{238}U \rightarrow ^{234}U$	^{230}Th	^{226}Ra	^{210}Pb	^{210}Po	^{228}Ra	^{232}Th
Air	1	0.5	0.5	500	30	1	1

Adapted from UNSCEAR (1993).

TABLE 1.9. Reference Annual Intakes of Some Terrestrial Radionuclides and Associated Lifetime Effective Doses for Adults.

	Ingestion		Inhalation	
Radionuclide	Intake (Bq)	Dose (μSv)	Intake (Bq)	Dose (μSv)
^{238}U	5.7	0.21	8.0	0.25
^{234}U	5.7	0.22	8.0	0.28
^{230}Th	2.9	0.22	4.0	0.20
^{226}Ra	22	4.84	4.0	0.01
^{210}Pb	36	31.0	4000	8.8
^{210}Po	127	26.7	400	0.40
^{232}Th	1.5	0.56	8.0	1.68
^{228}Ra	15	4.05	8.0	0.01
^{228}Th	1.5	0.10	8.0	0.53
^{235}U	0.24	0.01	0.40	0.01

The variability of activity concentrations in foods is clearly demonstrated in Table 1.10, which shows selected data on elevated levels (UNSCEAR, 1993). The reference values are exceeded many times. In the volcanic area of Minas Gerais in Brazil and in the mineral sands of Kerala in India there is evidence of excess activity in milk, meat and grain, leafy vegetables, roots, and fruits. In the granitic area of Guandong in China, elevated activity has been measured in foodstuffs such as rice and radishes. Perhaps the most pronounced increase over reference levels occurs in the arctic and sub-arctic regions of Scandinavian countries, where ^{210}Pb and ^{210}Po accumulate in the flesh of reindeer, which is an important part of the diet of the

TABLE 1.10. Elevated Activity Concentrations (in mBq·kg^{-1}) of Some Natural Radionuclides in Foods from Places with Excessive Radioactivity.

			Activity Concentration	
Food	Country	Radionuclide	Range	Arithmetic Mean
Cows' milk	Brazil	^{226}Ra	29–210	108
		^{210}Pb	5–60	45
Chicken meat	Brazil	^{226}Ra	37–163	86
		^{228}Ra	141–355	262
Beef	Brazil	^{226}Ra	30–59	44
		^{226}Ra	78–111	96
Pork	Brazil	^{226}Ra	7–22	13
		^{228}Ra	93–137	121
Reindeer meat	Sweden	^{210}Pb	400–700	550
		^{210}Po	–	11,000
Cereals	India	^{226}Ra	Up to 510	174
		^{228}Th	Up to 5,590	536
Corn	Brazil	^{226}Ra	70–229	118
		^{210}Pb	100–222	144
Rice	China	^{226}Ra		250
		^{210}Pb		570
Green vegetables	India	^{226}Ra	325–2,120	1,110
		^{228}Th	348–5,180	1,670
Carrots	Brazil	^{226}Ra	329–485	411
		^{210}Pb	218–318	255
Roots and tubers	India	^{226}Ra	477–4,780	1,490
		^{228}Th	70–32,400	21,700
Fruits	India	^{226}Ra	137–688	296
		^{228}Th	59–21,900	2,590

Based on UNSCEAR (1993).

TABLE 1.11. Activity Concentrations of Some Natural Terrestrial Radionuclides in Potable Waters of Various Sources (Ar.—Arithmetic, Geo.—Geometric).

Source	Country	Radio-nuclide	Activity Concentration mBq·L^{-1} Range	Ar. Mean	Geo. Mean
Bottled waters	Brazil	^{226}Ra	<10–130		27
		^{210}Pb	<50–190		77
	France	^{226}Ra	up to 2,700	60	
		^{238}U	up to 2,000	60	
		^{232}Th	–	<40	
	Germany	^{226}Ra	<1–1,800		25
		^{238}U	<1–140		4.4
		^{210}Pb	3.3–53		9.0
		^{210}Po	0.4–8.9		1.8
	Indonesia	^{226}Ra	<1–60	22	
	Portugal	^{226}Ra	<3–2,185		26.7
		^{210}Pb	2–392		18.5
Ground waters	Finland	^{226}Ra	up to 5,300	440	
		^{238}U	up to 74,000	4,200	
		^{210}Pb	up to 10,200	430	
		^{210}Po	up to 6,300	220	
	Sweden	^{226}Ra	2–2,460	42	13.7

Adapted from UNSCEAR (1993).

Lapps. It is known that reindeer feed mainly on lichens, which concentrate these radionuclides from the atmosphere. It is possible to calculate the effective dose associated with the known annual consumption of reindeer meat.

In addition to great variability of activity concentrations in foods, the contents of radionuclides in potable waters also show considerable variation (Table 1.11). As with foods, reference values in bottled mineral and groundwaters are exceeded by orders of magnitude. It is interesting to see that in Finland very high activities were found in wells drilled in bedrock throughout the south of the country in the Helsinki area. The consumption of these waters may lead to committed effective doses of about 550 μSv for adults, slightly lower for children, but about 50% higher for infants.

It has already been mentioned that the highest dose due to naturally occurring terrestrial radionuclides comes from the inhalation of radon decay products. The problem had not been fully recognized until the late 1970s when in some countries, notably in Sweden, Canada, and the USA, a high incidence of ordinary houses with very high radon concentrations were found. Long-term exposure to these elevated concentrations may substantially increase the individual risk of developing lung cancer. It has been estimated (NCRP, 1967; NCRP, 1989a) that the average annual dose equivalent to the lung bronchial epithelium of the USA public is about 25 mSv. This exposure may be causing some 5000 to 10,000 lung cancer deaths per year in the country.

Chemically, radon is a noble gas, behaving similarly to, for example, helium or neon; like them it is colorless and odorless. There are actually three natural isotopes of the radioactive element radon: ^{219}Rn (historically known as actinon), ^{220}Rn (thoron) and ^{222}Rn (radon). The first isotope of radon, ^{219}Rn, is of very little significance and may be ignored entirely, mainly because its parent, ^{235}U, is a relatively rare nuclide and also because the half-life of ^{219}Rn is very short (four seconds) so that it will usually decay before having a chance to escape from its place of origin. Although the abundance of ^{232}Th (the precursor of ^{220}Rn) in the earth's crust is slightly higher than the abundance of ^{238}U (the distant parent of ^{222}Rn) the average rate of production of these two radon isotopes in the ground is practically the same. The half-life of ^{220}Rn (fifty-five seconds), however, is much shorter than half-life of ^{222}Rn. Consequently, the amount of ^{220}Rn entering the environment is usually considerably less than the amount of ^{222}Rn. The most important of all radon isotopes is therefore ^{222}Rn, which comes from the most abundant uranium isotope, ^{238}U. The half-life of ^{222}Rn (3.82 days) is long enough for it to survive after diffusion and penetration into an enclosed space, where it can accumulate to quite high concentrations.

Radon ^{222}Rn is a naturally occurring decay product of ^{226}Ra, the fifth daughter of ^{238}U (see Figure 1.7) (NRC, 1991). As radon forms, some of its atoms leave the soil or rock and may enter surrounding soil, air, or water. Radon decays into a series of solid, relatively short-lived radionuclides that are collectively known as radon daughters, radon progeny, or radon decay products.

A comparison of alpha decay properties of ^{222}Rn and ^{220}Rn, and their short-lived decay products is summarized in Table 1.12 (UNSCEAR, 1993). In addition to alpha particles, the measurement of radon decay products is also possible through the detection of their gamma photons and beta particles. More details regarding these parameters of radon decay products are given in Table 1.13. The data presented are based on the ICRP Publication 38 (1983).

Although the contribution from the inhalation of ^{220}Rn decay products may not always be neglected, in most countries the exposure from radon comes mainly from ^{222}Rn and especially its short-lived daughters. This is why far more attention is paid to this radon isotope and its decay products.

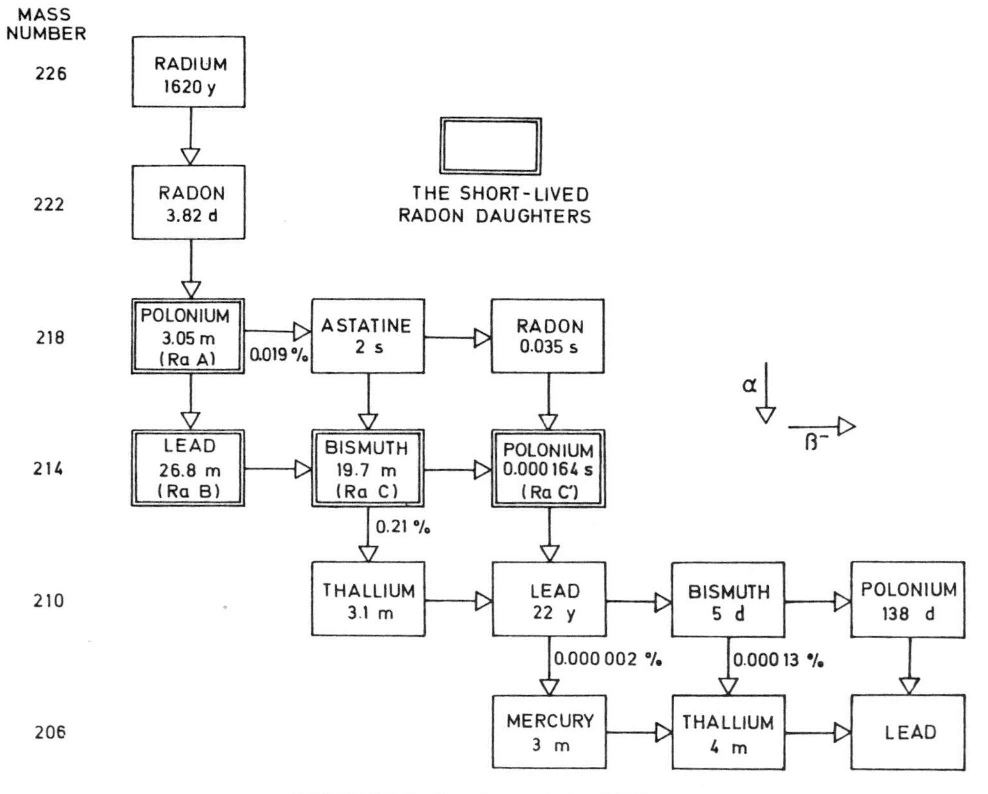

FIGURE 1.7. The decay chain of ^{222}Rn.

TABLE 1.12. Some of the Alpha Decay Parameters of ^{222}Rn and ^{220}Rn and Their Short-Lived Decay Products.

Radionuclide	Half-Life	E_α (MeV)	Intensity (%)
^{222}Rn:	3.824 d	5.49	100
^{218}Po	3.04 min	6.00	100
^{214}Pb	26.8 min	β, gamma	—
^{214}Bi	19.7 min	β, gamma	—
^{214}Po	163.7 μs	7.69	100
^{220}Rn:	55 s	6.29	100
^{216}Po	0.15 s	6.78	100
^{212}Pb	10.64 h	β, gamma	—
^{212}Bi	60.6 min	6.05	25
		6.09	10
^{212}Po	304 ns	8.78	100
^{208}Tl	3.10 min	β, gamma	—

Discussing the radon problem as a whole, one has to distinguish between the *outdoor radon concentrations* and *indoor radon concentrations* (UNSCEAR, 1988).

Radon is known to enter the atmosphere mainly by crossing the soil-air interface. There are a number of other secondary sources, such as the ocean, groundwater, natural gas, geothermal fluids, and coal combustion, however, which have also to be taken into considerations. The *outdoor radon concentrations* at ground level are usually governed by the source term, i.e., the exhalation rate, and by atmospheric dilution processes. Both of these factors are affected by the local meteorological conditions, which are also to a large extent responsible for the degree of radioactive equilibrium between radon and its daughters.

When the radon parent radionuclide, ^{226}Ra, decays in soil or rock, the resulting atoms of ^{222}Rn must first escape from the soil or rock particles to air-filled pores. They can then move through these pores and may finally reach the atmosphere. The fraction of radon that escapes into the pores of soil is called the *emanating power* or *emanating coefficient.* Its actual value depends on the composition of the soil and many other factors. The emanating power found in various measurements may lie between 1% and 80%.

Of course, not all radon atoms which escape from the soil particles will enter the free atmosphere. Only a fraction of the radon produced can reach the surface of the soil by the process of diffusion or convection through the air- or water-filled pores of the material.

It is generally known that the molecular diffusion process represents a spontaneous movement of the component such that a uniform concentration

TABLE 1.13. Further Radiation Parameters of ^{220}Rn and ^{222}Rn and Their Decay Products. In Case of Beta Radiation, the Average Energy Is Given.

Radionuclide	Historical Name	Principal Radiations, Intensities, and Energies
^{220}Rn	Thoron	Alpha: 100%, 6.288 MeV
^{216}Po	Thorium A	Alpha: 100%, 6.779 MeV
^{212}Pb	Thorium B	Beta: 5.2%, 45 keV; 84.9%, 94.4 keV; 9.9%, 173 keV Gamma: 44.6%, 239 keV; 3.4%, 300 keV
^{212}Bi	Thorium C	Alpha: 36%, 6.051 MeV Beta: 3.4%, 190 keV; 2.7%, 229 keV; 7.9%, 531 keV; 48.4%, 832 keV Gamma: 11.8%, 727 keV; 2%, 785 keV; 2.8%, 1.62 MeV
^{212}Po	Thorium C′	Alpha: 100%, 8.785 MeV
^{208}Tl	Thorium C″	Beta: 3.1%, 340 keV; 22.8%, 439 keV; 22%, 532 keV; 50.9%, 647 keV Gamma: 6.8%, 277 keV; 21.6%, 511 keV; 85.8%, 583 keV; 12%, 860 keV; 99.8%, 2.61 MeV
^{208}Pb	Thorium D	Stable
^{222}Rn	Radon	Alpha: 100%, 5.490 MeV
^{218}Po	Radium A	Alpha: 100%, 6.003 MeV
^{214}Pb	Radium B	Beta: 48.1%, 207 keV; 42.1%, 227 keV; 6.3%, 1.34 MeV Gamma: 7.5%, 242 keV; 19.2%, 295 keV; 37.1%, 352 keV
^{214}Bi	Radium C	Beta: 5.5%, 352 keV; 8.3%, 491 keV; 17.6%, 525 keV; 17.9%, 539 keV; 7.5%, 684 keV; 17.7%, 1.27 MeV Gamma: 46.1%, 609 keV; 4.9%, 768 keV; 15%, 1.12 MeV; 5.9%, 1.24 MeV; 15.9%, 1.765 MeV; 5%, 2.2 MeV
^{214}Po	Radium C′	Alpha: 100%, 7.687 MeV
^{210}Pb	Radium D	Beta: 81%, 15 keV
^{210}Bi	Radium E	Beta: 100%, 1.161 MeV
^{210}Po	Radium F	Alpha: 100%, 5.297 MeV
^{206}Pb	Radium G	Stable

of that component will be established in all parts of the assumed enclosure. As a result of thermal agitation, molecules of any gas are constantly moving in all directions. The number of molecules moving in any given direction at a particular point is proportional to the number of molecules present per unit volume. This results in a net transfer of molecules toward the region of their lower concentration.

On the other hand, the movement of ^{222}Rn atoms caused by forced convection depends on pressure differences created by meteorological conditions, which vary with time and which are very difficult to predict. The diffusion process usually dominates over convection as the mechanism by which radon enters the atmosphere from the surface of the earth.

Outdoor radon concentrations fluctuate depending on the place, time, height above the ground and actual meteorological situation. Since the only source of outdoor radon is the soil, and since radon has a rather short physical half-life, its concentration is usually constantly decreasing with height. The geographic location is also important; in such areas as islands and the Arctic, where there is less soil capable of emanating radon than over the continental temperate regions, outdoor radon concentrations are very low.

The outdoor radon concentrations are subject to both seasonal and diurnal changes. These variations are illustrated in Figures 1.8 and 1.9, respectively (Fisenne, 1982, 1984).

The typical average value of outdoor atmospheric radon concentrations for normal areas in the United States, after taking into account the above mentioned time fluctuations, was estimated to be about 9 $Bq \cdot m^{-3}$

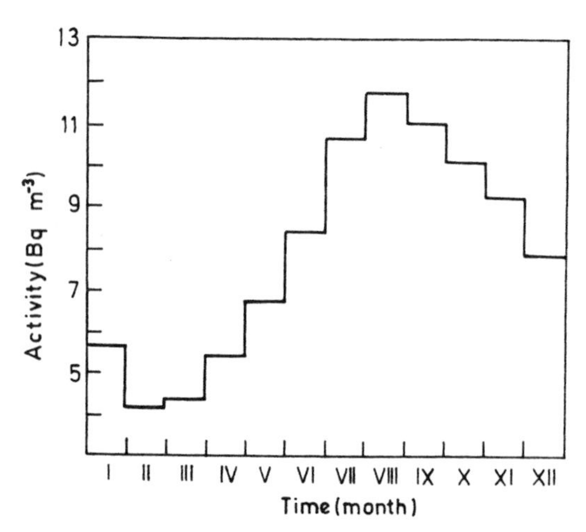

FIGURE 1.8. Time variations of outdoor radon concentrations measured at Chester, New Jersey (1977–1982 average values).

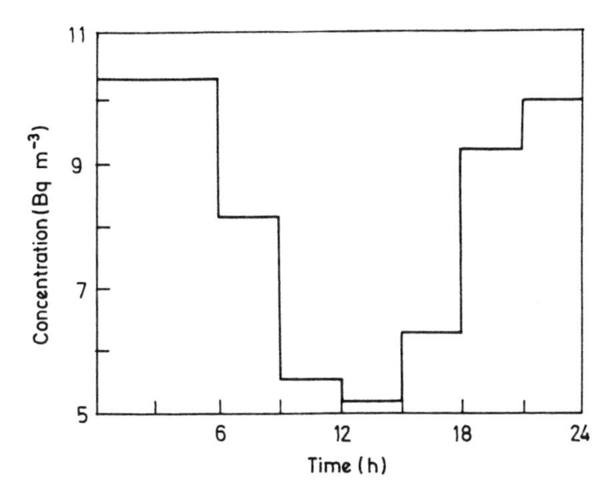

FIGURE 1.9. Diurnal changes of outdoor radon concentration at Chester, New Jersey (1977–1983 average of three-hour data).

(UNSCEAR, 1988). Different values of the equilibrium factor F have been obtained by measurements in various locations. A very low value of F has been observed in the mountainous regions of Taiwan, around 0.16 (Chen et al., 1992), while more common values for most places were about 0.8. For example, in Germany the equilibrium factor was found to be 0.77 (Jacobi, 1972) and in the USA about 0.87 (Cox et al., 1970). It seems that 0.8 for this factor is a representative value at the height of 1 m above ground.

For assessment of radiological hazards due to inhalation of radon and its decay products, the *indoor radon concentrations* are more important than the outdoor concentrations. Indoor radon comes from different sources, such as the soil or rock under the building, building materials, water supplies, natural gas, and outdoor air. One of the most important sources of indoor radon is the underlying soil, from which radon can be transported into a building either via diffusion or via the pressure-driven flow of air through the structural elements or through the openings in the structural elements. The actual indoor radon concentrations depend on two dominant factors: the entry or production rate from various sources, and the ventilation rate.

As has already been mentioned, exposure to radon presents the most significant element of human irradiation by natural radiation sources. The subsequent biological effects are due almost entirely to the inhalation of the short-lived radon decay products existing in indoor and outdoor air.

The current estimations of the concentrations of both ^{222}Rn and ^{220}Rn in the indoor and outdoor atmosphere are summarized in Table 1.14 (UNSCEAR, 1993). In this table the radon (gas) as well as its equilibrium

TABLE 1.14. Average World-Wide Concentrations of ^{222}Rn, ^{220}Rn and Their Decay Products, Together with Dose Coefficients and Resulting Annual Effective Doses.

		Concentration ($Bq \cdot m^{-3}$)		Effective Dose Coefficient (nSv per $Bq \cdot h \cdot m^{-3}$)		Annual Eff. Dose[a] (μSv)	
Radionuclide	Location	Gas	EEC[b]	Gas	EEC	Gas	Decay Prod.
^{222}Rn (radon)	Outdoors	10	8	0.17	9	3	130
	Indoors	40	16	0.17	9	48	1000
Total (rounded)							1200
^{220}Rn (thoron)	Outdoors	10	0.1	0.11	10	1.9	1.8
	Indoors	3	0.3	0.11	32	2.3	67
Total (rounded)							73

[a]Weighted for occupancy: 0.2 outdoors, 0.8 indoors.

[b]The values of the equilibrium factor have been taken to be 0.8 outdoors and 0.4 indoors for ^{232}Rn. Thoron EEC values are based on measurements.

Based on UNSCEAR (1993).

equivalent concentrations (EEC) are given together with the relevant dose conversion coefficients and resulting annual effective doses. The conventional indoor and outdoor occupancy factors of 0.8 and 0.2 are incorporated in the dose assessment. On the whole, the effective dose is an order of magnitude greater from ^{222}Rn and its decay products than the effective dose from ^{220}Rn and its products. From the data shown one can also see that the effective dose due to decay products is by far more significant than the contribution from the relevant gas component of these radionuclides.

1.1.2.3 COSMOGENIC RADIONUCLIDES

Cosmic rays generate a range of stable nuclides and radionuclides in the atmosphere, biosphere, and lithosphere by a variety of nuclear reactions. In these processes a dominant role is played by high-energy *primary cosmic rays,* although even the secondary particles released in their interactions are still very effective in the creation of *cosmogenic nuclides.*

It is well known that the primary cosmic rays are composed of electrons and the nuclei of light elements, principally hydrogen and also helium. The energy typical of these nuclei lies in the range of 1–10 GeV. The cascades of secondary particles are generated by the interaction of primary cosmic rays with air nuclei at the top of the terrestrial atmosphere and are found all the way to the ground and below. The primary cosmic ray fluence rate striking the top layers of the earth's atmosphere is about 0.02 $cm^{-2} \cdot s^{-1}$ at the equator and about 1 $cm^{-2} \cdot s^{-1}$ at the poles. The latitude effect is due to the screening of charged cosmic particles by the earth's geomagnetic field.

Some of generally important cosmogenic radionuclides with half-lives exceeding two weeks are listed in Table 1.15. A considerable portion of these radionuclides are produced by the reactions of cosmic rays with such atmospheric constituents as N, O, Ar, Kr, and Xe. The rest are formed by interactions with other elements in the biosphere and lithosphere (mainly O, Mg, Si, Fe, Al, Ca, K).

Cosmogenic radionuclides have received increasing emphasis and importance since higher-sensitivity techniques and methods of low-level counting have become available. The main applications of some atmospheric cosmogenic radionuclides are shown in Table 1.16 (Lal and Peters, 1967; Reedy et al., 1983).

Following its discovery by W. F. Libby in 1947, the cosmogenic ^{14}C has been used not only for archeological dating, but also for studying a wide variety of geophysical problems such as air-sea exchange, biological cycles, and large-scale ocean circulation. The cosmogenic ^{14}C is primarily produced in the earth's atmosphere by the capture of thermal neutrons with ^{14}N nuclei, i.e.,

$$n + {}^{14}N \rightarrow {}^{14}C + p + 0.625 \text{ MeV} \tag{1.34}$$

TABLE 1.15. Important Cosmogenic Radionuclides with Their Half-Lives and Production Target Elements.

Radionuclide	Half-Life	Atmospheric Targets	Other Targets
^{3}H	12.3 y	N, O	O, Mg, Si, Fe
^{7}Be	53 d	N, O	O, Mg, Si, Fe
^{10}Be	1.6×10^{6} y	N, O	O, Mg, Si, Fe
^{14}C	5730 y	N	O, Mg, Si, Fe
^{22}Na	2.6 y	Ar	Mg, Al, Si, Fe
^{26}Al	7.1×10^{5} y	Ar	Si, Al, Fe
^{32}S	87 d	Ar	Fe, Ca, K, Cl
^{36}Cl	3.0×10^{5} y	Ar	Fe, Ca, K, Cl
^{37}Ar	35 d	Ar	Fe, Ca, K
^{39}Ar	269 d	Ar	Fe, Ca, K
^{61}Kr	2.1×10^{5} y	Kr	Rb, Sr, Zr
^{129}I	1.6×10^{7} y	Xe	Te, Ba, La, Ce
^{40}K	1.3×10^{9} y		Fe
^{41}Ca	1.0×10^{5} y		Ca, Fe
^{46}Sc	84 d		Fe
^{48}V	16 d		Fe, Ti
^{53}Mn	3.7×10^{6} y		Fe
^{54}Mn	312 d		Fe
^{55}Fe	2.7 y		Fe
^{56}Co	79 d		Fe
^{59}Ni	7.6×10^{4} y		Ni, Fe
^{60}Fe	1.5×10^{6} y		Ni
^{60}Co	5.27 y		Co, Ni

TABLE 1.16. Principal Atmospheric Cosmogenic Radionuclides and Their Applications (Radionuclides Are Grouped in Order of Increasing Half-Lives).

Half-Life Group	Radionuclides	Application Fields
0.5 h–2.6 y	^{34m}Cl, ^{38}Cl, ^{39}Cl, ^{18}F, ^{31}Si, ^{38}S, ^{24}Na, ^{28}Mg	Cloud physics
days–years	^{32}P, ^{33}P, ^{7}Be, ^{35}S, ^{22}Na	Atmospheric structure, large-scale air circulation, and precipitation scavenging
>10 y	^{14}C	Archeology and paleobotany
	^{3}H, ^{32}Si, ^{39}Ar, ^{14}C, ^{36}Cl, ^{10}Be, ^{129}I	Air-sea exchange, geochemical and biological cycles, paleomagnetic reversal records and cosmic ray prehistory
	^{32}Si, ^{39}Ar, ^{14}C, ^{81}Kr, ^{36}Cl, ^{10}B	Hydrology and glaciology: chronology of groundwaters, lacustrine sediments, and glaciers

It has been found that most of the ^{14}C production takes place in the stratosphere. Radiocarbon then oxidizes to form $^{14}CO_2$, and this is taken up photosynthetically by plants, which in turn are eaten by animals. Radiocarbon dating is based on the known concentration of ^{14}C in the atmospheric CO_2: there are about 13.5 disintegrations per minute per gram of carbon (≈ 0.23 $Bq \cdot g^{-1}$).

Further applications of radiocarbon dating (see also Section 3.8) were possible owing to accelerator spectrometry introduced in the late 1970s. This novel technique allows for identification and counting of individual atoms. It is a very powerful method which is about 10^4–10^6 times more sensitive than earlier techniques relying on detection of the emitted radiation.

The estimated global average production rate of principal cosmogenic radionuclides including their volume concentration in air and their global inventory is listed in Table 1.17. Due to the decay and settling processes (the radionuclides with shorter half-lives usually decay before settling to the earth) there are considerable variations in the concentrations of some cosmogenic radionuclides with altitude.

Although more than twenty different cosmogenic radionuclides have been identified, only four of them may be considered as far as human exposure is concerned: ^3H, ^7Be, ^{14}C, and ^{22}Na. Table 1.18 (UNSCEAR, 1993) provides a summary of the annual intakes and the committed effective doses for adults, applying standard dose per unit intake coefficients.

The greatest contribution to human doses due to cosmogenic radionuclides comes from ^{14}C. Carbon dietary intake by adults is about 95 kg per

TABLE 1.17. Estimated Production Rate, Global Inventory and Air Activity of Some Atmospheric Cosmogenic Radionuclides.

Radionuclide	Production Rate of Atoms ($cm^{-2} \cdot s^{-1}$)	Global Inventory (kg)	Air Activity ($Bq \cdot m^{-3}$)
^3H	0.20–0.25	3.5	0.167
^7Be	0.08	0.032	0.017
^{10}Be	0.045–0.05	$(2.6–3.9) \times 10^5$	10^{-7}
^{14}C	2.5	$(6.8–7.7) \times 10^4$	0.0167
^{22}Na	8.6×10^{-5}	0.0019	1.7×10^{-7}
^{32}Si	1.6×10^{-4}	1.4	3.3×10^{-8}
^{32}P	8.1×10^{-4}	0.0004	3.3×10^{-4}
^{33}P	$(5.8–6.8) \times 10^{-4}$	0.0006	2.5×10^{-4}
^{35}S	1.4×10^{-3}	0.0045	2.5×10^{-4}
^{36}Cl	1.1×10^{-3}	1.45×10^4	5×10^{-10}
^{39}Ar	5.6×10^{-3}	23	
^{81}Kr	10^{-6}	16.2	

Based on data from Kathren (1984) and Lal and Peters (1967).

TABLE 1.18. Annual Intake by Cosmogenic Radionuclides and Annual Effective Doses to Adults.

Radionuclide	Intake ($Bq \cdot a^{-1}$)	Annual Effective Dose (μSv)
^{3}H	500	0.01
^{7}Be	1000	0.03
^{14}C	2000	12
^{22}Na	50	0.15

Based on UNSCEAR (1993).

year, which is equivalent to about 20 $kBq \cdot y^{-1}$ of ^{14}C on average entering the body. This annual intake corresponds to the committed effective dose of about 12 μSv.

The contribution from ^{22}Na is much smaller (only 0.15 μSv) and the remaining two radionuclides are even less significant as far as exposure from natural sources of radiation is concerned.

1.1.3 Radionuclides Released into the Environment

1.1.3.1 TRANSPORT OF RADIONUCLIDES

The utilization of nuclear energy for the generation of electricity, the applications of radionuclides in industry, technology, science, medicine, and consumer products as well as the tests of nuclear weapons result in the release of various radioactive materials into the environment. Assessment of the actual or potential radiological consequences of such releases of radionuclides into air and water, or their disposal in the ground, is a complex procedure. The major steps in the assessment of the impacts of radioactive material released in the environment are shown in Figure 1.10 (NCRP, 1984a) where the arrows indicate the links between individual compartments. The whole process begins at the source term (the release of radionuclides) and goes through the relevant pathways, which are related to the intake by individuals. The description of these pathways including the evaluation of the final consequences in terms of health effects is, in principle, possible by means of appropriate mathematical models. They have to take into account all the essential factors governing the behavior of radionuclides and their movement from one compartment to another. The mathematical models can be used to predict the transport, accumulation, and intake of radioactive materials released into the environment by persons.

The actual pathways are quite complex and not all details of the transport mechanisms are always fully understood. Usually, two general categories of environmental transport pathways are considered: (1) the pathway involving transport and entry of radionuclides into the foodstuffs through ground and surface waters and (2) the pathway between radionuclides introduced into the environment via the atmosphere. Generalized illustrations of these two ecological pathway types are shown in Figures 1.11 and 1.12.

Other pathways are also possible, but they are less important. These may include direct ingestion by animals or man and in some cases direct external exposures from the immersion of a person in a contaminated cloud.

The first serious studies of the transport of radioactive materials were initiated as a consequence of nuclear weapons explosions, during which huge amounts of various radionuclides were released into the environment. Later, primary attention was concentrated on radiation protection and ecological aspects of radioactive contamination caused by nuclear power

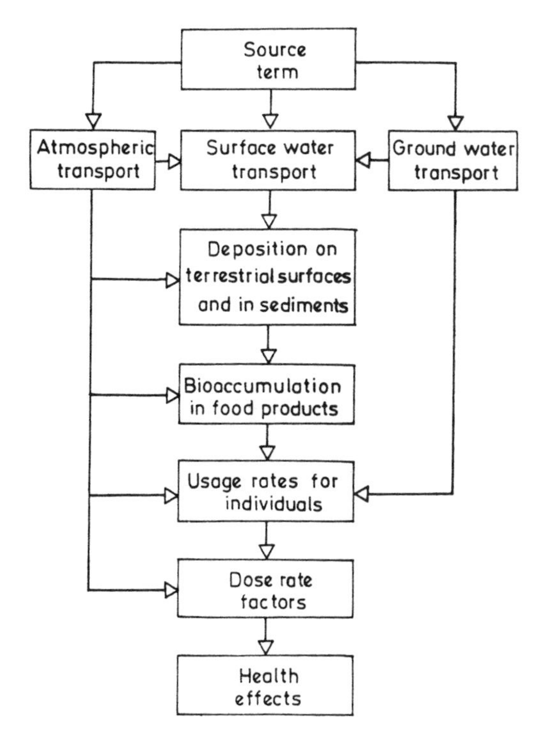

FIGURE 1.10. The basic pathways between the radioactive sources released into environment and resulting health hazards (with permission from the National Council on Radiation Protection and Measurements).

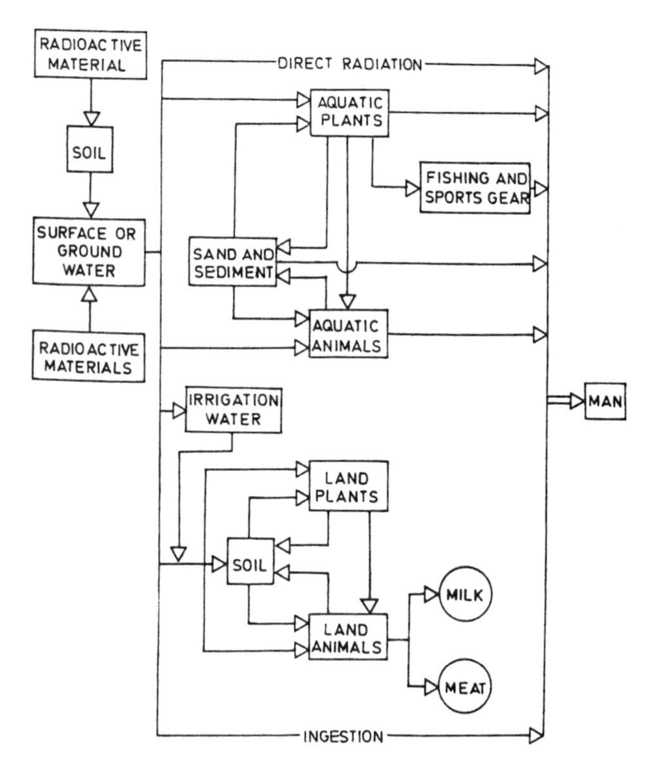

FIGURE 1.11. Pathways between radioactive materials released to surface and ground waters (directly or via soil) and man. Adapted from Kathren (1984).

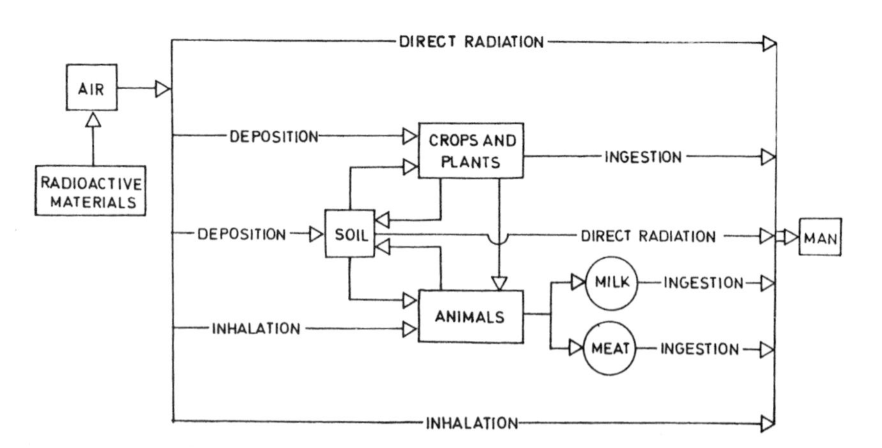

FIGURE 1.12. Pathways of radionuclides to man through air. Adapted from Kathren (1984).

plants. The problem of nuclear energy installations consists not only in released radioactivity, but also in storage and disposal of constantly produced radioactive wastes. During the last few years, the situation in this field has been affected by various reactor accidents and especially by the accident at the Chernobyl nuclear power plant. The Chernobyl catastrophe presented an opportunity to validate a number of existing hypotheses and models, but also revealed our lack of knowledge and lack of preparedness to deal with such accidents. Before Chernobyl, only two large-scale accidents were known: the first radioactive fallout over the Marshall Islands in 1954, and the accident at Kyshtym in the former USSR in 1957. The radioactive fallout occurred during the testing of the first American thermonuclear bomb on Bikini atoll which was heavily contaminated. The accident in Russia involved the explosion of a tank containing fission products, caused by the breakdown of a cooling system. The consequences: a contaminated area of 800 km^2 and the evacuation of more than 10,000 people from the affected region.

1.1.3.2 RADIOACTIVITY FROM NUCLEAR EXPLOSIONS

Most of what is now known about radioactivity from the explosions of nuclear weapons has been learned from studies and evaluations of the effects of test explosions that took place in various parts of the world since the first atomic bomb was detonated on a New Mexico desert in July 1945. The enormous destructive potential of nuclear explosions was for the first time revealed to the world public by the bombing of Hiroshima on 5 August, 1945 and Nagasaki on 9 August the same year (Eisenbud, 1987). In the following years, several hundreds of various nuclear weapons were tested, mainly by the USA and the USSR (first test in 1949) and also by some other countries such as Great Britain (1952), France (1960), the People's Republic of China (1964), and India (1974). Most nuclear tests have been conducted in various remote regions in the northern hemisphere.

The majority of nuclear tests in the atmosphere occurred before 1963. After 1963 the nuclear explosions were mainly carried out underground. In comparison with the atmospheric tests, the environmental consequences of the underground explosions are relatively small.

In large atmospheric nuclear explosions, most of the radioactive material is carried into the stratosphere, where radionuclides remain for some time and are later dispersed and deposited around the world. On average, the retention time may vary from less than a year to about five years, depending on the altitude and latitude. Smaller explosions carry the radioactive materials only to the troposphere, and fallout occurs within days or weeks (UNSCEAR, 1988).

TABLE 1.19. Numbers and Estimated Yields of Nuclear Explosions Carried Out in the Atmosphere Since 1945.

Years	Number of Tests	Estimated Yield in *Mt* of TNT	
		Fission	Total
1945–1951	26	0.8	0.8
1952–1954	31	37	60
1955–1956	44	14	31
1957–1958	128	40	81
1959–1960	3	0.1	0.1
1961–1962	128	102	340
1963	0	0.0	0.0
1964–1969	22	10.6	15.5
1970–1974	34	10.0	12.2
1975	0	0.0	0.0
1976–1980	7	2.9	4.8
1981–1993	0	0.0	0.0

The chronology of atmospheric nuclear explosions including their numbers and estimated fission as well as total yields in equivalent amounts of TNT is documented in Table 1.19 (UNSCEAR, 1988). Table 1.20 summarizes the yields of nuclear explosions that have been reported by the individual countries (UNSCEAR, 1982).

Each nuclear explosion produces a huge amount of different radionuclides. The actual activity and the type as well as the number of radionuclides released depend on the mode of production of the nuclear explosion. Two kinds of nuclear reactions can generate extremely large

TABLE 1.20. Estimated Yields of Nuclear Explosions Carried Out in the Atmosphere by Individual Countries.

Country	Period	Number of Tests	Estimated Yield in *Mt* of TNT	
			Fission	Total
USA	1945–1962	193	72	139
USSR	1949–1962	142	111	358
UK	1952–1953	21	11	17
France	1960–1974	45	11	12
China	1964–1980	22	13	21
Total	1945–1980	423	218	547

quantities of energy in a very short time interval: *fission reaction* and *fusion reaction*.

Fission reactions represent the splitting of a heavy nucleus into two lighter nuclei (or fission fragments) along with the release of about 200 MeV energy and the emission of two or three neutrons. To start the fission reaction, two things should be available: an initial source of neutrons, and a supercritical mass of a suitable fissionable material such as ^{233}U, ^{235}U, or ^{239}Pu. Only one of these nuclides, ^{235}U, can be found in nature (0.7% in natural uranium). The other two fissionable nuclides must be produced artificially—^{233}U from ^{232}Th and ^{239}Pu from ^{238}U (99.3% of natural uranium). Some of the physical parameters of these nuclides are listed in Table 1.21.

The fusion reaction is the reverse of fission: two light nuclei unite to form a heavier nucleus whose mass is slightly smaller than that of two light nuclei. The mass difference is converted into energy and released. The fusion reaction can only occur, however, if two nuclei are close enough together that the attractive nuclear forces exceed the electrostatic repulsion. This requires that the light nuclei possess sufficient energy, which can be obtained with the aid of an extremely high temperature (several million degrees). Such heat is easily achieved in a fission device, which may serve as the trigger for a so-called *thermonuclear* explosion, based on the fusion reaction.

The total radioactivity produced by nuclear explosions utilizing the fis-

TABLE 1.21. Some Physical Characteristics of Nuclides ^{233}U, ^{235}U, ^{238}U, and ^{239}Pu.

Nuclide	$T_{1/2}$ (y)	Alpha Emission (MeV)	Gamma Emission (keV)
^{233}U	1.592×10^5	4.729 (1.6%)	42 (0.06%)
		4.783 (13.3%)	55 (0.014%)
		4.824 (84.4%)	97 (0.022%)
^{235}U	7.037×10^5	4.209 (5.7%)	109 (1.5%)
		4.352 (17.0%)	144 (10.2%)
		4.392 (54.0%)	163 (4.7%)
		4.555 (4.5%)	186 (54.0%)
		4.597 (5.4%)	202 (1%), 205 (4.7%)
^{238}U	4.468×10^9	4.039 (0.2%)	
		4.149 (23.0%)	49 (0.07%)
		4.196 (77.0%)	
^{239}Pu	2.411×10^4	5.105 (11.7%)	38 (0.01%)
		5.143 (15.1%)	52 (0.027%)
		5.155 (73.0%)	129 (0.006%)

Based on Granier and Gambini (1990).

sion reaction is much higher than radioactivity following the detonation of thermonuclear devices (Table 1.22). A fission explosion equivalent to one kiloton of TNT can release radioactive material with an overall activity of about 6×10^{23} Bq. During the fission explosion more than 200 different radionuclides are produced; their half-lives are in the range of a fraction of a second to several million years.

The contamination of the environment caused by a fission nuclear explosion comes from both fission fragments and their decay products as well as from activation products produced by neutrons after interacting with some elements in the air and the ground. The main activation products related to fission explosions are ^{3}H, ^{14}C, ^{54}Mn, and ^{55}Fe.

In addition to the above mentioned radionuclides, fallout following a nuclear explosion always contains some amounts of transuranic radioactive elements which are characterized by long half-lives and very high toxicity.

As far as radiological hazards are concerned, the most important fallout constituents are ^{90}Sr and ^{137}Cs because of their long half-lives (twenty-eight years and thirty years, respectively) and relatively high uptake by biosystems. Strontium is chemically very similar to calcium, which is used for bone formation. This is why ^{90}Sr concentrates in bone. While radioactive strontium simulates calcium, the behavior of cesium is close to that of potassium, one of the most important biogenic elements.

The activity of released radionuclides is not distributed uniformly in space or time after an explosion. This can be illustrated by the strontium ground surface contamination as a function of latitude and time, Figures 1.13 and 1.14. The maxima shown in Figure 1.13 are due to the dynamics of stratospheric-tropospheric mixing (Budnitz et al., 1983). About half of ^{90}Sr falling out on land is found in the first 4 cm depth of soil. The time distribution of ^{90}Sr in New York City (Figure 1.14) shows the very obvious effects of the large-scale nuclear tests during the 1960s.

TABLE 1.22. Estimated Activity of the Principal Radionuclides Produced per Mt *of a Fission Explosion.*

Radionuclide	Half-Life	Activity (Bq)
^{89}Sr	50.5 d	7.4×10^{17}
^{90}Sr	28.8 y	3.7×10^{15}
^{95}Zr	65 d	9.25×10^{17}
^{103}Ru	40 d	6.85×10^{17}
^{106}Ru	1 y	1.07×10^{16}
^{131}I	8 d	4.63×10^{18}
^{137}Cs	30.2 y	5.92×10^{15}
^{144}Ce	284 d	1.37×10^{17}

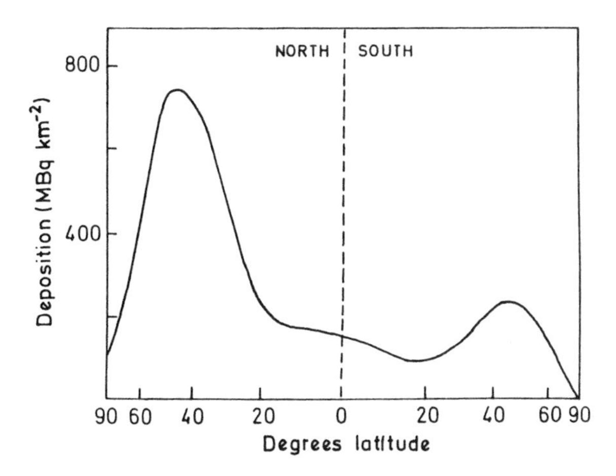

FIGURE 1.13. Latitudinal distribution of ^{90}Sr in late 1959. Adapted from UNSCEAR (1982).

Much of the attention paid to ^{90}Sr since it was identified shortly after World War II was due to its recognition as the most toxic radionuclide in fallout. From available data the global inventory of ^{90}Sr can now be estimated to be about 3.5×10^{17} Bq on the earth's surface of the northern hemisphere and approximately 10^{17} Bq on the earth's surface of the southern hemisphere. It is now assumed that practically all the ^{90}Sr injected into the atmosphere during the period of intensive nuclear explosions in the 1960s was deposited on the earth's surface by 1970: The total activity of ^{90}Sr is

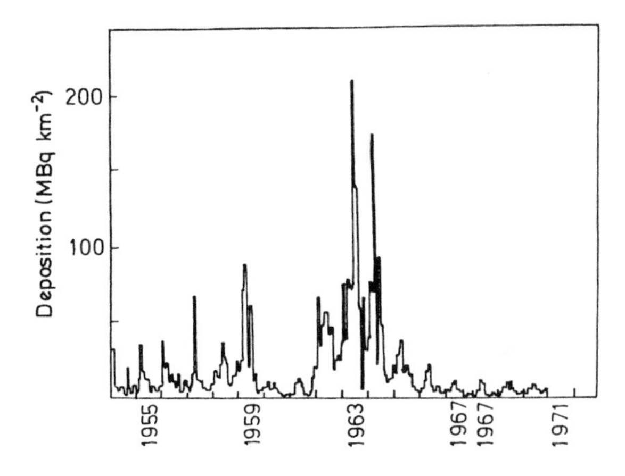

FIGURE 1.14. Time distribution of ^{90}Sr as measured in New York City. Adapted from Budnitz et al. (1983).

decreasing by its decay at a rate of 2.5% per year. The occasional tests carried out later by France and China have not essentially affected the present global inventory of this controversial radionuclide.

The biological hazards of radiostrontium depend on its concentration in food and on the food composition and consumption habits of the population. The situation may differ considerably from country to country and from one region to another. This can be illustrated by the ^{90}Sr intakes of people from New York City and San Francisco (Table 1.23). The significant difference between these two cities is believed to be due mainly to different annual rainfall in the regions that supply food to the New York and San Francisco areas.

The other strontium isotope, ^{89}Sr, with its relatively short half-life (50.5 days), can contribute to the committed effective dose for only a short period of time following a nuclear explosion.

Another widely studied and monitored radionuclide present in fallout is ^{137}Cs; it has a half-life of thirty years, which is comparable to that of ^{90}Sr (twenty-eight years). Although ^{137}Cs is tightly bound by soil and thus is not readily incorporated metabolically into vegetation, it can contaminate foodstuff through foliar absorption (Eisenbud, 1987). In 1968, yearly intake of ^{137}Cs through the consumption of various foods typical for the Chicago area was about 455 Bq.

In the case of ^{137}Cs, one has to consider both of its contributions to the effective dose: (1) the component from ingestion and inhalation and (2) the component from the external gamma exposure. According to UNSCEAR data (1982), the average committed effective dose equivalent from ingesting ^{137}Cs in the northern hemisphere can be estimated to be about 170 μSv. At present, the external radiation doses due to radionuclides from fallout are caused almost exclusively by ^{137}Cs.

Long-lived plutonium isotopes, ^{239}Pu and ^{240}Pu, are also an important part of radioactive fallout. They originate from the volatilization of both unfissioned plutonium and plutonium produced through neutron capture by ^{239}Pu. Since these radionuclides cannot be distinguished by alpha spectrometry, usually they are referred to together as $^{239/240}$Pu or ^{239,240}Pu. The global deposition of $^{239/240}$Pu was assessed to be about 12×10^{15} Bq, of which approximately 75% was deposited in the northern hemisphere (Kathren, 1984). Most of the plutonium injected into the atmosphere by nuclear explosions has already fallen out on the earth's surface. The cumulative depositions on soil in such areas as New York were reported to be in the range of 100 MBq·km^{-2}. The radiological impact of plutonium is caused mainly by its concentration in the bone (about 50% of the total body burden), liver ($\sim$40%), and also in the lymph nodes.

TABLE 1.23. Typical Concentrations in Diet and Intakes of ^{90}Sr in Two American Cities.

Diet Category	Consumption (kg/y)	New York City Concentration (mBq/kg)	New York City Intake (Bq/y)	San Francisco Concentration (mBq/kg)	San Francisco Intake (Bq/y)
Dairy products	200	118	23.7	37	7.4
Fresh vegetables	48	326	15.6	89	4.3
Canned vegetables	22	200	4.4	107	2.4
Root vegetables	10	125	1.3	111	1.4
Potatoes	38	85	3.3	78	2.9
Dry beans	3	588	1.8	292	2.0
Fresh fruit	59	96	5.6	48	2.8
Canned fruit	11	41	0.4	30	0.3
Fruit juice	28	63	1.8	52	1.5
Bakery products	44	111	4.8	70	3.1
Flour	34	167	5.7	130	4.4
Grain products	11	222	2.6	107	1.2
Macaroni	3	89	0.3	85	0.3
Rice	3	22	0.1	30	0.1
Meat	79	15	1.3	15	1.1
Poultry	20	11	0.2	11	0.2
Eggs	15	22	0.4	22	0.3
Fresh fish	8	7	0.04	4	0.04
Shell fish	1	7	<0.04	26	0.04
Yearly intake			73.2 Bq		35.8 Bq

Adapted from Eisenbud (1987).

A number of radionuclides are produced by neutron activation of elements present in the explosive device, as well as in elements of the atmosphere and the ground. The most significant of the activation products are ^{3}H and ^{14}C, both of which are also important cosmogenic radionuclides. Their relatively large amounts produced by weapons tests has changed the long-standing balance of these radionuclides in our environment. In the case of ^{14}C this can be demonstrated by Figure 1.15, where the excess concentrations of ^{14}C in the troposphere and on the surface of the oceans (UNSCEAR, 1977) are shown. The distortion of the natural level of ^{14}C by nuclear explosions, however, is partly compensated by increased fossil fuel burning.

The radiation doses from nuclear explosions are delivered over a long period of time at changing dose rates, which depend on such factors as the time of detonation and the nature of the radioactive fallout. Accurate assessment of the total annual doses from fallout is not possible, since we do not have all the information needed. The dose commitment, or so-called lifetime dose, is estimated by the use of a series of approximations and simplistic models that are subject to many uncertainties. The mean dose equivalent commitments, as estimated for the United States, are presented in Table 1.24 (NCRP, 1987). With the exception of ^{14}C, most of the doses shown have already been received and only a few percent of these doses are yet to be received. The total dose commitment from ^{14}C will reach about 1.4 mSv and will be delivered at a decreasing rate over many generations (as a matter of fact, over the next few thousand years). The resulting total annual effective dose based on these estimations will not be higher than about 0.01 mSv now. Comparing this number with the world average effective dose from all natural sources of radiation, 2.4 mSv, one may easily conclude that radio-

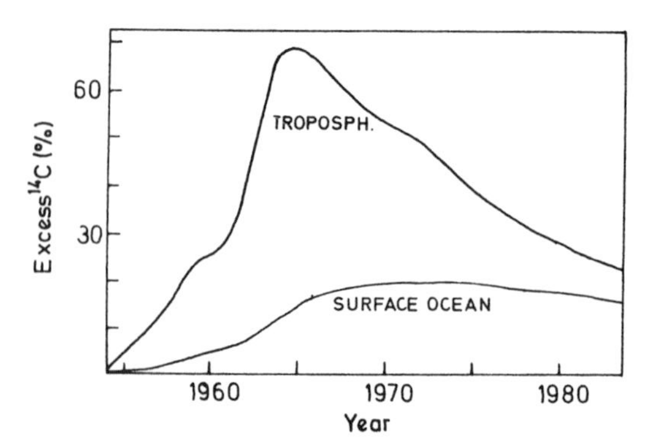

FIGURE 1.15. Excess atmospheric ^{14}C produced by nuclear explosions. Adapted from UNSCEAR (1977) and Kathren (1984).

TABLE 1.24. Mean Dose Equivalent Commitments to the Year 2000 in the USA due to the Nuclear Weapons Testing through 1970.

Source of Radiation	Dose Equivalent Commitment (μSv)
External	
Whole body	750
Internal	
^{90}Sr	
Marrow	450
Endosteal	650
^{137}Cs, whole body	150
$^{239,\,240}Pu$	
Lung	400
Bone	20
^{3}H, whole body	20
^{14}C, whole body	20

Published with permission of the National Council on Radiation Protection and Measurements, 1987.

active fallout from the detonated nuclear explosions is no longer a significant source of population exposure.

As far as the collective effective dose equivalent commitment due to all atmospheric nuclear tests is concerned, it was estimated in the UNSCEAR 1982 Report (1982) that this dose is about 3×10^7 man·Sv, an estimate that is still valid.

1.1.3.3 NUCLEAR POWER PRODUCTION

It is now slightly more than fifty years since Enrico Fermi succeeded in controlling the first chain reaction of nuclear fission in Chicago on December 2, 1942. Nuclear power production is still based completely on the use of nuclear fission as the heat source for the generation of electricity.

At the end of 1992, the 419 reactors operating in thirty countries had an installed capacity of 330 GW. This means that about 16% of the world's electricity is generated by nuclear reactors. Projections to the year 2000, although they may now look somewhat speculative, amount to around 450 GW.

The installed nuclear electrical energy capacity in the years 1979–1992 is illustrated in Figure 1.16.

The complete list of all countries producing electricity by nuclear power is shown in Table 1.25, where the latest data regarding the reactors in opera-

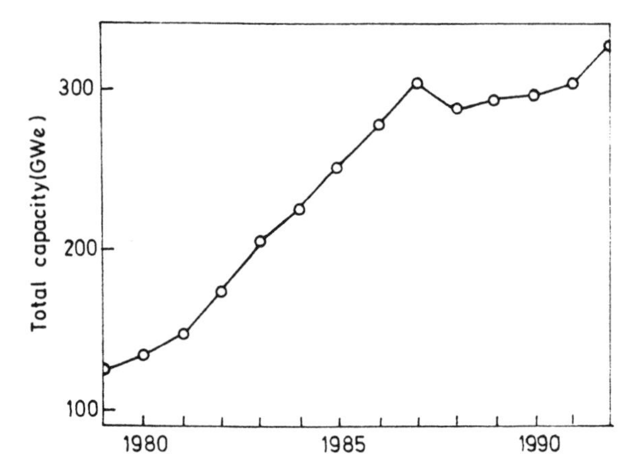

FIGURE 1.16. The installed nuclear electrical energy capacity in the years 1979–1992.

tion, reactors under construction, and nuclear electricity generation are summarized.

From this table we can see that the highest proportion of nuclear electricity in 1992 was in Lithuania, whose 78.2% put France into the second place. Splitting the former USSR figures in 1992 shows how nuclear power (8% of all electricity produced before its collapse) was unevenly concentrated. Apart from Lithuania, only Ukraine with about 29.4% and Russia with nearly 12% now possess nuclear capacity of any significant size. The USA, although its nuclear power plants cover only about one-fifth of the country's total electricity production, accounts for nearly one-third of the world electricity generation by nuclear power.

Among all types of power reactors, pressurized water-moderated and -cooled reactors (PWR) have a dominant position. The percentage of individual nuclear reactors in operation, under construction, and planned is shown in Figure 1.17.

Nuclear power production, by virtue of its nature, is inevitably associated with the release of radioactive materials into the environment. The reactor itself is only one part of a complex *nuclear fuel cycle,* which includes several steps:

(1) Uranium mining and uranium milling
(2) Enrichment of the isotopic content of ^{235}U for some types of reactors
(3) Fabrication of fuel elements
(4) Production of energy in the reactors
(5) Reprocessing of spent fuel

TABLE 1.25. *Electricity Generated by Nuclear Power in 1992.*

	Reactors in Operation		Reactors under Construction		Nuclear Electricity Generation	
Country	Units	MWe	Units	MWe	TWh	%
Argentina	2	1,005	1	745	7.08	14.8
Belgium	7	5,749	0	–	40.09	59.9
Brazil	1	657	1	1,309	1.75	0.8
Bulgaria	6	3,760	0	–	11.5	32.5
Canada	20	14,523	2	1,870	81.78	16.4
China	0	–	3	2,220	–	–
Czech Rep.	4	1,760	2	2,028	12.25	20.7
Finland	4	2,400	0	–	18.2	28.9
France	56	58,257	6	8,655	321.7	72.8
Germany	21	22,416	0	–	158.8	34
Hungary	4	1,760	0	–	13.98	44.6
India	8	1,500	8	2,410	6.33	2.1
Japan	41	14,771	12	12,124	214	34.7
Korea	9	7,616	5	4,700	56.53	43.2
Lithuania	2	2,500	0	–	14.64	78.2
Mexico	1	675	1	675	3.92	3.2
Netherlands	2	539	0	–	3.21	5.4
Pakistan	1	137	1	310	0.55	1.2
Russia	28	20,242	3	3,000	119.6	11.8
Slovakia	4	1,728	4	1,760	11.05	49.5
Slovenia &						20.9
Croatia	1	632	0	–	3.77	17.4
South Africa	2	1,930	0	–	9.29	6.2
Spain	9	7,400	0	–	55.73	35.5
Sweden	12	10,145	0	–	61.0	43.3
Switzerland	5	3,065	0	–	22.23	38.7
Taiwan	6	5,144	0	–	32.5	25.7
Ukraine	14	12,808	3	3,000	73.75	29.4
UK	37	16,609	1	1,258	48.44	18.1
USA	112	106,391	6	7,372	606.3	21.7
Total	419	326,119	59	53,436	2,009.97	

Adapted from Howles (1993).

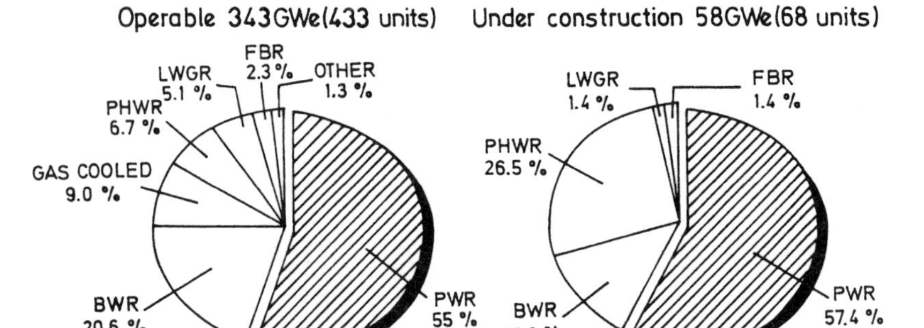

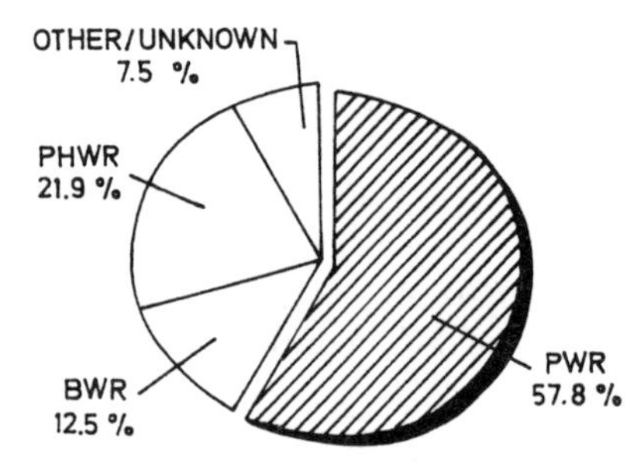

FIGURE 1.17. Reactors operable, under construction and planned as of 31 August, 1992 (PWR—pressurized water moderated and cooled reactor, BWR—boiling water moderated and cooled reactor, PHWR—pressurized heavy water moderated and cooled reactor, LWGR—light water cooled graphite moderated reactor, FBR—fast breeder reactor).

(6) Transportation of nuclear materials between fuel cycle installations
(7) Storage and disposal of radioactive wastes

Each of these steps is responsible for a certain amount of released radioactivity, although in the case of transportation, under normal conditions (assuming no accidents), the leakage of radioactive materials into the environment is insignificant. The problems related to the radioactive wastes and nuclear reactor accidents will be treated separately later.

Under normal circumstances, in the routine operation of nuclear installations, small quantities of various radionuclides are emitted in effluents, which disperse in the environment and result in low-level exposures to the population. Only in the unlikely event of certain types of accidents can large amounts of radioactive materials be released.

Almost all the *artificial radionuclides* associated with nuclear energy production are present in the irradiated nuclear fuel; only a small portion of them are produced by neutron activation of structural and cladding materials. The release of some *natural radionuclides,* namely uranium and its decay products, is due mainly to the mining and milling of uranium ores. This increase in the concentration of natural radionuclides falls into the category of so-called *technologically enhanced radioactivity.* It is widely recognized that some of man's specific industrial activities, particularly those directed toward energy production, can significantly change the level of naturally occurring concentrations of certain radionuclides to which people are exposed.

Uranium mining operations involve the removal of uranium ore principally from underground or surface (open pit) mines, although solution mining is also widespread. The extracted ore contains uranium and its decay products at concentrations between a tenth to a few percent U_3O_8. Exploitable deposits have on average a concentration of 0.1–0.5%, i.e., about 1–5 kg of natural uranium per metric ton. Such concentrations are several thousand times higher than the concentrations of these radionuclides present in the normal natural terrestrial environment.

Large deposits of uranium are found in central Africa and around the gold-mining areas of South Africa, in Canada's Great Bear Lake region, and also in Australia. Ores with lower concentrations of uranium have been mined extensively on the Colorado Plateau in the USA since the 1940s. Some deposits of low grade ores can also be found in west-central France, in the western mountains of the Czech Republic and at many other locations.

The U.S. uranium mining industry went through its boom years and peaked between 1960 and 1980, with a minor boom appearing in 1968, as can be seen from Figure 1.18 where the average grade of uranium ore mined is also shown. The production of uranium in the USA in 1985 was approximately the same as that in 1953.

The primary radiological concern of *uranium mining* is related to the exposure of miners. In addition to external doses from gamma emitters (radionuclides of the uranium and thorium series), inhalation of radon and especially its decay products have particular effects on miners.

Radiation and mining have been closely associated for over 400 years. Georgius Agricola, in his book written in 1556 recognized that the miners of Schneeberg (Germany) and Jáchymov (Joachimsthal), now in the Czech Republic, had an increased risk of death from lung diseases, called Bergkrankheit or mountain sickness. He also noticed that the disease was caused by something in the air. This disease was first identified as malignant in 1879 and finally recognized as lung cancer in 1921 (Beckman, 1988). The possibility that the lung cancer of miners might have been caused by radon was

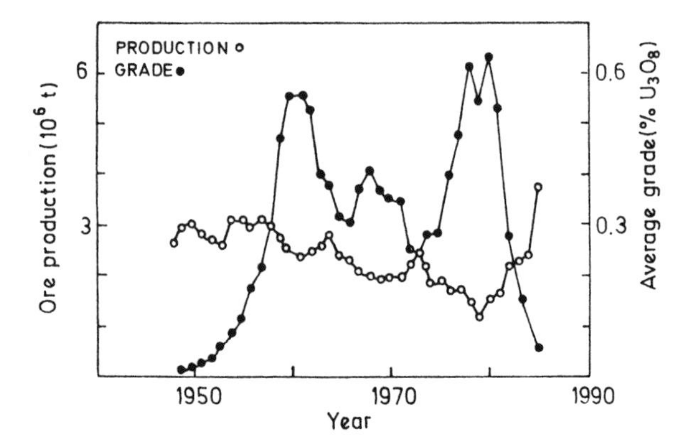

FIGURE 1.18. Underground uranium ore production and ore grade in the USA mines. Adapted from Beckman (1988).

suggested in the early 1920s. In 1951 W. Bale of the University of Rochester pointed out that the major portion of the radiation dose to the lungs came rather from radon decay products and not from radon. This is why in the mine environment radon daughter monitoring is more important than the measurement of radon concentrations. The results of radon daughter measurements are usually given in terms of *working levels* (WL). Since we already know that one working level corresponds to 1.3×10^5 MeV of potential alpha energy per liter of air, this potential energy concentration is equivalent to that of 37 Bq·m^{-3} of radon in equilibrium with its four short-lived products. The exposures of miners to radon decay products are measured in *working level hours* (WLH) or *working level months* (WLM). A working level hour represents an exposure to 1 WL for one hour, or 0.5 WL for two hours, etc. A working level month is the product of the number of months worked and the working level.

A working level month actually expresses the cumulative exposure received during the working hours in one month. Consequently, an exposure of 1 WLM may be defined as the inhalation of a radon daughter concentration of one WL for 170 (some regulatory agencies cite 173) hours.

Before intensive ventilation and other appropriate measures had been introduced, radon concentrations in uranium mines were very high; in the Schneeberg and Jáchymov mines concentrations of around several hundred thousand were no exception. A similar situation also observed in the U.S. mines, although in some mines in Utah and Colorado radon concentrations showed levels as high as 2 MBq·m^{-3}. In the late 1960s, both in the USA and the other countries, strict measures were implemented in order to limit the exposure to miners to approximately 4 WLM per year, corresponding to the concentration of radon decay products of about 0.3 WL. The limit of 4 WLM may now be in contradiction with the radiation protection philosophy of the International Commission on Radiological Protection and its latest recommendations (ICRP, 1991). Actual exposures of underground miners are now estimated to average about 1 to 2 WLM per year (NCRP, 1984b; UNSCEAR, 1988). These exposures correspond to about 10 to 20 mSv of annual effective dose. The exposures of the surface miners are assumed to be lower, possibly around 5 mSv.

The mine effluents include the following basic radioactive wastes: liquids, solids (both particulates and tailings), and gases. The liquid wastes are mainly water from mine drainage and also water used in drilling operations. The waste rock and remaining low-grade ore are considered solid waste. Most of the radon is removed via the ventilation system to the outside atmosphere, but part of it can contaminate underground and surface water, which may affect nearby sources of drinking water.

After uranium ore has been mined, several milling operations follow, including crushing, leaching, extraction, and conversion processes, at the end of which is the uranium in a form suitable for fabrication of fuel elements. The mill wastes produced (also called tailings) always contain such radionuclides as ^{230}Th and ^{226}Ra as well as their decay products, notably radon and its daughters. Airborne releases from the uranium mills contain both particulates and gases. The liquid mill effluents include various waste solutions from the leaching, grinding, extraction, and washing operations.

At a typical mill, the tailings contain more than 99% of the original uranium ore (Nelson, 1988). The solid tailings contain practically all the ^{230}Th and ^{226}Ra originally present in the ore. These tailings are discharged as a slurry that, on average, is about equal parts by weight, solid and liquid. The slurry is then discharged to an impoundment where the tailings settle and the water portion of the liquid evaporates. The radon exhalation from the liquid-covered tailings is minimal and is much less from wet uncovered tailings than from dried tailings.

The amount of radon released into the atmosphere from an ore stockpile or a tailings impoundment depends heavily on such factors as the physical configuration of the source, ore grade, emanating fraction (the portion of radon that escapes the ore or tailings particles), porosity, moisture content, temperature and barometric pressure. These parameters can vary considerably within the same source as well as between similar sources at different locations.

At present, tailings are usually kept in open, uncontained piles or behind engineered dams or dikes with solid or water cover. The UNSCEAR (1988) assumes, however, that some further engineering will be carried out to minimize the release of radionuclides from the abandoned piles. One of the possible solutions, as suggested by Kathren (1984), to the problem of how to control releases from tailings would be to return the tailings to the mine shaft for disposal.

The uranium ore concentrate produced at the mills is further processed and purified and converted to uranium tetrafluoride (UF_4), if it is to be enriched in the isotope ^{235}U before being transformed into uranium oxide or metal and finally fabricated into fuel elements. Releases of radionuclides into the environment from the conversion, enrichment, and *fuel fabrication* processes are generally very small. The individual doses in the vicinity of fuel fabrication facilities, however, may reach several tens of mSv per year for members of the public (UNSCEAR, 1988). The main contribution to this exposure arises from the inhalation of the isotopes of uranium.

The most important feature of the whole nuclear fuel cycle is the *reactor,* which contains huge amounts of various radionuclides. Almost all the electric energy generated now by nuclear power plants is produced in thermal reactors, the majority of which are either boiling water reactors (BWR) or pressurized water reactors (PWR).

During the operation of a nuclear reactor, in addition to radioactive fission products formed within the fuel, induced radioactivity is produced in structural and cladding materials as well as in the coolant circuit. The actual amount of radioactive materials discharged from reactors depends on the reactor type, its design and the system used for the treatment of escaping radionuclides. The most important radionuclides released into the atmosphere are fission noble gases (krypton and xenon), activation gases (^{14}C, ^{14}N, ^{35}S, ^{41}Ar), tritium, iodine and particulates. Radionuclides discharged into the aquatic environment usually include tritium, fission products, and activated corrosion products.

The characteristics of individual types and categories of effluents from reactor operations, as summarized by UNSCEAR (1988) are as follows:

(1) *Fission noble gases* include radioactive isotopes of krypton (at least nine) and xenon (eleven) formed during fission. Many of them have half-lives too short to be of any significance or health hazards. The most important are ^{85}Kr, ^{88}Kr, ^{133m}Xe, ^{133}Xe, ^{135}Xe and ^{138}Xe, having half-lives of 10.8 years, 2.3 days, 5.3 days, 9.2 days, and 17 minutes, respectively.

(2) *Activation gases are* formed especially during gas-cooled reactor operation. These are primarily ^{41}Ar, produced by the activation of stable argon in air, and ^{35}S coming from sulphur and chlorine impurities in the graphite core. The other radioactive gas, ^{14}N (half-life 7 seconds, high-energy gamma emitter—6.1 and 7.1 MeV), is generated in the coolant water of the boiling water reactors. After production, ^{14}N is transferred

in the steam to the turbine buildings where it may cause direct external exposures.

(3) *Tritium* is produced by both neutron activation and ternary fission. The proportion and its activities depend mainly on the type of reactor. Tritium is released into the atmosphere as well as into the hydrosphere. A typical discharge from a PWR normalized over a five-year period amounts to about 6 TBq per year per one GW. The same type of reactor may be responsible for a release of about 27 TBq $(GW \cdot y)^{-1}$ of ^{3}H to the hydrosphere.

(4) *Radioactive carbon* discharges are of interest because of its long half-life (5.73×10^{3} years) and contribution to the collective dose commitments (lifetime doses). Radionuclide ^{14}C is produced primarily by the activation of ^{17}O, which is present in the coolant and moderator, and also by the reaction (n,p) with nitrogen impurities in fuel and elsewhere. The National Council on Radiation Protection and Measurements (NCRP, 1985) has estimated the production rate of ^{14}C in PWRs to be in the range of 2 to 3 TBq $(GW \cdot y)^{-1}$, and for BWRs to be between 3 and 4 TBq $(GW \cdot y)^{-1}$.

(5) *Iodine* isotopes are produced in the fission process. From the radiological point of view important radioactive isotopes of this volatile element are ^{129}I (half-life 1.6×10^{7} years), ^{131}I (half-life 8.04 days), ^{132}I (half-life 2.3 hours), ^{133}I (half-life 21 hours)., ^{134}I (half-life 53 minutes), and ^{133}I (half-life 6.6 hours). The radionuclides of special interest are ^{129}I (long-term consequences) and ^{131}I (short-term effects). Typical iodine isotope releases from U.S. nuclear power plants using PWRs and BWRs in terms of normalized activity given in GBq $(GW \cdot y)^{-1}$ are shown in Table 1.26.

(6) *Particulates in airborne effluents* include radionuclides in particulate form, which can arise directly or as decay products of fission noble gases or can come from the corrosion of materials in the primary coolant circuit. These aerosols are generated because of leaks in the primary circuit or because of maintenance work involving active compo-

TABLE 1.26. Isotopic Composition of Iodine Discharged from Reactors in the United States in 1982. The Data Presented for Individual Radionuclides Are Given in GBq $(GW \cdot y)^{-1}$.

Reactor Type	^{131}I	^{132}I	^{133}I	^{134}I	^{135}I
PWR	2.6	0.05	6.50	0.00067	0.41
BWR	13.56	0.46	56.27	0.80	137.03

nents removed from the primary circuit. Radionuclides identified in the releases of particulate activity into the atmosphere are as follows: ^{7}Be, ^{22}Na, ^{51}Cr, ^{54}Mn, ^{59}Fe, ^{57}Co, ^{58}Co, ^{60}Co, ^{63}Ni, ^{65}Zn, ^{76}As, ^{88}Ru, ^{89}Sr, ^{90}Sr, ^{91}Sr, ^{95}Zr, ^{97}Zr, ^{95}Nb, ^{99}Mo, ^{99m}Tc, ^{103}Ru, ^{105}Ru, ^{106}Ru, ^{108m}Ag, ^{110m}Ag, ^{113}Sn, ^{115}Cd, ^{122}Sb, ^{124}Sb, ^{125}Sb, ^{123m}Sn, ^{123m}Te, ^{134}Cs, ^{137}Cs, ^{139}Ce, ^{140}Ba, ^{140}La, ^{141}Ce, ^{144}Ce, and ^{182}Ta. Radiation doses from particulates in airborne effluents are due mainly to their transfer through foodchains to man, as well as via external exposures to deposited radionuclides and inhalation of radioactive particulates.

(7) *Liquid effluents* consist of tritium and essentially the same other radionuclides as those mentioned in the case of radioactive particulate releases to the environment atmosphere. The majority of doses due to liquid effluents come especially from ^{137}Cs, ^{134}Cs, ^{131}I, and ^{60}Co.

During operation of a nuclear power plant, various *radioactive wastes* are generated in a number of ways. Although some of them are originally in a liquid form and may be stored in this form at the site, these wastes are usually solidified before disposal. Radioactive wastes are generally classified into three categories: *high-level wastes* (HLW), *intermediate-level wastes* (ILW), and *low-level wastes* (LLW). The categorization is a rather broad and is usually made according to the activity concentration (see Table 1.27).

High-level wastes include primarily the spent fuel elements and the solidified waste products from reprocessing. They are characterized by high

TABLE 1.27. One Possible Classification Scheme for the Categorization of Radioactive Wastes.

	Physical State		
Category of Waste	Gaseous (Bq/L)	Liquid (Bq/L)	Solid (Bq/m^3)
Low-Level:			
Alpha	$<3.7 \times 10^{-6}$	$<3.7 \times 10^{7}$	$<1.3 \times 10^{13}$
Beta/photon	$<3.7 \times 10^{-3}$	$<3.7 \times 10^{7}$	$<1.3 \times 10^{13}$
Intermediate-Level:			
Alpha	From 3.7×10^{-6} to 3.7	From 3.7×10^{7} to 3.7×10^{10}	From 1.3×10^{13} to 1.3×10^{15}
Beta/photon	From 3.7×10^{-3} to 3700	From 3.7×10^{7} to 3.7×10^{10}	From 1.3×10^{13} to 1.3×10^{15}
High-Level:			
Alpha	>3.7	$>3.7 \times 10^{10}$	$>1.3 \times 10^{15}$
Beta/photon	$>3.7 \times 10^{3}$	$>3.7 \times 10^{10}$	$>1.3 \times 10^{15}$

Adapted from Kathren (1984, p. 172).

activity concentrations of both actinides and fission products and are significantly heat-generating.

Intermediate-level wastes are defined to some extent by exclusion from the other two waste categories (high-level and low-level). This category contains either actinides or long-lived beta and/or gamma radionuclides in amounts that are not negligible, or substantial activity concentrations of beta/gamma emitters with shorter half-lives. Intermediate-level wastes are not supposed to be significantly heat-generating. In principle, this category would be applied to wastes that can be safely released into the environment after some treatment and under controlled conditions, or to wastes that may require shielding for protection of personnel against external radiation. In general, however, the category of intermediate-level radioactive waste is not usually used.

Low-level wastes include primarily reasonably short-lived beta/gamma radionuclides in low-to-moderate activity concentrations. They may contain actinides or long-lived beta/gamma radionuclides but only in very small quantities. The term *low-level waste* is quite broad and is usually used for all other wastes except high-level wastes.

Low-level waste is often defined as radioactive waste that is neither spent fuel nor high-level waste from the reprocessing of spent fuel, transuranic (TRU) waste, thorium and uranium mill tailings. Low-level wastes may be further categorized according to concentrations of some specific radionuclides (Knittel, 1989).

There is a slightly different approach as to the classification of LLW, for example, between the International Atomic Energy Agency (IAEA) and the U.S. Nuclear Regulatory Commission (USNRC). The IAEA recommends that two categories of low-level beta-gamma waste be considered with regard to disposal, depending on whether or not the waste contains significant amounts of transuranic alpha-emitting radionuclides (IAEA, 1981). Low-level wastes containing significant alpha activity are by the IAEA termed "low-level, long-lived." The IAEA recommends that these wastes be immobilized and contained in structurally sound packages. It does not recommend them, however, for burial at shallow depths. Solidified LLW that contain insignificant amounts of alpha activity are in the "low-level, short-lived" category and are considered by the IAEA as being acceptable for shallow land burial.

The USNRC, on the other hand, defines three classes of low-level radioactive wastes (USNRC, 1982). According to this categorization, Class A waste contains relatively low concentrations of radionuclides and need not be structurally stable after disposal. It must be in disposal units separate from the wastes of higher activity—Class B and Class C. Wastes of Class B contain radionuclide concentrations higher than those in Class A. They must be in a form or in a container that will ensure physical stability after final closure of the disposal facility. Class C waste includes the highest concentrations of radionuclides covered by the USNRC regulations. This type of LLW must not only meet the stability requirements of Class B waste, but special measures are required at the disposal facility to provide protection for inadvertent intruders for at least 500 years.

The production of radioactive wastes from nuclear installations has to be considered realistically and to be put into perspective. This can be demonstrated by giving some data relevant for the UK (Wilkinson, 1992). British estimates suggest that the quantity of radioactive waste produced from an individual's lifetime supply of electricity from nuclear power would be 16 L of LLW, 4 L of ILW, and only about 0.14 L of HLW. These figures include waste generated during power plant operation, as well as during fuel reprocessing.

Totally, some 45,000 m^3 of radioactive waste is produced in the UK each year. This is quite a small portion, as a matter of fact only about 1%, of the 4 million tons of all toxic industrial wastes. Most of the radioactive waste, about 40,000 m^3, is LLW, such as paper towels and discarded clothing. The strategy for dealing with this type of waste is to press them into compacts, and to grout the compacts into containers. The containers are usually placed in shallow concrete vaults, which are then capped with clay. This disposal procedure does not present any significant problems.

Much of the remaining 5000 m^3 is ILW, such as cladding material from fuel elements. This waste is encapsulated in cement in 500 L stainless steel drums, which are stored prior to disposal in a deep repository.

The balance of less than 100 m^3 produced per year is HLW, but it contains more than 95% of the radionuclides present in spent fuel. This high-level waste, which is now stored as a concentrated liquid, is being converted into glass in the new vitrification plant at Sellafield. Waste in this form will be stored for at least fifty years in order to allow the reduction of heat generated by fission products before final disposal in a deep repository.

There are special and very strict requirements regarding the repositories. They have to be designed to meet very high standards to prevent the escape of radioactivity into the environment. The current British regulations require that the effective dose to any member of the public over a very long timescale (no less than 10,000 years) should be less than 0.1 $mSv{\cdot}y^{-1}$. It is worthy to note that the dose of 0.1 mSv per year is less than 5% of the annual dose due to all natural sources of radiation. Actually, 0.1 $mSv{\cdot}y^{-1}$ is very much less than the variations in the natural background dose between different parts of the UK.

The operation of nuclear power plants results in the exposure of some members of the public to releases of radioactive materials to the atmosphere

TABLE 1.28. Summary of Annual Collective Effective Dose Equivalents to Regional Populations Due to Radioactive Effluents in Air and Water Pathways from Fuel Cycle Facilities.

Facility	Annual Collective Eff. Dose Equiv. (man·Sv)	Basis of Estimates
Mining		
Open pit – air	0.01	Model mine
Open pit – water	0.002	
Underground – air	0.1	Model mine
Underground – water	0.21	
Milling	0.62	Airborne effluent from model mine
Conversion		
Wet	0.004	Plant airborne effluent in 1980
Dry	0.029	Plant airborne effluent in 1980
Enrichment	0.00002–0.004	3 Plants, airborne effluents in 1980
Fabrication	0.0001–0.007	7 Plants, airborne effluents in 1980
Nuclear power plant		
Air	0.00003–0.13	47 Plants in 1980
Water	0–0.4	47 Plants in 1980
Low-level waste storage	<0.04	Maxey Flats, KY facility (estimate)

Published with permission of the National Council on Radiation Protection and Measurements, 1987.

or hydrosphere, and to direct gamma radiation emitted from those facilities. Table 1.28 summarizes the radiological impact of the individual parts of the fuel cycle in terms of the annual collective dose equivalents to regional populations (NCRP, 1987).

1.1.3.4 STORAGE AND DISPOSAL OF RADIOACTIVE WASTES

The storage and disposal of radioactive waste has become a very sensitive issue in these environmentally conscious times. A few decades ago, all kinds of wastes including most radioactive wastes were disposed of with little or no regard for their long-term safety and health hazards. The modern approach to waste management means that once we dispose of waste, we should not expect to suffer the consequences of its eventual return to our environment. Since the early 1970s, all types of waste have come under much stricter regulatory control, radioactive wastes more so than any other type. It may be noted here that radioactive wastes are losing their toxicity while they decay. Although, depending on the half-life, it sometimes takes a very long time, the radioactivity (at least in principle) eventually disappears completely or decays to innocuous levels. The other toxic (non-radioactive) wastes, however, currently disposed of into the environment (e.g., those containing arsenic, cadmium, and mercury) do not decay and will remain toxic practically forever.

In addition to the control, movement, and conditioning, the *storage* and *disposal* of *low-level radioactive wastes* are the most important aspects of waste management. Here, one must clearly distinguish between storage of wastes and their disposal. Wastes are stored for a certain period of time, after which they will be retrieved, or at least they are intended to be retrieved at some time in the future. On the other hand, when waste is disposed of, it is practically abandoned and there is no intention of retrieval.

The problem of HLW has not been solved completely. So far, attention is mainly concentrated on their reliable isolation for very long periods of time, during which they will remain potentially hazardous. In addition to the storage of HLW in deep repositories, other options and possibilities are also considered. These include extraterrestrial disposal (canisters containing HLW are launched by rockets into space) and transmutation of radioactive elements into stable elements.

Recent studies (Skalberg and Liljenzin, 1993) have shown that partitioning and transmutation (P-T) of radioactive waste nuclides is technically feasible but much research and development remains to be done before it can be regarded as mature. There currently seems to be no economic gain from P-T technology as compared with direct disposal of spent fuel, and it promises only insignificant dose savings compared with the current disposal of HLW. Nevertheless, such concepts will continue to be evaluated further, and in the longer term the picture may look different.

It has already been mentioned that, as far as the total produced amounts are concerned, the vast majority of radioactive wastes belong to the category of LLW. These wastes are generated mainly from the operation of the nuclear fuel cycle (both commercial and military) and also from various applications of radionuclides in medicine, industry, research, and many other fields, including use in consumer products. In the USA, for example, of the 130,000 m^3 of low-level waste generated in 1988, about 70% originated in the weapons program, 20% in the commercial nuclear fuel cycle, 7% in industry, and 3% from research and medical applications.

The choice of the appropriate method for disposal of low-level radioactive wastes depends mainly on chemical and physical form, activity concentration, total activity, and some other factors.

In general, the following disposal procedures are usually applied:

(1) Radioactive waste containing short-lived radionuclides is kept until it decays sufficiently so it may be considered harmelss.
(2) Radionuclide material that mixes well with air or water is diluted to a very low concentration and then dispersed so that the concentrations of released radionuclides in the environment become lower than limits set by the radiation safety authorities.
(3) Long-lived wastes are concentrated and isolated from the environment.

Before 1970, a total of 3.7×10^{15} Bq of LLW produced in the USA weapons program was also disposed of by ocean dumping, and prior to 1983, 4.8×10^{16} Bq was also injected along with grout into the hydrofractured shale formation under the Oak Ridge National Laboratory. These disposal practices have since been discontinued, and such methods of disposal are no longer used.

At present, for disposal of LLW one of the following burial disposal systems may be used:

(1) Shallow burial in trenches, tunnels, concrete bunkers, or caissons
(2) Deep burial in a purpose-built or modified existing mine, usually referred to as a repository
(3) Deep burial in boreholes drilled from the surface

Most operating disposal facilities are of the first type (Knittel, 1989). At present, only three commercial shallow-land burial sites are used for LLW in the USA. Radioactive waste sent to these disposal sites comes from the nuclear fuel cycle, as well as from other practices involving the production and the use of radionuclides. Typical amounts of individual radionuclides in solid waste from PWR and BWR are given in Table 1.29. The activity of radionuclides present in the non-fuel-cycle LLW sent to the disposal sites in one year in the USA is shown in Table 1.30.

Reactor solid wastes include spent resins from the ion exchangers used to purify the coolant, evaporator bottom wastes, and compactable trash (e.g., paper, rags, disposable protective gloves and clothing, air and water filters) and contain a mixture of fission and activation products. To non-fuel-cycle wastes (institutional wastes, i.e., wastes produced by hospitals, medical clinics, universities, and research centers), on the other hand, belong laboratory materials, contaminated animal carcasses, protective clothing, and also liquid scintillation "cocktails" containing such low-energy beta emitters as ^{3}H and ^{14}C.

TABLE 1.29. Activity of the Most Important Radionuclides in LLW from Pressurized Water and Boiling Water Reactors (BWR Is Equipped with Charcoal Absorber System).

	Activity [$Bq \cdot y^{-1} \cdot (GWe)^{-1}$]	
Radionuclide	PWR	BWR
^{134}Cs	1.04×10^{14}	4.81×10^{12}
^{137}Cs	9.99×10^{13}	4.44×10^{12}
^{60}Co	1.04×10^{13}	7.40×10^{12}
^{55}Fe	7.40×10^{12}	2.96×10^{13}
^{58}Co	4.44×10^{12}	4.81×10^{12}
^{54}Mn	8.14×10^{11}	4.81×10^{11}
^{144}Ce	1.55×10^{11}	1.89×10^{11}
^{90}Sr	9.99×10^{10}	3.33×10^{12}
^{89}Sr	7.40×10^{10}	1.81×10^{12}
^{59}Fe	6.66×10^{10}	2.44×10^{11}
^{106}Ru	6.29×10^{10}	–
^{95}Zr	2.37×10^{10}	4.07×10^{10}
Total	2.26×10^{14}	5.92×10^{13}

Adapted from Kathren (1984).

TABLE 1.30. Radionuclides in Non-Nuclear-Fuel-Cycle Sent to the Commercial Disposal Sites per Year in the USA.

Radionuclide	Half-Life	Activity (Bq)
^{3}H	12.3 y	1.33×10^{15}
^{14}C	5730 y	3.96×10^{14}
^{35}P	14.3 d	2.32×10^{14}
^{32}S	87.4 d	9.03×10^{13}
^{51}Cr	27.7 d	7.00×10^{13}
^{67}Ga	78.3 h	4.88×10^{12}
^{99m}Tc	6.0 h	1.39×10^{15}
^{125}I	60.2 d	4.71×10^{14}
^{131}I	8.04 d	2.69×10^{14}
Other		5.73×10^{14}
Total (approximately)		5×10^{15}

Adapted from Anderson et al. (1978).

The management of low-level radioactive wastes includes many steps, ranging from segregation, treatment, conditioning, packaging, and shipment to disposal (see Figure 1.19) (Tang, 1988). Treatment of radioactive *gaseous waste* is usually accomplished through filtration units fixed with high-efficiency particulate air filters to remove particulate materials, and charcoal absorbers to eliminate elements such as iodine. After their use the filter frame and charcoal materials are then treated as solid wastes. *Liquid wastes,* after processing and conditioning, are separated into an effluent for recycle or controlled discharge and concentrated stream for solidification. Also, these wastes are eventually converted into a solid form by packaging, transport, and disposal.

In the near future we will be more and more occupied with radioactive wastes coming from decommissioned nuclear facilities. At the end of their lifetime, nuclear reactors have to be dismantled and the accumulated radioactivity disposed of. The wastes related to the decommissioning of nuclear facilities include neutron-activated wastes, surface-contaminated wastes, and such contaminated materials as concrete, steel, resins, etc. The neutron-activated wastes are mainly confined to the reactor pressure vessel and its internal components, which have been irradiated during the reactor operation by neutrons of very high fluence. These components contain significant concentrations of long-lived radionuclides such as ^{94}Nb with a half-life of 20,000 years. Only part of these wastes will be disposed of as LLW in the shallow burial sites. Most of them will, by necessity, be disposed of in geologic repositories.

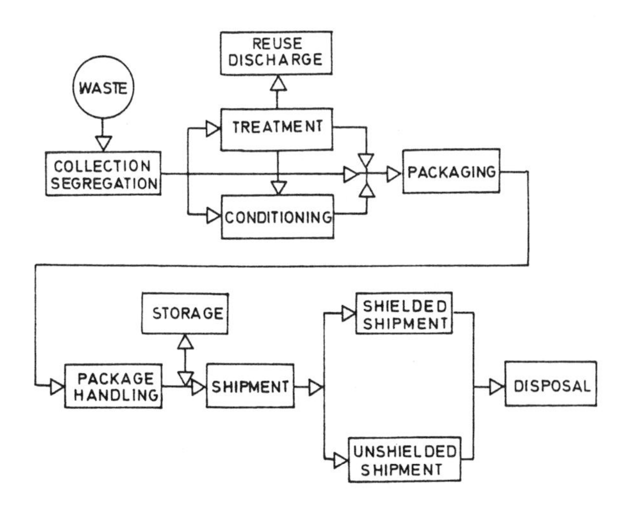

FIGURE 1.19. Illustration of the waste management steps for low-level radioactive waste.

An integral part of any waste management operation has to be an efficient and reliable monitoring program that conducts regular measurements of radioactivity and radiation levels at all locations where a potential for population exposure exists. The monitoring program must include especially the following objectives:

- Measure and record radiation exposures to the population in the vicinity of the disposal facilities.
- Verify regularly the proper operation of waste treatment facilities.
- Provide warning if any unexpected increase in radiation or radionuclide concentration level occurs.
- Ensure preoperational background radioactivity and radiation levels and data on wildlife, vegetation, or fish populations.
- Be prepared for handling emergency situations and accidents (including adequate monitoring).

The above-mentioned requirements and objectives are accomplished by measuring ambient radiation levels, as well as radioactivity in liquid and gaseous effluents, cooling ponds and receiving streams, groundwater, air samples, milk, crops, and local foodstuffs. In some cases, measurements should be carried out and recorded continuously. Other measurements require periodic sampling and evaluation at a low-level laboratory. Environmental monitoring at a low-level radioactive waste disposal facility and its vicinity may also include some other measurements and evaluations needed to confirm that the disposal system is performing as expected.

1.1.3.5 NUCLEAR ACCIDENTS

During the more than fifty years of the nuclear industry, there have been several accidents involving reactors and critical assemblies, which have resulted in radioactive contamination of the site and the environment. The actual and potential risk especially involves such accidents where reactor core was affected and damaged. So far, about fourteen reactor accidents of this category have occurred in various countries with different consequences (see Table 1.31) (Eisenbud, 1987). Four of these accidents are of major significance from an environmental point of view: Windscale (1957), Idaho Falls (1961), Three Mile Island (1979), and in particular Chernobyl (1986).

The first of these accidents, which occurred in October 1957 at Windscale (now called Sellafield) in northwest England, resulted in the first release of radioactive materials from a reactor accident. The amount of various fission products released to the surrounding countryside was esti-

TABLE 1.31. Reactor Accidents That Involved Core Damage.

Year	Location	Name of Reactor	Type of Reactor	Extent of Contamination
1952	Canada	NRX	Experimental	None
1955	Idaho, USA	EBR-1	Experimental	Trace
1957	United Kingdom	Windscale	Military production reactor	7.4×10^{14} Bq of ^{131}I
1957	Idaho, USA	HTRE-3	Experimental	Slight
1958	Canada	NRU	Research reactor	None
1959	California, USA	SRE	Experimental	Slight
1960	Pennsylvania, USA	WTR	Research	None measured
1961	Idaho, USA	SL-1	Experimental	3.7×10^{14} Bq of ^{131}I
1963	Tennessee, USA	Orr	Research	Trace
1966	Detroit, USA	Fermi	Experimental power	No release outside plant
1969	France	St. Laurent	Power	Little, if any
1969	Switzerland	Lucens	Experimental	None
1979	Pennsylvania, USA	TMI-II	Power	Slight
1986	USSR	Chernobyl-4	Power	Extensive

mated originally by Dunster et al. (1958) and later analyzed by Clark (1974). These estimations of activities of principal fission radionuclides are summarized in Table 1.32. The first evidence that an accident had occurred was the increase in measured beta activity of atmospheric aerosols monitored by an instrument installed about 1 km from the reactor. The measured con-

TABLE 1.32. Main Fission Products Released from Windscale No. 1, a Graphite Moderated Reactor during the 1957 Accident.

	Estimated Releases (Bq)	
Radionuclide	Dunster et al. (1958)	Clarke (1974)
^{131}I	7.4×10^{14}	6.0×10^{14}
^{137}Cs	2.2×10^{13}	4.6×10^{13}
^{89}Sr	3.0×10^{12}	5.1×10^{12}
^{90}Sr	3.3×10^{11}	2.2×10^{11}

centration was about ten times higher than normal background activity concentration, mainly from radon and thoron decay products.

The radiological impact of this accident appeared largely through milk contaminated by ^{131}I and its inhalation. The highest concentration of ^{131}I in milk was about 37 $mBq \cdot L^{-1}$, while the average concentration of this radionuclide in air during the period of the accident was approximately 170 $mBq \cdot L^{-1}$.

The second major nuclear accident was the explosion of the Army Low-Power Reactor (SL-1) situated near Idaho Falls, Idaho, in January 1961. The SL-1 was a direct-cycle boiling-water power reactor fueled with enriched uranium and designed to operate at a level of e MW(th). The explosion resulted in the death of three persons and destruction of the reactor. It was estimated that about 5–10% of the total fission product inventory escaped from the reactor core, which at the time of the accident contained about 37 PBq of long-lived radionuclides. Fortunately, only about 0.01% of the total activity escaped from the reactor building. High radiation levels (about 5 $Gy \cdot h^{-1}$) were measured in the reactor building following the accident. Farther from the building the dose rate was sharply reduced so that at a distance of about 1 km it was less than about 20 $\mu Gy \cdot h^{-1}$. It was assumed that about 3 TBq of ^{131}I escaped into the environment. The air concentration of ^{131}I was found to be below approximately 3 $Bq \cdot m^{-3}$. Practically all fission products that escaped from the reactor building remained within a few hundred meters of the reactor.

The accident in Three Mile Island Unit 2 (TMI-2) was, in fact, the first major nuclear accident at a power plant. It occurred early in the morning of March 28, 1979, at the 880 MWe pressurized water reactor near Harrisburg, Pennsylvania. Although radiation exposure to the plant personnel and the general public was insignificant, the financial consequences of the accident were tremendous. The TMI mishap had a negative impact on public opinion all over the world. This led to a freeze in the construction of many plants and a curtailment or reduction in nuclear energy development in a number of countries. Although during the accident, the reactor core was partially melted, owing to its design, which provided a safe containment for radioactive releases, the exposure of personnel and the nearby population was minimal. The most credible estimation of the collective effective dose would be on the level of about 3.3 man·Sv. On the other hand, the decontamination of the plant posed many problems.

The worst accident in the history of nuclear energy, with very serious consequences, occurred early in the morning on 26 April, 1986 in Unit 4 of the Chernobyl nuclear power station near Kiev in the Ukraine (on the territory of the former USSR). A water-cooled, pressure tube, graphite-moderated, 1000 MWe reactor was destroyed and huge amounts of radionuclides were released.

TABLE 1.33. Core Inventory and Estimate of Total Radioactivity Released into the Atmosphere (the Inventory Values Are Corrected to 6 May, 1986).

Radionuclide	Half-Life	Inventory (Bq)	Release (%)
^{85}Kr	10.72 y	3.3×10^{16}	Up to 100
^{133}Xe	5.21 y	1.7×10^{18}	Up to 100
^{131}I	8.04 d	1.3×10^{18}	20
^{132}Te	3.26 d	3.2×10^{17}	15
^{137}Cs	30.0 y	2.9×10^{17}	13
^{134}Cs	2.06 y	1.9×10^{17}	10
^{89}Sr	50.5 d	2.0×10^{18}	4
^{90}Sr	29.12 y	2.0×10^{17}	4
^{95}Zr	64.0 d	4.4×10^{18}	3
^{99}Mo	2.75 d	4.8×10^{18}	2
^{103}Ru	39.3 d	4.1×10^{18}	3
^{106}Ru	368 d	2.1×10^{18}	3
^{140}Ba	12.7 d	2.9×10^{18}	6
^{141}Ce	32.5 d	4.4×10^{18}	2
^{144}Ce	284 d	3.2×10^{18}	3
^{239}Np	2.36 d	1.4×10^{17}	3
^{238}Pu	87.74 y	1.0×10^{15}	3
^{239}Pu	24,065 y	8.0×10^{14}	3
^{240}Pu	6,537 y	1.0×10^{15}	3
^{241}Pu	14.4 y	1.7×10^{17}	3
^{242}Cm	163 d	2.6×10^{16}	3

The accident was caused by an attempt by the reactor operators to perform an experiment that required conditions different from those of routine operation and that permitted the reactor to become unstable. This accident occurred during shutdown of the reactor for routine maintenance. In order to perform the experiment, the operators deliberately violated many of the safety rules. The results were explosions, the first of which occurred at about 1:24 A.M. local time. More explosions were caused by the production of hydrogen and carbon monoxide, which mixed with atmospheric oxygen. These explosions destroyed parts of the reactor, including its cooling system and the containment building, and allowed fission products and other radionuclides to escape into the environment. This release of large amounts of radioactive material did not occur as a single massive event. About 25% of the total released materials escaped during the first day of the accident; the rest of the radioactivity escaped over the following nine-day period. The estimated percentages of various radionuclides released into the environment are given in Table 1.33 (IAEA, 1986; UNSCEAR, 1988). The time reconstruction of the total releases is shown in Figure 1.20.

The release-rate diagram may be subdivided into four stages (UNSCEAR, 1988):

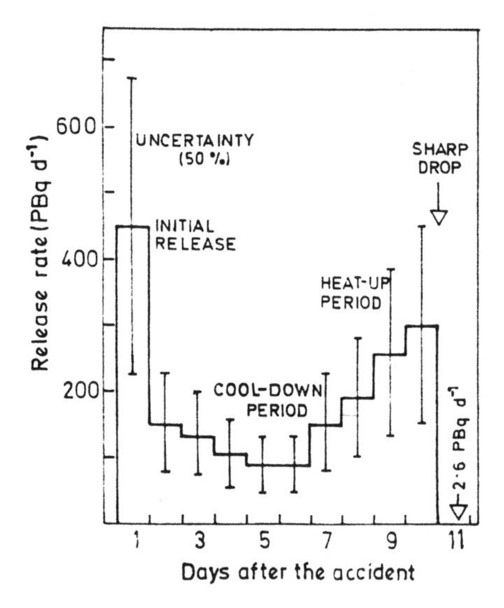

FIGURE 1.20. Daily release rate of the radioactive materials, excluding noble gases, during the Chernobyl accident.

(1) The initial release occurring on the first day of the accident—During this stage, the mechanical discharge of radioactive materials was caused directly by the explosion in the reactor core.
(2) A period of five days, during which the release rate fell to a minimum of about six times lower than the release on the first day—This decline was the result of the measures taken to fight the graphite fire. During this operation about 5000 tons of boron carbide, dolomite, clay, and lead were dropped from helicopters.
(3) A period of the following four days, during which the release rate increased again, reaching about 70% of the initial release rate—This increase was caused by heating of the fuel in the core, where the temperature exceeded 2000°C due to residual heat release.
(4) A sudden reduction in the release rate nine days after the accident—The release rate fell to less than 1% of the initial rate and continued declining thereafter.

Long-range atmospheric transport spread the released radionuclides throughout the entire northern hemisphere. The initial radioactive contaminations reached Japan on 2 May, China on 4 May, India on 5 May, and Canada and the USA on 5–6 May. It is worthwhile to note that no airborne activity from Chernobyl has been reported in the southern hemisphere.

The consequences of the accident were tremendous: loss of lives; overexposure of personnel and members of the firefighting crews; increase in exposure of large segments of the population both in the former USSR and in other countries, especially in Europe; contamination of large areas; economic losses (material costs of control, resettlement, and decontamination); and a negative impact on public opinion, which strengthened the opposition against the nuclear industry.

The immediate medical consequences can be summarized as follows (Goldman and Catlin, 1988; UNSCEAR, 1988):

(1) Thirty-one workers and emergency personnel exposed on the reactor site have died. Two of them died immediately from trauma and burns. The remaining twenty-nine died over the next two months from radiation effects complicated by severe thermal and radiation burns.
(2) Excessive exposure to radiation incurred during firefighting or rescue operations resulted in the hospitalization of more than 200 on-site personnel. No acute radiation sickness was found in the off-site population.
(3) About 24,000 persons (from about 130,000 citizens who were evacuated from a 30-km zone around the reactor because of high ambient radiation levels) received average doses of about 450 mGy.

On the basis of external dose rate measurements and analysis of environmental samples, including air, soil, and water samples, the total amounts of radionuclides deposited at various distances from Chernobyl were estimated. The proportions of the core inventory deposited were assumed to be as follows:

- on-site: 0.3–0.5%
- within the 20-km radius: 1.5–2%
- beyond the 20-km radius: 1–1.5%

The *main pathways* and radionuclides contributing to the population exposures are

(1) External irradiation from radionuclides deposited on the ground—Most of these doses can be attributed to ^{137}Cs. Its total release is estimated to have been between 30 to 100 PBq (3×10^{16} to 10^{17} Bq), of which about one-third was deposited on the territory of the former USSR, one-third in other European countries, and the remaining third over the rest of the northern hemisphere.
(2) The dietary ingestion of radionuclides through the contaminated foodstuffs, primarily ^{131}I in milk and leafy vegetables during the first month and, after that, ^{134}Cs and ^{137}Cs in foods.

Two *secondary pathways* have also been considered: (1) external exposure from radioactive materials present in the cloud and (2) inhalation of radionuclides during passage of the radioactive cloud. Both these pathways contributed to the total exposure only for the relatively short period after the accident (before the airborne material had been deposited or dispersed).

It has been found that, for the evaluation of doses received via ingestion, it is enough to take into consideration only basic food items, such as milk products, grain products, leafy vegetables, other vegetables and fruit, and meat. Those five categories proved to be sufficient to account for the food ingestion of most individuals. Radionuclide uptakes in other foods, e.g., mushrooms and fishes, some of which show increased ability to concentrate particular radionuclides, are generally of minor importance.

After the accident, extensive national monitoring programs were undertaken in order to evaluate the extent and degree of contamination by released radionuclides. In addition, it was also necessary to determine the need for possible countermeasures. The results obtained from measurements of environmental radiation levels and radionuclide concentrations in diet and air, as well as in the human body, provided a solid basis for assessment of radiation exposures in terms of relevant radiation protection quantities.

Radionuclides in air were measured, mainly using filter sampling followed by gamma spectrometry. This method made it possible to detect such radionuclides as ^{99}Mo, ^{99m}Tc, ^{103}Ru, ^{127}Sb, ^{129}Te, ^{132}Te, ^{131}I, ^{132}I, ^{133}I, ^{134}Cs, ^{136}Cs, ^{137}Cs, ^{140}Ba, and ^{140}La. After the decay of some short-lived radionuclides emitting interferring gamma lines, certain additional radionuclides, e.g., ^{95}Nb, ^{106}Ru, ^{110m}Ag, ^{125}Sb, ^{129m}Te, ^{141}Ce, and ^{144}Ce, could also be identified and evaluated.

The main interest has centered on ^{134}Cs and ^{137}Cs. Since ^{134}Cs did not occur in the nuclear weapon debris existing in the environment prior to the Chernobyl accident, the concentration ratio of $^{134}Cs/^{137}Cs$ gave a convenient means of recognizing the Chernobyl material (Pattenden, 1991). As an example, the ^{137}Cs concentrations in the air at Chilton near Harwell (England) from 1982 to 1989 are shown in Figure 1.21. The presence of ^{134}Cs, ^{137}Cs, ^{131}I and other fission products were also clearly identified both in air and rainwater in Taiwan, 8000 km away from Chernobyl (Chung, 1989).

A number of different radionuclides present in air were determined by beta or alpha spectrometry. It was possible to monitor the following radionuclides: ^{89}Sr, ^{90}Sr, ^{85}Kr, ^{3}H, ^{238}Pu, $^{239,240}Pu$, and ^{242}Cr.

The deposition of radioactive materials is closely related to rainfall, which occurred very sporadically throughout Europe during the passage of the contaminated air. This is why the deposition pattern in some European

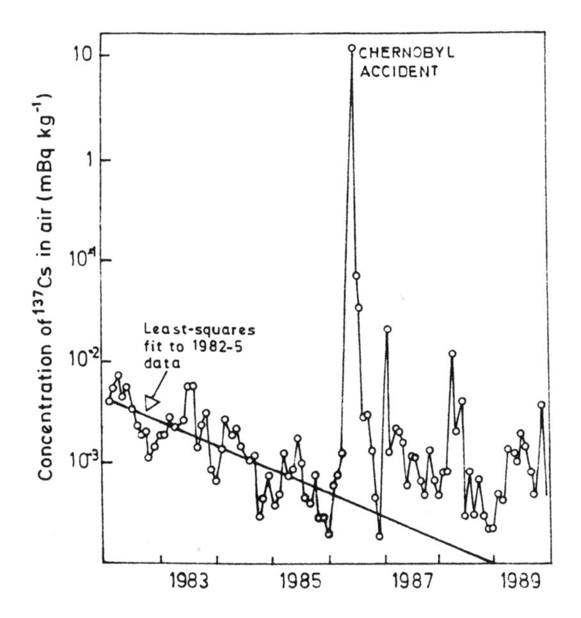

FIGURE 1.21. The air concentration of ^{137}Cs measured in Chilton during 1982–1989.

and other countries was very irregular. The highest surface contamination of ^{137}Cs outside the former USSR was measured in Sweden north of its capital Stockholm, where the deposition amounted to more than 85 $kBq \cdot m^{-2}$. Depositions of ^{137}Cs and also ^{134}Cs, ^{103}Ru, ^{106}Ru, and ^{131}I in the former USSR and some countries in North, Central, and South Europe, West and East Asia, and North America are shown in Table 1.34.

Deposited radioactive materials resulted in external irradiation, which in the first months after the accident was caused by a number of such short-lived radionuclides as ^{132}Te, ^{132}I, ^{131}I, ^{140}Ba, ^{140}La, ^{103}Ru, and ^{106}Ru. In the long term, however, external exposure will be due primarily to ^{134}Cs and ^{137}Cs.

Since ingestion of contaminated foods is an important pathway contributing to radiation doses from ^{131}I and ^{137}Cs, all countries paid special attention to monitoring the concentrations of these radionuclides in various foodstuffs. The short-lived ^{131}I is rapidly transferred to man through the consumption of milk and leafy vegetables and is a signifcant source of exposure in the first days and weeks following its deposition. On the other hand, ^{137}Cs is also effective in other basic foods, such as cereals, root vegetables, fruit, and meat, which are produced during longer growing periods.

By way of illustration, the measurements of the ^{137}Cs concentrations in milk carried out in Germany and Finland are shown in Figure 1.22. It is in-

TABLE 1.34. Deposition of Selected Radionuclides in Some Countries (Numbers in Parentheses Are Inferred Values).

	Deposition Density ($kBq \cdot m^{-2}$)				
Country	^{137}Cs	^{134}Cs	^{103}Ru	^{106}Ru	^{131}I
USSR (northwest of Chernobyl)	39	21	41	8.8	590
Finland	15	7.6	19	12	100
Sweden (Stockholm region)	31	17	9.9	3.7	160
Austria	23	(12)	31	(6.3)	120
Czech Republic (western part)	2.3	1.3	4	(0.72)	(26)
Slovakia (eastern part)	2.8	1.3	6.1	(0.85)	(30)
Germany (Dresden)	6.1	2.9	14	(1.8)	(45)
Germany (Munich)	16	8.0	20	4.8	100
Poland	5.2	2.6	(13)	(1.6)	38
Belgium	0.84	0.4	1.4	0.4	5.2
Ireland	3.4	1.7	4.9	1.3	10
United Kingdom (England)	0.1	0.05	0.18	0.06	0.8
Italy (north)	6.0	3.0	14	3.8	25
Portugal	0.02	(0.01)	(0.04)	(0.012)	0.07
Israel	(0.4)	(0.2)	(1.6)		(0.07)
China	(0.015)	(0.075)	(0.21)	(0.044)	(0.29)
Japan	0.18	0.087	(0.45)	(0.090)	1.6
Canada	0.030	0.015	(0.04)	(0.016)	0.10
USA	0.026	0.013	(0.062)	(0.008)	0.15

Adapted after UNSCEAR (1988).

teresting to compare the concentrations of ^{137}Cs in milk and in meat. This can be illustrated using results from Finland (Figure 1.23).

After the Chernobyl accident, world-wide measurements were made of the concentrations of some radionuclides in the human body or organ doses due to the internal contamination. Extensive monitoring focused, in particular, on the evaluation of ^{131}I in the thyroid and ^{137}Cs in the body. The contribution of ^{131}I to the thyroid in terms of the average dose equivalent for adults in various countries is presented in Figure 1.24. It should be noted on this occasion that doses due to ^{131}I are generally higher in infants than in adults because the main pathway is through milk consumption.

The actual radiological impact of the accident can be assessed on the basis of the values of effective dose equivalent commitment and collective

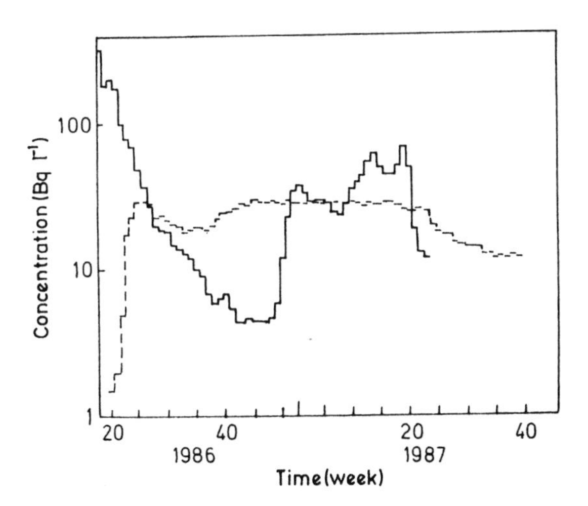

FIGURE 1.22. Weekly monitoring results of ^{137}Cs concentrations in milk from the Federal Republic of Germany (diary farm in southeast Bavaria) and Finland (country-wide average). Adapted from UNSCEAR (1988).

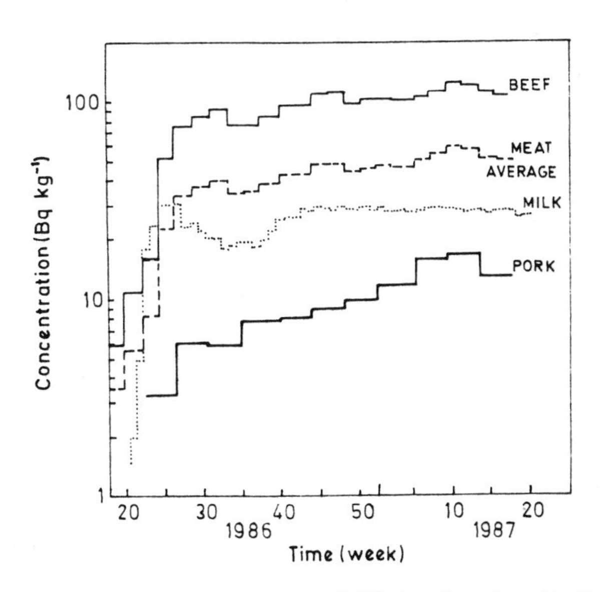

FIGURE 1.23. Average concentrations of ^{137}Cs in milk and meat in Finland.

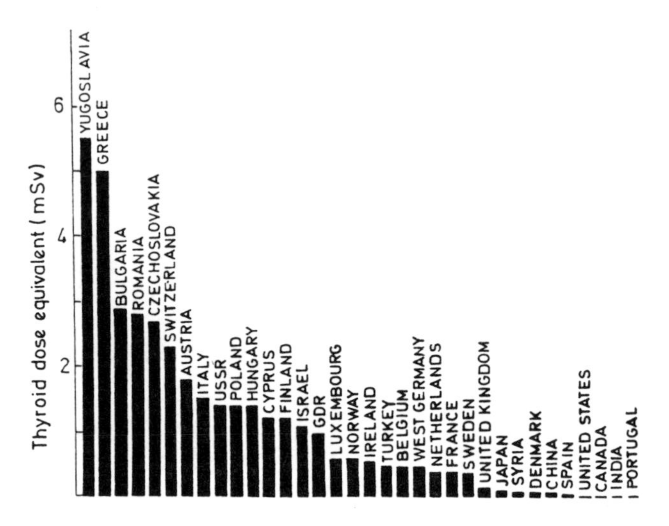

FIGURE 1.24. The average adult thyroid dose equivalents from the Chernobyl accident. From UNSCEAR (1988).

effective dose equivalent commitment. Most of the doses were received through external and internal exposures during the first year following the accident. The country-wide average first-year effective dose equivalents (including committed doses from radionuclide ingestion and inhalation in this period of time) of individuals in various countries are shown in Figure 1.25. The first-year committed effective dose equivalents resulted primarily from the ingestion pathway with dominant contributions from ^{131}I, ^{134}Cs, and ^{137}Cs (contributions between about 60–80%, the rest was mainly due to the external irradiation).

The effective dose equivalent commitments from all radionuclides released in the accident averaged over large regions are evaluated in Table 1.35. The estimated values range from 1.2 mSv in southeastern Europe, 970 μSv in Scandinavia, 940 μSv in central Europe, 820 μSv in the former USSR and 510 μSv in Mediterranean countries to 20 μSv or less in other regions (UNSCEAR, 1988). These results are graphically illustrated in Figure 1.26.

The radiological impact of the Chernobyl accident may be estimated using the collective effective dose equivalent commitments. The contributions of main radionuclides to the total collective dose equivalent commitment, together with their releases and dose factors, are summarized in Table 1.36.

The estimations of potential health impacts found in various studies may differ considerably. This difference depends on applied input data and risk factors, which have recently been updated by the ICRP (1991). Neverthe-

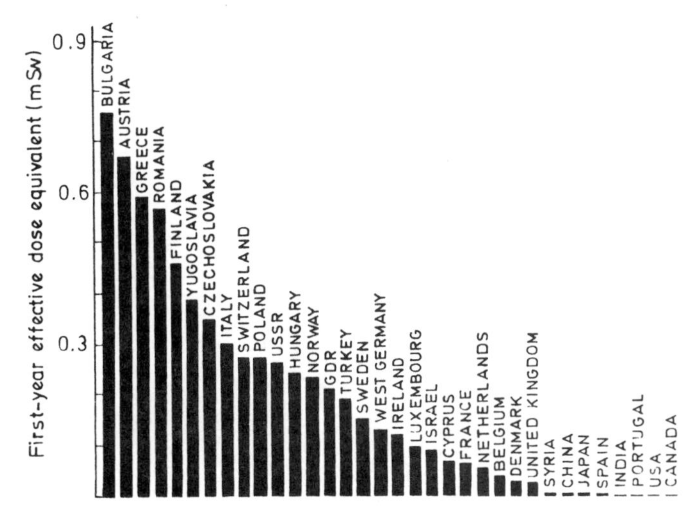

FIGURE 1.25. The average first-year effective dose equivalents received by adult individuals from the Chernobyl accident. From UNSCEAR (1988).

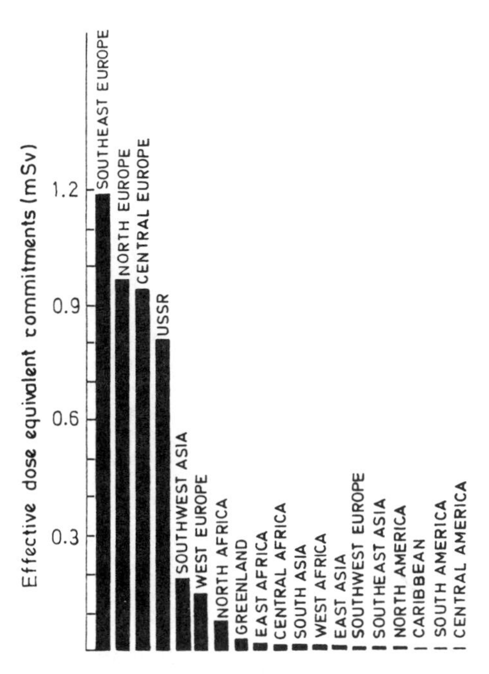

FIGURE 1.26. The regional average effective dose equivalent commitments from the Chernobyl accident. From UNSCEAR (1988).

TABLE 1.35. The Deposition of ^{137}Cs and Average Effective Dose Equivalent Commitments in Various Regions.

Region	Population-Weighted Deposition of ^{137}Cs ($kBq \cdot m^{-3}$)	Effective Dose Equivalent Commitment (μSv)		
		First Year	After First Year	Total
USSR	5.1	260	560	820
North Europe	7.0	210	760	970
Central Europe	6.1	270	670	940
West Europe	1.0	48	110	160
Southeast Europe	7.4	390	810	1200
Southwest Europe	0.03	3.7	3.4	7
West Asia	3.2	160	350	510
East Asia	0.1	5.6	13	19
North America	0.03	1.5	3.2	5

From UNSCEAR (1988).
Note: *North Europe*—Denmark, Finland, Norway, Sweden; *Central Europe*—Austria, Czechoslovakia (now the Czech Republic, and Slovakia), Germany, Hungary, Poland, Romania, Switzerland; *West Europe*—Belgium, France, Ireland, Luxembourg, Netherlands, United Kingdom; *Southeast Europe*—Bulgaria, Greece, Italy, Yugoslavia (its former territory, before splitting.; *Southwest Europe*—Portugal, Spain; *West Asia*—Cyprus, Israel, Syria, Turkey; *East Asia*—China, India, Japan; *North America*—Canada, United States.

TABLE 1.36. The Contribution of the Most Important Radionuclides to the Total Collective Effective Dose Equivalent Commitment (CEDEC), Including Release Data and Dose Factors.

Radionuclide	Release (PBq)	Dose Factor ($man \cdot Sv \cdot PBq^{-1}$)	CEDEC ($man \cdot Sv$)
^{137}Cs	70	6,100	430,000
^{134}Cs	35	3,400	120,000
^{131}I	330	110	37,000
^{14}C	0.005	110,000	550
^{133}Xe	1,700	0.05	85
^{85}Kr	33	0.21	7
^{129}I	0.00003	170,000	5
^{3}H	2	0.4	1

Based on UNSCEAR (1988).

TABLE 1.37. The Estimations of the Possible Consequences Caused by the Chernobyl Accident.

Region	Population (10^6)	Fatal Cancer	
		Natural (10^3)	Radiation Induced (10^3)
USSR evacuees	0.135	17	0.41
European USSR	75	9,400	11
Asian USSR	225	28,000	2.5
Europe (other)	400	72,000	13
Asia (other)	2,600	450,000	0.6
USA	226	41,000	0.02
Northern hemisphere	3,500	~600,000	28

Adapted from Goldman and Catlin (1988).

less, all assessments show that the expected consequences in terms of additional fatal cancer cases or severe mental retardation and genetic disorder rates will be quite minor. On the global scale, most of them will be within the fluctuations of these health effects due to toxic chemical pollutants and causes other than radiation.

The number of fatal cancers attributed to the Chernobyl accident could possibly increase by as much as 0.004% over the next seventy years (Goldman and Catlin, 1988). About 50% of this increase is projected to occur within the former Soviet territory. More data about the projected Chernobyl health impact is presented in Table 1.37 (Goldman and Catlin, 1988).

1.1.3.6 THE MEDICAL USE OF RADIONUCLIDES

Medical applications of radiation and radionuclides can be divided into three conventional groups, although sometimes other categorization may be used:

- diagnostic examination using external X-ray beams
- radiation therapy based on sealed sources, including external beam therapy or intracavitary therapy (brachytherapy)
- nuclear medicine covering the use of unsealed (open) sources mainly for diagnostic examination, but also to a certain limited extent for therapy treatment

The use of sealed sources does not present a significant hazard as far as actual or potential radioactive contamination is concerned. These sources are properly capsuled, and there are strict requirements regarding tolerated leakage of radioactivity outside the capsule in which the radionuclides are confined. The leakage has to be regularly checked, for example, by wipe tests. Furthermore, storage and eventual disposal of sealed sources can be done without special difficulties.

Regarding radiation protection and possible contamination hazards, the situation is more complicated with the application of unsealed sources, which, in principle, may be in various physical and chemical forms. In *nuclear medicine* special radionuclide-labeled chemical substances, usually called *radiopharmaceuticals,* are administered to humans for the purpose of diagnosis or treatment of both benign and malignant diseases and also for research purposes. The radiopharmaceutical must be in a suitable form to deliver radioactive material to particular parts or organs in the body.

Diagnostic nuclear medicine examinations are based on the *tracer principle,* invented shortly after 1910 by G. Hevesy. He found that radioactive elements had chemical properties identical to their non-radioactive forms, and therefore, they could be used to trace the behavior of some substances under particular conditions. Radioactive tracers are extremely useful for the assessment and study of the dynamic state of body constituents. With the tracers and imaging systems, the movement of labeled molecules could be followed virtually from the very beginning of their entry into the body (regional blood flow, organ function, and other *in vivo* biological processes) to their excretion.

The half-life of radionuclides used in diagnostic nuclear medicine must be suitable to the nature of the particular examinations on the one hand, yet sufficiently long to be fixed for a necessary time interval in the investigated organ on the other hand, but as short as possible in order to limit the dose received by the patient. The length of time during which the radiopharmaceutical remains within the body depends on both the physical half-life $T_{p1/2}$ and biological half-life $T_{B1/2}$ of the radionuclide involved. The resultant *effective half-life* $T_{eff1/2}$ is given as follows:

$$\frac{1}{T_{eff1/2}} = \frac{1}{T_{P1/2}} + \frac{1}{T_{B1/2}} \tag{1.35}$$

One of the first applications of unsealed radioactive sources in medicine was the use of ^{131}I (discovered in the late 1930s) for examinations and investigations of abnormalities in thyroid function. This method made iodine metabolism studies very efficient and reliable.

Imaging methods make it possible to monitor organ or tissue function and to present the results in the form of suitable functional or biochemical pictures of the relevant parts of the body. These pictures (images) can enhance the information about the structure of the selected organs obtained by other imaging methods, such as computerized tomography (CT) or nuclear magnetic resonance imaging (MRI), which can provide information about abnormalities in anatomy. The nuclear examinations, however, can often reveal disease before the anatomical changes show up on a CT or MRI scan. Since some biochemical problems occur before anatomical changes, nuclear medicine has the potential to diagnose certain diseases at an earlier stage than was possible in the past using conventional X-ray imaging methods.

The imaging procedures rely on the detection and measurement of organ activity, usually by gamma cameras directed at the body from the outside. These systems register photons emitted by a radionuclide inside the examined organs or regions of interest and are able to interpret this information in the form of the distribution pattern of radioactivity in the body.

In addition to imaging procedures, diagnostic nuclear medicine also includes the so-called *in vitro* studies. These procedures do not involve the administration of radioactive material to the patient. Such applications require only a small sample of blood or tissue that contains the drug or chemical substance to be measured. The blood sample is then subjected to specific tests using radioactive reagents to quantify the unknown amount of substance present. Most of the *in vitro* tests are based on the *radioimmunoassay* principle. Radioimmunoassay includes a number of different sensitive techniques for measuring antigen or antibody titers with the use of radioactively labeled reagents.

Therapeutic procedures in nuclear medicine are, at present, quite limited, in most cases, to the treatment of thyroid disease with ^{131}I and occasionally to the treatment of polycytemia with ^{32}P as sodium phosphate. The purpose of these procedures is to selectively destroy tumors by beta radiation, delivering the dose locally. It is likely that therapeutic applications of radiopharmaceuticals will increase in the future because of new developments in immunotherapy and in methods utilizing unique metabolic characteristics of both benign and malignant cancers.

Nuclear medicine exposures may be grouped into the following five categories:

(1) Exposure of the patients
(2) Exposure of the hospital personnel
(3) Exposure occurring during the transport of radiopharmaceuticals
(4) Exposure during the manufacture and production of radiopharmaceuticals
(5) Exposure from radioactive wastes

Patient exposures depend on the type of examination and the ad-

TABLE 1.38. Average Activities of the Radiopharmaceuticals Used for Some Common Nuclear Medicine Examinations.

Organ	Radiopharmaceutical	Average Activity (MBq)		
		Germany	Sweden	UK
Thyroid	^{99m}Tc–pertechnetate	37	81	75
	^{123}I–iodine	3.7	1.5	
	^{131}I–iodine	1.9	3	
Liver/spleen	^{99m}Tc–colloid	167	100	90
	^{198}Au–colloid	5.6	18.5	
Renal	^{131}I–Hippuran	1.5	4.5	28
	^{99m}Tc–DTPA	370	163	248
	^{99m}Tc–glucoheptonate	370	370	
	^{99m}Tc–DMSA	111	177	102
Bone	^{99m}Tc–phosphate	555	409	520
Cardiac	^{99m}Tc–erythrocytes	740	540	658
	^{201}Tl–chloride	74	65	68
Lung	^{99m}Tc–microspheres	167	80	88
Brain	^{99m}Tc–DTPA	463	601	536
	^{99m}Tc–pertechnetate	463	403	590
	^{99m}Tc–glucoheptonate		403	570

Adapted from UNSCEAR (1988).

ministered activity of the radiopharmaceutical used. The administered amounts of the radiopharmaceuticals in some common nuclear medicine examinations in various countries are given in Table 1.38. There are some differences between individual countries as far as the average activity used for the same type of examinations. For example, the range of actual thyroid doses related to the ^{131}I scans may lie between 1 and 2 Gy per examination. Developments during recent years have tended to minimize the patient exposure through the introduction of radionuclides with shorter half-lives. The shorter half-life also reduces the problem of radioactive wastes.

For some radiopharmaceuticals used frequently in nuclear medicine, absorbed doses per unit of administered activity for certain organs of interest, including gonads, are given in Table 1.39 (NCRP, 1989b).

The estimation of possible radiological hazards related to nuclear medicine diagnostic examinations can be derived from the effective dose equivalent and collective effective dose equivalent. Some data about the frequency of examinations and effective dose equivalent, as well as the collective effective dose equivalent for the USA in 1982, are given in Table 1.40 (NCRP, 1989b). The per capita annual effective dose equivalent in the United States is higher than that reported for most other developed countries, primarily due to the increased frequency of examinations relative to other countries. These doses in some other countries are as follows: 20 μSv in Australia in 1980, 50 μSv in Denmark in 1985, 60 μSv in Sweden in 1983, and 17 μSv in the UK in 1982 (NCRP, 1989b).

Occupational exposures from nuclear medicine procedures contribute

TABLE 1.39. Absorbed Dose to Organs from Some Representative Radiopharmaceuticals per Unit of Administered Activity.

	Organ Dose ($mGy \cdot MBq^{-1}$)			
Radiopharmaceutical	Ovary	Testes	Whole Body	Other
^{3}H inulin	0.00011	0.00011	0.00013	Kidney 0.0009
^{11}C monoxide (inhal.)	0.0024	0.0024	0.0030	Heart 0.023
^{14}C aminopyrine (oral)	–	–	0.0029	–
^{14}C inulin	0.010	0.010	0.0011	Kidney 0.0077 Bladder 0.15
^{52}Fe ion or citrate	8.0	4–32	8.0	Spleen 7–41 Liver 7–41 Kidney 18
^{57}Co bleomycin	0.28	0.02	0.027	Bladder 0.3 Liver 0.1
^{57}Co vitamin B_{12} (oral)	1.0	0.1	4.0	Liver 25 Kidney 4
^{58}Co vitamin B_{12} (oral)	3.5	0.55	5.0	Liver 100 Kidney 10
^{66}Ga citrate	0.17	0.15	0.18	Colon 0.96 Spleen 0.41
^{67}Ga citrate	0.076	0.065	0.07	Colon 0.19
^{68}Ga citrate	0.013	0.011	0.014	Small intestine 0.0057 Colon 0.046
^{85}Sr ion	1.0	1.0	1.5	Skeleton 4.0 Marrow 4.0
^{99m}Tc pertechnetate	0.006	0.0024	0.004	Stomach 0.07 Thyroid 0.04
^{99m}Tc DMSA	0.005	0.002	0.004	Kidney 0.2
^{99m}Tc glucoheptonate	0.003	0.003	0.003	Kidney 0.06 Marrow 0.003 Bladder 0.03

TABLE 1.39. (continued).

	Organ Dose ($mGy \cdot MBq^{-1}$)			
Radiopharmaceutical	Ovary	Testes	Whole Body	Other
^{99m}Tc phosphate	0.01	0.005	0.003	Kidney 0.01 Skeleton 0.015 Bladder 0.06 Marrow 0.008
^{99m}Tc sulphur colloid	0.0015	0.0003	0.005	Liver 0.09 Spleen 0.05 Marrow 0.007
^{111}In blemycin	0.045	0.03	0.05	Liver 0.5 Spleen 0.2 Marrow 0.1
^{124}I ionic	0.07	0.04	0.27	Thyroid 340
^{125}I fibronogen	0.053	0.044	0.06	Spleen 0.26 Lungs 0.17 Marrow 0.11
^{126}I ionic	0.043	0.032	0.37	Thyroid 660
^{130}I ionic	0.07	0.04	0.084	Thyroid 49
^{131}I ionic	0.04	0.026	0.26	Thyroid: Adult 530 Newborn 3200–8600 1 year 2200–3900
^{127}Xe gas (inhal.)	0.0004	0.0003	0.00035	Lungs 0.007 Bronchus 0.01
^{201}Tl ion	0.10	0.20	0.045	Kidney 0.3 Heart 0.15 Thyroid 0.20

Adapted from NCRP (1989b).

approximately 2% to the collective doses received by radiation workers. On average, the annual individual occupational doses range from 0.3 to 2.0 mSv. Most of these doses are from ^{99m}Tc, which is usually administered by injection. This may lead to relatively high extremity doses especially in the case when syringes are not properly shielded. The patient represents an additional source of exposure to medical and paramedical personnel and sometimes to members of his family also.

This is illustrated in Table 1.41 where absorbed dose rates on the surface of the body, as well as at certain distances from the patient's body, are given.

TABLE 1.40. Annual Effective Dose Equivalents (H_E), Number of Some Common Nuclear Medicine Examinations and Collective Effective Dose Equivalents for the USA in 1982.

Examination	H_E (mSv)	Number of Examinations (× 10³)	Collective Effective Dose Equivalent (man·Sv)
Brain	6.5	813	5,300
Hepatobiliary	3.7	180	700
Liver	3.4	1,424	3,400
Bone	4.4	1,811	8,000
Pulmonary	1.5	1,203	1,800
Thyroid	7.5	530	4,000
Renal	3.1	236	700
Tumor	12.2	121	1,500
Cardiovascular	7.1	961	6,800
Total			32,100
Per capita			140,000

Adapted from NCRP (1989b).

It can be seen that at some time following the administration of a radiopharmaceutical, dose rates drop. The decrease in dose rate depends mainly on the half-life of the radionuclide involved.

The environmental impact of the releases of radiopharmaceutical wastes is not very significant since most radionuclides used in nuclear medicine are characterized by relatively short half-lives. Only small quantities of

TABLE 1.41. Typical Absorbed Dose Rates from Commonly Used Radiopharmaceuticals and Their Time as Well as Distance Profiles.

Type of Scintigraphy and Radiopharmaceutical Used	Typical Range of Activity (MBq)	Dose Rate ($nGy \cdot h^{-1}$ per MBq)					
		Immediately			After		
		Close	0.3 m	1 m	Close	0.3 m	1 m
Bone, ^{99m}Tc MDP	150–600	27	13	4	13	7	2
Liver, ^{99m}Tc colloid	10–250	27	13	4	20	10	3
Blood pool determination, ^{99m}Tc RBC	550–740	27	13	4	20	10	3
Myocard, ^{201}Tl	50–110	36	18	6	36	18	6

Adapted from ICRP (1987).

radiopharmaceutical wastes are treated and disposed of as low-level radioactive wastes. Nuclear medicine patients are administered on the order of 10^8 Bq of radiopharmaceuticals of which a large proportion is excreted in the urine. Small portions of radioactivity are also eliminated in sweat and saliva and can be distributed anywhere that is accessible to the touch, e.g., sinks, faucet handles, door knobs. Most radiopharmaceutical wastes, however, go into the environment through patient excretions via sewers from hospitals and homes (in the case of out-patients). Solid and liquid wastes from hospital and research laboratories are partly disposed of directly into the sewer or along with conventional wastes (as long as activity concentrations and total activity are within the limits set by regulatory authoritiess), or they may be kept for decay before disposal. On the other hand, the widespread use of radioimmunoassay tests and other research methods involving labeled organic compounds may cause some concern because of long-lived ^{3}H and ^{14}C, as well as liquid scintillators usually based on toluene or xylene. For these wastes incineration is sometimes recommended since their disposal into the public sewer system is potentially hazardous and environmentally no longer tolerated. The combustion of solid or liquid wastes can considerably reduce their volume, and the resulting ash can be treated more safely and easily than the bulk of the original radioactive wastes.

Small amounts of radioactive wastes are produced during the manufacturing and preparation of radiopharmaceuticals. In addition, disposal of spent radionuclide generators containing some residual radioactive impurities may sometimes be of environmental concern.

The majority of radionuclides used as radiopharmaceuticals are artificially produced either in a nuclear reactor or in a particle accelerator. The radionuclides produced can be applied directly, or their short-lived radioactive decay products, obtained from *radionuclide generators,* can be used. Such a generator is, in fact, a system that contains a long-lived parent radionuclide that decays to a suitable short-lived daughter, and this is used in preparing the desired radiopharmaceuticals (Kowalsky and Perry, 1987). Because parent and daughter are not isotopes of the same nuclide, chemical separation and isolation of the daughter from the parent is possible. After separation, new daughter radioactive atoms are produced in the radionuclide generator. In this way, a generator provides a fresh supply of short-lived daughter radionuclides when needed until the parent activity is depleted. The useful life of a radionuclide generator thus depends on the half-life of the parent radionuclide. The most common radionuclide generators are listed in Table 1.42.

The impurities present in radiopharmaceuticals may slightly affect the dose received by a patient and also may cause some problems as to the dis-

TABLE 1.42. Properties of Some Radionuclide Generators.

Parent Radionuclide	Parent $T_{1/2}$	Daughter Radionuclide	Daughter $T_{1/2}$
^{99}Mo	67 h	^{99m}Tc	6 h
^{113}Sn	115 d	^{113m}In	1.7 h
^{132}Te	3.2 d	^{132}I	2.3 h
^{68}Ge	270 d	^{68}Ga	68 m
^{87}Y	3.3 d	^{87m}Sr	2.8 h
^{81}Rb	4.5 h	^{81m}Kr	13 s

Adapted from Kowalsky and Perry (1987).

posal of both radiopharmaceuticals and depleted generators. Usually, the contamination of radiopharmaceuticals arises during their production process, either because of activation in the target of radionuclides other than those intended or because of an unavoidable side reaction in the relevant nuclear process (ICRP, 1987). Table 1.43 contains a list of the most frequent radioactive impurities in radionuclides used in nuclear medicine.

The general trend in diagnostic nuclear medicine is toward an increase in the number and frequency of examinations per unit population. Persons undergoing these examinations are usually much older than the average patient having diagnostic X-ray examinations. In the USA, for example, three-fourths of all nuclear medicine examinations are performed on patients over

TABLE 1.43. Radioactive Impurities Found in Some Radionuclides Used in Preparations of Radiopharmaceuticals.

Main Radionuclide	Radionuclide Impurities
^{47}Ca	^{47}Sc (daughter)
^{52}Fe	^{52m}Mn (daughter), ^{55}Fe
^{59}Fe	^{55}Fe, ^{60}Co
^{57}Co	^{56}Co, ^{58}Co, ^{60}Co
^{58}Co	^{57}Co, ^{60}Co
^{81m}Kr	^{81}Rb (parent)
^{99m}Tc	^{99}Mo (parent), ^{99}Tc (daughter), ^{131}I
^{123}I	^{124}I, ^{125}I, ^{126}I
^{125}I	^{126}I
^{111}In	^{113}Sn
^{198}Au	^{199}Au
^{197}Hg	^{197m}Hg, ^{203}Hg
^{201}Tl	^{200}Tl, ^{202}Tl, ^{203}Pb

Based on ICRP (1987).

the age of forty-five and more than one-third on persons older than sixty-four years. It seems that there are not so many possibilities to reduce patients' doses, but the replacement of ^{131}I- by ^{123}I- or ^{99m}Tc-labeled radiopharmaceuticals in thyroid studies can considerably decrease the final dose. A reduction of the absorbed dose can also be obtained by a better choice of the equipment and through a number of simple precautions.

1.1.3.7 OTHER SOURCES OF RADIOACTIVITY

In addition to the nuclear industry and nuclear-related activities and applications, there are some other practices and fields, essentially *non-nuclear,* where radioactive materials are used or unintentionally generated. The radioactivity of such sources represents direct or potential impact on the level of the radioactive contamination of the environment.

Certain radionuclides are used in a number of *consumer products* and some special applications, for example, luminescent paints, batteries for pacemakers, eliminators of static electricity, smoke and fire detectors, ceramics and U-Th alloys, stabilizers of the electric discharges in gauges, and analyzers of gases using ionization, as well as many different tracer techniques (they will be discussed in detail later). These may be considered possible contributors to the additional ambient radioactivity. Other examples may also include tobacco products containing ^{210}Pb and ^{210}Po, combustible fuels and building materials accommodating members of the uranium and thorium decay series, and gas mantles, camera lenses, and welding rods containing thorium.

The manufacture and use of *luminescent paints* is the oldest and most widespread of all applications of the radionuclides. These paints were especially used for radioluminescent dials and markers of various kinds. The luminescent paints take advantage of the permanent state of excitation of a fluorescent compound maintained because of the radiation emitted by a radioactive salt mixed with it. Alpha and beta emitters, such as ^{226}Ra, ^{90}Sr, ^{147}Pm, and ^{3}H, have been used in these applications. Because of the high toxicity of ^{226}Ra, it has been largely replaced by the pure beta radionuclides ^{147}Pm and ^{3}H, which are characterized by considerably lower radiotoxicity than alpha emitters. There are, however, still many millions of timepieces and alarm clocks, thousands of aircraft and marine instrument dials and other similar devices employed in night activities (compasses, etc.) or safety indicators (direction arrows, exit signs)—all containing radium-based luminescent paints. The activity of radium used in a typical wrist-watch is between 3 to 10 kBq, while dials of aircraft and marine board instruments may contain radium activities even ten times higher.

Radium-bearing consumer products may cause radiological hazards, which include:

(1) External exposure by emitted gamma radiation – For example, the dose rate to which the wrist is subjected from an older type radium watch may reach about 7 $nGy \cdot h^{-1} \cdot Bq^{-1}$. This would result in an annual dose equivalent of 24 mSv if the watch were worn continuously (Kathren, 1984).

(2) Release of radon, which is continuously produced at the rate of approximately 1 $mBq \cdot h^{-1}$ per unit activity of radium (1 Bq) – Since neither timepieces nor instrument dials are tightly sealed, leakage of radium may also be expected.

(3) Radioactive contamination of the environment through the scrapped or abandoned watches or instruments after their useful lifetime – The contamination may also be the result of breakage of the radium-containing component.

Radionuclide batteries have been used successfully as sources of electrical energy not only in scientific and communication instruments aboard space satellites, but also in pacemakers. The heat produced by the absorption of the radiation emitted by an appropriately chosen radionuclide can be converted into a current generator using the Peltier effect. The most suitable radionuclide for these applications seems to be ^{238}Pu, an alpha emitter with a half-life of 87.8 years. A typical plutonium battery used in pacemakers contains about 150 mg of ^{238}Pu, i.e., nearly 100 GBq. The lifetime of such a battery is usually longer than ten years. The battery must be designed and constructed in such a way as to resist the progressive increase of its internal pressure due to the release of helium. In addition to this problem, the use of radionuclide batteries in pacemakers always results in a small exposure by gamma photons emitted in the process of ^{238}Pu decay. The risks associated with the plutonium batteries have led to further development and perfection of long-life chemical batteries so that this type of battery is no longer used in pacemakers.

The ionization of air by alpha or beta particles emitted by a conveniently chosen radionuclide can be used for the *neutralization of static electricity.* For this purpose suitable radionuclides are ^{241}Am, ^{238}Pu, ^{210}Po, ^{239}Pu (alpha emitters), and ^{3}H, ^{85}Kr, ^{204}Tl (beta emitters) in the form of rods or ribbons placed in the proximity of charged material. Attention should be paid to the installation and maintenance of these sources to prevent leakage of radioactivity to the environment and ensure their adequate disposal.

Smoke and fire detectors represent another example of the use of a radionuclide, such as ^{241}Am or ^{238}Pu, in combination with a small open ionization chamber. The source, placed inside the chamber, will produce ionization, and this results in a constant current. When smoke particles enter the chamber, they form centers of condensation, which, due to recombination, causes a decrease in the ionization current. The drop in electrical current then signals fire. The activities of ^{241}Am used in these devices is in the range from about 30 kBq (recent models) to more than 100 GBq (earlier models). The possible hazards related to the use of smoke detectors is partly connected to their fate at the end of their useful life and dispersion of radioactivity in the case of fire.

With the almost full replacement of gas for lighting purposes by electricity, the *gas mantles* containing thorium are now no longer a big problem. These mantles are now used only for camping lanterns and in places without electricity. The dose to campers and other users of gas mantles is estimated to be less than 10 nSv per year (Eisenbud, 1987).

Most of the additional radioactivity, however, comes from some *industrial processes* that bring to the surface of the earth materials characterized by the above average concentrations of the naturally occurring radionuclides. Such industrial activities include geothermal work, phosphate mining and the production of fertilizers from phosphate rock, coal mining and coal burning, and non-uranium ore mining and processing. Since the hazards from radioactivity in these industrial processes is usually much smaller than the hazard from other chemical substances, neither radiation nor radioactive contamination are systematically monitored and evaluated (UNSCEAR, 1988).

Although there are other alternatives now available, *energy production from coal* is still dominant in many countries. It is assumed that, on the average, about 70% of the coal is burned in electric power stations, 20% in coke ovens, and about 10% in houses for cooking and heating. Coal, like the majority of materials in nature, contains trace quantities of natural radionuclides, namely ^{40}K, ^{238}U, and ^{232}Th and their decay products. In the process of burning, the activities of these radionuclides are *redistributed* from underground into our environment. The content of radionuclides in coal varies substantially; for example, in coal mined in the USA the concentrations of ^{238}U range from 1 to as high as 1000 Bq·kg^{-1}. The UNSCEAR assumes that the average activity concentrations of ^{40}K, ^{238}U, and ^{232}Th in coal are 50, 20, and 20 Bq·kg^{-1}, respectively.

Generally speaking, radiation exposure due to coal comes from all stages of its fuel cycle, which includes

- coal mining
- various uses of coal
- use of fly ash

Coal mining results in small additional exposures of both coal miners and members of the public mainly through *radon and its decay products.* While the miners are subjected to increased concentrations of radon directly, other people are exposed to radon present in the exhaust air of coal mines. The activity of radon released per year from coal mining all over the world is estimated to be 30–800 TBq, which results in collective effective dose equivalent commitments of about 0.5 to 10 man·Sv. The exposures of miners rarely exceed 1 mSv of effective dose equivalent per year. The upper estimate of the total collective effective dose equivalent of all world miners in 1980 was about 2000 man·Sv (UNSCEAR, 1988).

The following figures show that the radiological impact of the *uses of coal* (electrical energy production, carbonization, heating in dwellings) is much higher than the impact of coal mining. A coal-fired power plant produces about four times more fly ash than bottom ash. Average concentrations of individual radionuclides in fly ash released from these power plants are as follows: ^{40}K–265 Bq·kg^{-1}, ^{238}U–200 Bq·kg^{-1}, ^{226}Ra–240 Bq·kg^{-1}, ^{210}Pb–930 Bq·kg^{-1}, ^{210}Pb–1700 Bq·kg^{-1}, ^{232}Th–70 Bq·kg^{-1}, ^{238}Th–110 Bq·kg^{-1}, and ^{228}Ra–130 Bq·kg^{-1}. The actual activities of natural radionuclides discharged by coal power plants to the atmosphere depend on many factors, one of which is related to the fraction of the fly ash released. Old power plants usually release about 10% of the fly ash produced, while the modern plants equipped with sophisticated retention devices release only 0.5% of the fly ash. The combustion of 3×10^9 kg of coal, required to produce 1 GWe·y of electric energy, may result in releases of individual radionuclides in a wide range: ^{40}K from 350 to 200,000 MBq per GW·y, ^{238}U from 70 to 23,000 MBq per GW·y, ^{226}Ra from 70 to 23,000 MBq per GW·y, ^{210}Pb from 100 to 81,000 MBq per GW·y, ^{210}Po from 300 to 74,000 GBq per GW·y, ^{232}Th from 40 to 34,000 MBq per GW·y, ^{238}Th from 40 to 14,000 MBq per GW·y depending on the activity concentration in coal, the efficiency of the filtering system and some other factors.

The dominant pathways through which the populations living around coal-fired power plants are exposed to enhanced levels of natural radionuclides are inhalation during the cloud passage and external irradiation and ingestion following deposition of radionuclides on the ground. The collective effective dose equivalent commitments received via these pathways, normalized to the generation of 1 GW·y of electric energy, are presented in Table 1.44 (UNSCEAR, 1988).

The *domestic use of coal*—for cooking and heating—may result in high collective dose commitments because the chimneys of most residential buildings are usually low and without any filtration systems, and the density of population in the vicinity of the sources of emission is generally high.

It is estimated that about 280 million tons of coal ash are produced annually in all coal-fired power plants in the world. The average activity concentrations of ^{40}K, ^{238}U and ^{232}Th in coal ash are estimated to be about 400,

TABLE 1.44. Estimates of Collective Effective Dose Equivalent Commitments per Unit Energy Generated Resulting from Atmospheric Releases from a Typical "Old" and "Modern" (in parentheses) Coal-Fired Power Plant.

Radionuclide	Inhalation during the Cloud Passage	Exposure due to Deposited Activity	
		Internal	External
^{238}U	15 (0.75)	2 (0.1)	
^{234}U	17 (0.9)	2 (0.1)	
^{230}Th	60 (3)	3 (0.15)	
^{226}Ra	2.3 (0.07)	3 (0.15)	12 (0.6)
^{222}Rn & daughters	0.1 (0.1)	20 (20)	
^{210}Pb, ^{210}Po	14 (0.7)	66 (3.3)	
^{232}Th	290 (14)	1 (0.05)	
^{228}Ra & daughters	60 (3)	6.7 (0.3)	18 (0.9)
^{220}Rn & daughters	–	30 (1.5)	

Based on UNSCEAR (1988).

150, and 150 $Bq \cdot kg^{-1}$, respectively. When we compare this with the activity concentration in normal soil, the average content of ^{40}K is similar, but the concentrations of ^{238}U and ^{232}Th are about six times higher.

Coal ash is used especially in the manufacture of cement and concrete, as well as in some other building materials. This may result in an increase in the indoor doses due to external irradiation and especially through the inhalation of radon decay products.

Other types of energy production, such as *geothermal energy* and *combustion of oil and natural gas,* may also, to a certain extent, contribute to environmental radioactivity. *Geothermal energy* is extracted through hot steam or water derived from high-temperature rocks located deep inside the earth. Most of the radionuclides found in geothermal fluids belong to the uranium decay chain. Although solid radioactive elements could slightly pollute water, only radon released directly into the atmosphere or from water to buildings may cause some problems (Sabol et al., 1993).

While geothermal energy plays an insignificant role so far, *oil* has a dominant position and is widely used in many applications, such as road transport, the production of electricity, and domestic heating. Ash from oil-fired power plants, which are not usually equipped with filtration systems, contains natural radionuclides in concentrations similar to those in coal ash.

Like oil, natural gas is also a very important source of energy used in domestic heating, the generation of electricity, and in a number of applications in industry. The only radionuclide present in natural gas that is of interest in radiation protection is radon. Its concentration in gas may vary widely around a typical value of 1000 $Bq \cdot m^{-3}$ (UNSCEAR, 1988).

Radiation exposures attributable to discharges of some natural radionuclides from various systems producing electricity are summarized in Table 1.45. The exposures are given in terms of collective effective dose equivalent commitments normalized to the production of 1 GW·y of electrical energy. We can see that by far the highest radiation doses come from the use of coal.

Another source of additional radioactivity is related to the *use of phosphate rock*—the main source of phosphorus for fertilizers. The concentrations of natural radionuclides in phosphate rock depend on the type and origin of the rock. Most phosphate rock is of sedimentary origin (about 85%); the rest is of volcanic and biological origin. An activity content of ^{232}Th and ^{40}K in phosphate rocks of all types is similar to the concentrations normally found in soil. On the other hand, the concentrations of ^{238}U and members of its decay chain tend to be elevated in phosphate deposits of sedimentary origin. A typical content of ^{238}U in phosphate rock of sedimentary origin is about 1500 $Bq \cdot kg^{-1}$.

The radiation doses related to phosphate rock include both occupational and public exposures. Occupational exposures come mainly from the mining, processing, and transportation of phosphate rock, as well as from the transportation and use of phosphate fertilizers. Exposures of members of the general public are principally due to: (1) effluent discharges of materials containing radionuclides of the ^{238}U decay chain into the environment from phosphate rock mining and processing, (2) the use of phosphate fertilizers, and (3) the use of various by-products and wastes.

TABLE 1.45. Collective Effective Dose Equivalent Commitment to Members of the Public Attributable to Various Plants for the Production of Electricity.

Type of Plant	Collective Effective Dose Equivalent Commitment per Year of Practice (man·Sv)	Normalized Collective Effective Dose Equivalent Commitment (man·Sv per GW·y)
Coal-fired	2000	4
Oil-fired	100	0.5
Natural gas	3	0.03
Geothermal	3	2
Peat	–	2

TABLE 1.46. Estimates of Collective Effective Dose Equivalent Commitment to the World Population Arising from One Year of Exploitation of Phosphate Rock.

	Collective Effective Dose Equivalent Commitment (man·Sv)	
Source of Exposure	Members the Public	Workers (External Exposure)
Phosphate industrial operations	60	20
Use of phosphate fertilizers	10,000	50
Use of by-products and waste	300,000	Not estimated

From UNSCEAR (1988).

The total collective effective dose equivalent commitment attributable to the use of phosphate rock is estimated to be about 300,000 man·Sv. This exposure is essentially due to the use of some by-products, especially gypsum in dwellings. The hazards from occupational exposures are much smaller than the impact of doses received by members of the public. The relevant estimates by the UNSCEAR are listed in Table 1.46.

Non-uranium ore mining and processing also give rise to small additional exposures, which may sometimes be significant for miners, but have very little impact on the doses to members of the public. This is illustrated in Table 1.47 where estimates of collective effective dose equivalent commitments arising from one year of practice for most industrial activities are given for comparison. Individual exposures due to these various activities are generally small in comparison to the overall doses from natural radiation background. The total annual contribution from all industrial activities is less than 5×10^5 man·Sv, which corresponds to a per capita effective dose equivalent commitment equal to about 100 μSv.

1.1.4 Radiotracers in Science and Technology

In the preceding parts of this chapter, mainly two basic sources of low-level radioactivity have been analyzed:

- radionuclides occurring in nature as *naturally represented components*
- radionuclides penetrating into the environment in an undesirable way in consequence of *human activity*

In addition to these two basic groups of low-level sources, there is a third one, where radioactivity is intentionally introduced into the system. This tracer methodology is represented by the introduction of one or more radionuclides into the experimental system as *radiotracers*. By determining the distribution of radiotracers in the investigated system, an illustration of the course of the studied phenomena and/or structures of the investigated parts of the system is obtained. Radiotracer methodology is widely used, especially in the life sciences, but also in various other fields, e.g., in industrial engineering and technology. It represents not only an extremely powerful tool, but, in connection with low level analysis, it is also extremely sensitive.

Either a radioisotope of the investigated element with a sufficient half-life is used as a radiotracer (for example, in solving problems with heavy metals ^{109}Cd, ^{115m}Cd, ^{203}Hg, etc.), or compounds are labeled with radionuclides. During radiolabeling, the stable isotopes in molecules of the labeled substance are exchanged for radioactive ones (for example, radioactive ^{14}C is substituted for the stable ^{12}C). This substitution is achieved by chemical synthesis from a certain simple basic labeled compound (e.g., ^{14}C-labeled barium carbonate), biosynthesis (e.g., by cultivating a certain plant in a $^{14}CO_2$ atmosphere), or catalytic exchange reactions (mainly labeling with ^{3}H or ^{125}I). The preparation of labeled compounds is connected to

TABLE 1.47. Estimates of Collective Effective Dose Equivalent Commitments per Year of Practice for Various Industrial Activities Including Non-Uranium Mining and Processes.

Source of Exposure	Collective Effective Dose Equivalent Commitment (man·Sv)	
	Public	Workers
Coal combustion:		
Power plants	2,000	60
Homes	2,000–40,000	
Coal mining	0.5–10	2,000
Use of coal ash in the building industry	500,000	
Geothermal energy production	3	
Oil combustion in power plants	100	
Natural gas combustion in power plants	3	
Phosphate industrial operations	60	20
Use of phosphate fertilizers	10,000	
Use of phosphategypsum in houses	300,000	
Non-uranium mining	Small	20,000

From UNSCEAR (1988).

a series of specific problems, which are analyzed in detail in specialized monographs (Buncel and Jones, 1987; Jones, 1988).

If a substance labeled with a radionuclide is used in a tracer experiment, it is desirable that it should be radiochemically and radionuclide pure (i.e., without impurities of other substances or other radionuclides), as well as being stable under certain storage and experimental conditions (the bindings of the entire molecule, as well as of the incorporated radioindicator, must be stable), and it should have sufficient specific activity. The achievement of a certain value of specific activity is given primarily by the procedure used in the preparation of the labeled compound, and also by the requirements on the location of the labels in the molecule. This can be either specific of a given position (e.g., deoxy[5-^{3}H]cytidine or L-[methyl-^{14}C]carnitine hydrochloride) or without a defined position (e.g., D[U-^{14}C] glucose). The position of radionuclides in the molecules of the labeled compound are given in square brackets, as above.

The fate of the same radiotracer in a chosen experimental system is quite different according to the labeled substance and/or its application. Our study of the rat exposure to heavy metals using their radioisotopes offers an intuitive example of such low-level counting. The binding of mercury to soluble proteins from different rat tissues (Table 1.48) was examined following intraperitoneal ([^{203}Hg]chloride) or intravenous (methyl-[^{203}Hg]chloride) administration (Tykva et al., 1987a). It is evident from Table 1.48 that, in both cases, a large amount of mercury is tightly bound to protein, but the values in kidney differ considerably. Also, the specific activity of the applied compound has to be taken into consideration because usually its total amount used in the experiment is decisive for the course of the investigated effect.

Radiolabeling should always be considered from these viewpoints in a complex manner. Thus, for example, the labeling with ^{3}H of a certain compound permits attainment of an average of three orders of magnitude higher specific activity in comparison with ^{14}C, with a less expensive catalyzed exchange, in comparison with the usually considerably more difficult and time-consuming multistep synthetic preparation in ^{14}C-labeling. However, in contrast to this the ^{14}C-labeling displays a greater stability of the bond during storage; moreover, a substantially smaller ratio of the masses ^{14}C/^{12}C than ^{3}H/^{1}H removes possible distortion of the reaction kinetics analyzed by radiotracer techniques (isotope effects). Thus, the conditions for measuring low activities are frequently created by the limitation of the accessible specific activity of the applied radiotracer.

The low activity of radioindicators may also be due to a number of other causes. The starting active substance itself, containing the radioindicator and introduced into the experimental system, represents a source of a low

TABLE 1.48. Distribution of ^{203}Hg in Soluble Tissue Proteins (A $^{203}HgCl_2$; B Methyl$^{203}HgCl$) (S.D. $\leq$ 3%).

	10^{-2} % of Appl. Activity per mg Proteins					
			Protein-Bound			
	Total after Extraction				Mercaptoethanol Added	
	A	B	A	B	A	B
Liver	0.072	0.197	0.070	0.078	–	–
Kidney	11.20	2.694	6.600	0.817	0.02	0.005
Plasma	0.028	0.094	0.023	–	–	–
Skin	0.219	–	0.172	–	–	–
Hair	0.023	0.784	–	–	0.0003	0.005

Not measured indicated by –, Tykva et al. (1987a).

activity even before use. It consists in the controls of its radiochemical purity, the presence of side-products contaminated with the radionuclide and formed during its preparation, or products of radioanalysis taking place during storage. In the system itself the measurement of low activity can take place, both owing to the presence of small amounts of radionuclide-labeled investigated substances or their metabolites in the measured site (in consequence of the mechanism of the process followed), and also after using the low activity of the applied entering compound. This low entering activity is caused not only by the low achieved specific activity of the applied compound, but also by its limited quantity (for example, in order to preserve physiological conditions). Sometimes the low activity of the radioindicator is used to exclude radiation damage to the system.

From the total arrangement of the radiotracer technology used, it is evident that the levels of low radioactivities are usually higher after an intentional introduction of radioactivity into the system than in environmental radioactivities. Nonetheless, the possibility of measuring low activity extends the use of tracer techniques considerably. However, the measurement of the low activity of radiotracers has to be interpreted prudently in view of the character of the experimental system, for example, with respect to the variability of the biological material, the possibility of artifacts formation (for example, sorption), etc. Very frequently the determination of the low activity of the radioindicator in itself is not sufficient without additional analyses that characterize its chemical bond in the active compound more closely, for example by radiochromatography (Roberts, 1979).

Thus, radiotracer methodology enriches the theme of low levels of radionuclides by the insertion of detectors themselves into sophisticated PC-monitored and evaluating devices (Tykva et al., 1988), which are used for imaging of different effects. Frequently, the determination of low levels is also combined with separation micromethods, e.g., gel electrophoresis (Tykva and Votruba, 1972), thin-layer chromatography (Tykva and Votruba, 1974), gas chromatography (Tykva and Šeda, 1975), or two-dimensional combination of chromatography and electrophoresis (Tykva and Franěk, 1977).

Useful methodical approaches offer multilabeling, using more than one radiotracer simultaneously in the experimental system. This makes possible:

- The fate of more substances labeled individually with different radiotracers can be investigated in the same experimental system (Tykva et al., 1992).
- The responsibility of the individual parts of the molecule of the

investigated substance can be studied using their labeling with different radiotracers (Tykva and Bennettová, 1993).

- The dependencies of the investigated effects on the used amount of the applied substance and/or on the time interval of application and/or the way of application can be followed by the use of the appropriate substance labeled independently with more radiotracers.

Some examples of such radiolabeling we have summarized previously (Tykva, 1974). The measuring arrangement is briefly described in Section 2.1, from which it follows that multilabeling further increases the demands on low-level methodology (Tykva, 1994).

Sometimes other analytical methods, indispensable for a complex evaluation of the investigated system, are also applied simultaneously, for example gas chromatography in combination with mass spectrometry (GC-MS method), supplied by a library of reference spectra.

The use of the techniques of measurement of low-level radioactivities in tracer methodology is very widespread. In spite of the development of a number of highly sensitive non-nuclear analytical methods during the 1980s, radioindicator methods remain irreplaceable for the solution of a number of goals of present fundamental and applied research, mainly in the chemical, biological, and medical fields (for example, radiochemistry, biochemistry, physiology, toxicology, morphogenesis, pharmacology, analysis of solid substance surfaces, and a number of other fields).

The basic reason for the application of radiotracer methodology is its high detection sensitivity. Thus, in our experimental arrangements we followed the detection sensitivity of fractions obtained by separations by high-pressure liquid chromatography–HPLC: the sensitivity of the activity measurement of individual fractions labeled with radiotracer ^{14}C was two orders of magnitude higher than in the detection of the same non-labeled compounds by means of standard UV absorption under the same chromatographic arrangement.

Figure 1.27 illustrates a record from the outlet of an HPLC chromatographic column (Tykva and Bennettová, 1993). From the measured peaks it is evident that the detection of two radioactive fractions is not measurable under standard HPLC detection conditions using DAD detector (the specific activity of the applied [benzene-U-^{14}C]carbamate juvenoid was 2.294 GBq/mmol at a molecular weight of 363.4; the highest radiopeak represents the applied juvenoid). This example demonstrates clearly the advantage of the radiotracer methodology, which permits – in view of the high detection sensitivity – researchers to follow compounds that are not detect-

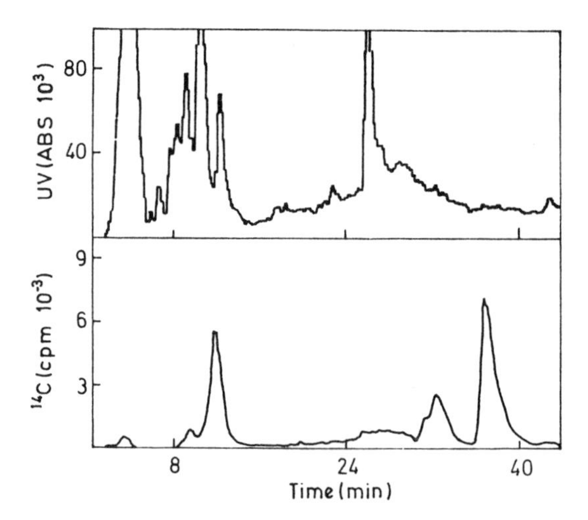

FIGURE 1.27. UV and radioactivity detection after HPLC of an aliquot of the whole-body extract from three flies on day seven after individual topical applications of 185 kBq of ^{14}C-labeled carbamate juvenoid on each *Sarcophaga bullata*, Tykva and Bennettová (1993).

able by sensitive non-nuclear methods (the detection sensitivity was two orders of magnitude higher by radio-HPLC).

The fundamental advantage of the measurement of low-level radioactivity is the increasing of detection sensitivity of radiotracers, which is the main advantage of tracer methodology.

In addition to tracer methodology, where radionuclides are introduced into the experimental system, analogous low-level measurements require activation analyses as well. In these, the content or the distribution of the inactive element is determined by transmutation of its nucleus, either with a neutron (neutron activation analysis – NAA) or with an accelerated charged particle (e.g., proton induced X-ray emission – PIXE), and the activity of the radionuclide formed in this way is then measured. For more detailed data, see Valkovič (1980). These procedures are used predominantly for the determination of trace elements in biological material (see Section 3.3.1). At this point they are mentioned only because of their similarity to tracer methodology.

Finally, the procedures and/or arrangements elaborated for the determination of low levels of radionuclides also find application in various experiments in physical research without radiotracing, especially in the investigation of very rare phenomena, such as the double beta decay of ^{78}Ge, one of the rarest processes in nature (Heusser et al., 1989) or the search on two-neutrino and neutrinoless double beta decay of ^{136}Xe in the underground laboratory (Bellotti, 1992). On the other hand, sometimes measuring devices for physical research such as a multiwire proportional chamber (Stanislaus et al., 1992) are also used in low-level counting. The importance of the fundamental knowledge on the determination of low activity is thus extended indirectly to the methods of further scientific disciplines.

1.1.5 Low-Level Methodology in Space Research

Besides radionuclides in the earth and on its surface, including the atmosphere, radionuclides in the heliosphere should be considered as a source of low levels to be measured.

In the 1970s the accessibility of various samples from the moon permitted scientists to carry out their spectrometric low-level analysis, thus indicating the possibilities for investigating further *planets* of the solar system in the future. *Meteorites* are another source of cosmic samples with low activity (Surkov, 1974).

Solar and galactic cosmic radiation falling on the moon causes nuclear transmutations, leading to the formation of cosmogenic radionuclides. Determination of their low-level concentrations and their ratios helps in investigations on the chronology of lunar processes, as well as in the study on the solar system and its history. The low-level specific activities of primordial radionuclides give valuable information about the development of the moon.

The results of the low-level gamma spectrometry (Tykva, 1975) of the lunar regolite obtained by the Russian expedition Luna-20 are summarized in Table 1.49 (the accuracy was given primarily by a very small amount available). A similar relatively high $^{22}Na/^{26}Al$ ratio was found in other lunar samples, except those collected by the expedition Apollo-17 after an increased flow of solar protons during solar eruptions in 1972 (O'Kelley et al., 1974). The values of uranium and thorium given in Table 1.49 are close to those exhibited by surface soils of the Taurus-Littrow region (Silver, 1974) or in fines and rocks from the Descartes region (Nunes et al., 1973). However, our values are considerably lower than in samples collected in the

TABLE 1.49. Specific Activities of Cosmogenic and Primordial Radionuclides Collected on the Surface of the Moon by Luna-20.

^{22}Na ($Bq \cdot kg^{-1}$)	^{26}Al ($Bq \cdot kg^{-1}$)	K[a] (%)	U (ppm)	Th (ppm)
1.27 ± 0.45	1.40 ± 0.42	0.08 ± 0.02	0.28 ± 0.09	1.01 ± 0.20

[a]Estimated from ^{40}K activity, Tykva (1975).

Fra Mauro formation (Tatsumoto, 1972) and most samples from the Lunar mare regions (O'Kelley et al., 1971). It may be concluded from the above comparisons that the development of the different parts of the surface of the moon was considerably different.

Another possibility is to analyze planetary gamma-ray emission spectra by orbital remote sensing. In this way, the mean concentration of as many as fifteen elements in the upper tenths of a meter of the surface can be observed. Thus for about 15% of the lunar surface the maps of the chemical composition of a number of key elements have been obtained (Yadav and Arnold, 1990).

In addition to the activity determination of radionuclides in cosmic materials, the low-level methodology in space research has another important application. It is its use to analyze cosmic radiation, which includes, as a rule, low *fluences of different particles.* The analyses of these fluences contribute to the elucidation of various space phenomena, either galactic (Simpson, 1983) or within the heliosphere as, for example, solar events (Plaga and Kirsten, 1991). These possibilities comprise a very broad extent of problems of investigations of the universe, and they are, therefore, the subject of regular international meetings. Satellites provide very good experimental conditions since they are outside the influence of the atmosphere. Several silicon spectrometric systems for such space laboratories have been prepared to identify particles without fluxes of charged particles using time-of-flight resp. $\Delta E/E$ spectroscopy based on measurement of the time interval that it takes an analyzed particle to pass a given length resp. on measurement of energy ΔE_i lost in the individual input thin detectors of a telescope and the total particle energy E. So, in the project Intershock investigated the fluence rates of different light nuclei in certain sites of space (Tykva et al., 1987b), the phenomena taking place on the sun was investigated.

A further area of application of low-level dose methods is the measurement of the instantaneous low-level dose to which a cosmonaut is subjected during the changing conditions of the satellite flight. In this manner we have determined the course of an exposure of the crew within short time intervals at the cosmic station Mir (Dachev et al., 1992).

Thus, since the 1970s, space research affords an extensive source of requirements on the development of highly sensitive detection methods for ionizing radiation.

1.2 PROPERTIES OF EMITTED RADIATION AND ITS INTERACTION WITH MATTER

All radionuclides release energy in the form of *ionizing radiation,* which can be divided into two groups—*directly* ionizing radiation and *indirectly* ionizing radiation. The first category includes *charged particulate radiation* such as electrons (both beta particles—positive or negative—and energetic electrons produced by other processes), as well as heavy charged particles, e.g., protons, alpha particles, fission products, and products of some other nuclear reactions. The second category of ionizing radiation comprises *uncharged radiation,* which includes two main types: *electromagnetic radiation* (gamma radiation, X-rays) and *neutrons.*

Toxicity and radiological impact of individual radionuclides depend, to a large extent, on the radiation emitted in the process of their disintegration. In addition to health effects, applications of radionuclides, as well as their measurements, are also related and based on the parameters of the radiation resulting from the decay of radioactive nuclei. This radiation forms a radiation field whose properties reflect the source involved. The radiation source can usually be quantified by means of the detection of particles or electromagnetic quantities producing the characteristic field around the source.

All detection and spectrometric methods rely on the interaction processes accompanying the passage of radiation through a suitable medium. There is a principal difference between the interaction of directly and indirectly ionizing radiation. While the charged particles lose their energy in matter more or less continuously, the uncharged radiation interacts with matter with a certain probability, and energy losses in one interaction event are relatively large.

1.2.1 Quantities Characterizing a Radiation Field

Every radiation source will give rise to a radiation field characterized in the first place by the parameters of the source and also by the properties of the surrounding material. Both radiation and radioactivity measurements require various degrees of specification of the radiation field at the point of interest in free space or in matter. Many quantities and units can be used in the description of a radiation field. The selection and definition of appropriate quantities for radiation and radionuclides have been a primary task of the International Commission on Radiation Units and Measurements (ICRU). Its basic report on radiation quantities and units (ICRP, 1980) represents the fundamental guidelines that have been widely adopted and used in all radiation or radioactivity related areas. The ICRP is now preparing a revised form of its report devoted to radiometry, interaction coefficients, dosimetry, and radioactivity (Allisy et al., 1992). Our approach in introducing quantities and units relevant to both radiation fields and to the interaction of radiation with matter relies mainly on the above-mentioned materials.

We may use two types of quantities to describe a radiation field produced by one or more sources: (1) the first class refers to the number of particles, and (2) the second type is related to the energy carried or transported by these particles.

The most general quantities associated with the radiation field are *particle number (N)* and *radiant energy (R)*. The full specification of the radiation field, however, requires more details on the nature and energy of the particles, as well as their spatial, directional, and temporal distributions. More detailed description of the radiation field can be achieved by introducing other quantities through certain successive differentiations of N and R. Consequently, such quantities may be considered as point functions with regard to each variable of differentiation. This means that these quantities relate to a particular value of this variable. In this way we can define such quantities as *fluence* (Φ) or *energy fluence* (Ψ).

The *particle number, N,* is the number of particles that are present, emitted, transferred, or received. This is a dimensionless quantity, and its unit is 1.

The *radiant energy, R,* represents the energy (excluding rest energy) of particles that are present, emitted, transferred, or received. The basic unit of radiant energy is J (joule).

Both particle number and radiant energy are defined in very general terms and can only be used with additional information such as type of particles, geometry, time, etc. It is obvious that for mono-energetic particles, each of energy E, the radiant energy is equal to $N \cdot E$. As long as the distribution in energy of the particle number, i.e.,

$$N_E = \frac{\mathrm{d}N}{\mathrm{d}E} \tag{1.36}$$

is known, we can write for the radiant energy the following relation:

$$R = \int_E E N_E \mathrm{d}E \tag{1.37}$$

The particle number (N) and the radiant energy (R) are often used in the derivation of the other, more specific quantities, which are defined as differential quotients. The numerator is the differential of the actual or expected value of N or of R, and the denominator is usually the differential of time, energy, area, direction, or a product of them.

The most important quantities for the specification of the radiation field are *fluence, fluence rate, energy fluence,* and *energy fluence rate.*

The *fluence* (Φ) is defined as the quotient dN over da, where dN is the number of *particles* incident on a sphere of cross-sectional area da, i.e.,

$$\Phi = \frac{dN}{da} \tag{1.38}$$

The unit of fluence is m^{-2}.

The *energy fluence* (Ψ) represents the quotient of dR over da, where dR is the radiant energy incident on a sphere of cross-sectional area da,

$$\Psi = \frac{dR}{da} \tag{1.39}$$

The SI unit of the energy fluence is $J \cdot m^{-2}$. In principle, such units as $MeV \cdot m^{-2}$ or $keV \cdot cm^{-2}$ can also be used.

The definition of the fluence as well as the energy fluence is based on a sphere. Its cross-sectional area da must be perpendicular to the direction of each particle. This definition applies equally well to mono-directional or multidirectional particles.

Both quantities are applicable in the common situation in which radiation interactions are independent of the direction of the incoming particles. In other situations, however, other quantities, involving the differential solid angle (dΩ) associated with a specific direction (Ω), may be required.

In practice, the distribution in energy of the fluence is commonly used; therefore, the short name *spectric fluence* has been recently recommended for this quantity (Allisy et al., 1992).

The *spectric fluence* (Φ_E) is introduced as the quotient of dΦ over dE, where dΦ is the fluence of particles of energy between E and E + dE, i.e.,

$$\Phi_E = \frac{d\Phi}{dE} = \frac{d^2N}{da \cdot dE} \tag{1.40}$$

The unit of spectric fluence is $J^{-1} \cdot m^{-2}$.

A further important quantity is the *fluence rate* (φ), which is defined as

$$\varphi = \frac{d\Phi}{dt} = \frac{d^2N}{da \cdot dt} \tag{1.41}$$

where dΦ is the increment of particle fluence in the time interval dt. The unit of the fluence rate is $m^{-2} \cdot s^{-1}$.

A quantity similar to the previous one is the *energy fluence rate* (ψ) which is the quotient of dΨ over dt

$$\psi = \frac{\mathrm{d}\Psi}{\mathrm{d}t} = \frac{\mathrm{d}^2R}{\mathrm{d}a \cdot \mathrm{d}t} \tag{1.42}$$

where dΨ is the increment of the energy fluence in the time interval dt. The SI unit of the energy fluence rate is $\mathrm{W \cdot m^{-2}}$.

In some cases, when the field is characterized by polyenergetic radiation, we may refer to the *mean* ($\bar{E}$) or, as it is sometimes called, *effective energy* (E_{ef}), weighted by influence or by energy fluence

$$\bar{E} = \frac{\int_0^{E_{max}} E\Phi_E \mathrm{d}E}{\int_0^{E_{max}} \Phi_E \mathrm{d}E} \tag{1.43}$$

or

$$\bar{E} = \frac{\int_0^{E_{max}} E\Psi_E \mathrm{d}E}{\int_0^{E_{max}} \Psi_E \mathrm{d}E} \tag{1.44}$$

Since in general $\Psi_E/\Psi \neq \Phi_E/\Phi$ (Greening, 1985), the above-mentioned Equations (1.43) and (1.44) lead to different values of mean energy E. This means that, in expressing the mean energy of any radiation, it is necessary to name the relevant quantity whose differential energy distribution was used in the determination of the mean value.

1.2.2 Interaction of Charged Particles

All charged particles during passage through matter lose their energy in a distinctly different manner than uncharged particles. For example, a photon or a neutron may pass through a slab of material without any interaction, but they can lose a substantial portion of their energy in one or a few interaction processes. On the other hand, a charged particle surrounded by a coulomb electric field will interact with the medium primarily by means of this field and the negative charge of the orbital electrons of the atoms located along the track of the charged particle. The energy transfer during these individual interactions represents only a minute fraction of the kinetic energy of the charged particles. It is sometimes practical to think of the charged particle as losing its energy gradually in a friction-like process, often referred to as the *continuous slowing-down approximation* or CSDA (Attix, 1986).

The main *interaction coefficients* applied for the interaction of charged particles with matter are the *stopping power*, the *linear energy transfer*, and the *range* or *pathlength*.

The *linear stopping power* (S_l) is defined by the expression

$$S_l = \frac{\mathrm{d}E}{\mathrm{d}l} \tag{1.45}$$

where $\mathrm{d}E$ is the energy lost by a charged particle in traversing a distance $\mathrm{d}l$ in the material. The SI unit of the linear stopping power is $\mathrm{J \cdot m^{-1}}$; the units based on eV are also used, e.g., $\mathrm{keV \cdot \mu m^{-1}}$.

The energy loss of charged particles has, in principle, two main components: (1) the first components take into account *collision* losses (interactions through coulomb forces), and (2) the second contribution is due to radiative losses (during acceleration of a charged particle electromagnetic radiation is emitted). In the case of high-energy charged particles, some nuclear reactions can occur as well. For particles emitted by radionuclides we may ignore this type of energy loss. The total linear stopping power ($S_{l,tot}$) can then be expressed in the form

$$S_{l,tot} = \left(\frac{\mathrm{d}E}{\mathrm{d}l}\right)_{col} + \left(\frac{\mathrm{d}E}{\mathrm{d}l}\right)_{rad} \tag{1.46}$$

where $(\mathrm{d}E/\mathrm{d}l)_{col}$ and $(\mathrm{d}E/\mathrm{d}l)_{rad}$ represent the linear collision stopping power and the linear radiative stopping power, respectively. Generally speaking, the energy spent in collision interactions results in ionization and excitation, while the energy spent in radiative processes is carried away from the charged particle path by the bremsstrahlung photons.

The quotient $(\mathrm{d}E/\mathrm{d}l)$ is also known as the *specific energy loss* or the *rate of energy loss*.

In practice, instead of linear stopping power, more often the *mass stopping power* is used. The *total mass stopping power* (S_l/ϱ) of a material having density ϱ for charged particles is the quotient of $\mathrm{d}E$ over $\varrho \mathrm{d}l$, i.e.,

$$\frac{S_l}{\varrho} = \frac{1}{\varrho}\frac{\mathrm{d}E}{\mathrm{d}l} \tag{1.47}$$

where dE is the energy lost by a charged particle in its passage through a distance dl in a material of density ϱ. The basic unit of the mass stopping power is $J \cdot m^2 \cdot kg^{-1}$; the energy E may also be expressed in eV and hence S/ϱ can be given in $eV \cdot m^2 \cdot kg^{-1}$.

As long as the velocity of a charged particle remains large compared with the velocities of the orbital electrons in the atoms of the absorbing medium, the collision mass stopping power can be expressed by a classical Bethe formula which can be written in the following form (Knoll, 1989)

$$\frac{1}{\varrho}\frac{dE}{dl} = \frac{4\pi e^4 z^2}{m_0 v^2} NB \qquad (1.48)$$

where

$$B = Z\left[\ln\frac{2m_0 v^2}{I} - \ln\left(1 - \frac{v^2}{c^2}\right) - \frac{v^2}{c^2}\right] \qquad (1.49)$$

In the above expressions, the symbols used have the following meaning:

- ze – the charge of the primary charged particles (e is the charge of the electron)
- m_0 – the electron rest mass
- v – the velocity of the charged particle considered
- c – the velocity of light
- N – the absorber atomic density
- I – the average excitation and ionization potential of the material

Heavy charged particles, such as alpha particles or protons, interact essentially with electrons by two possible mechanisms, namely by elastic and inelastic coulomb scattering. Because of the big difference in mass between the heavy charged particle and the electrons, the track of the incident charged particle remains unaffected and is more or less a straight line. The collision stopping power tends to rise significantly with the decreasing energy of the charged particle. The increase in stopping power that occurs when the energy of the particle is approaching zero is responsible for the so-called Bragg peak, which can be observed near the end of a charged particle's track.

The dependence of the mass collision stopping power on the energy of the charged particle is illustrated in Figure 1.28, where four different absorbing media are considered: water, Al, Cu, and Pb.

The passage of heavy charged particles through matter differs substantially from that of light charged particles, such as electrons and positrons. The track of these particles is no longer straight, and they are light enough

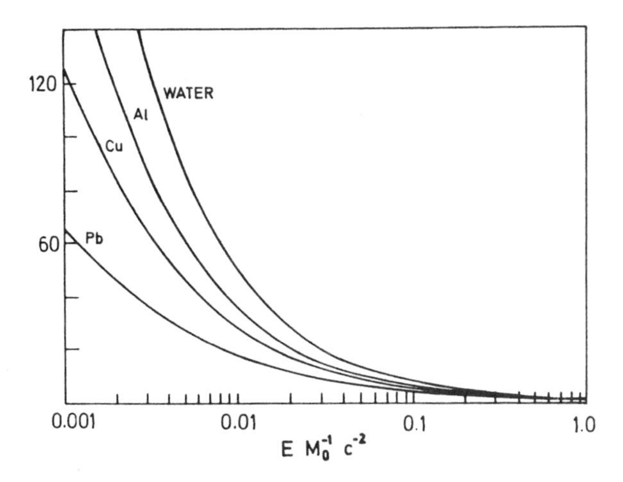

FIGURE 1.28. The mass collision stopping power for singly charged heavy particles, as a function of their kinetic energy E normalized by the rest mass M_0c^2, i.e., E/M_0c^2. For protons, for example, this ratio is equal to 1 at $E = 938$ MeV. In the case of a charged particle with a charge ze, the ordinate has to be multiplied by z^2. Adapted from Attix (1986). From *Introduction to Radiological Physics and Radiation Dosimetry* by F. H. Attix. © Copyright 1986, John Wiley & Sons, Inc. Reprinted by permission of John Wiley & Sons, Inc.

to generate significant bremsstrahlung, which depends on the inverse square of the particle mass for equal velocities. The radiative energy losses of an electron are very roughly proportional to its kinetic energy.

The *mass radiative stopping power* (in MeV·cm^2·g^{-1}) of electrons or positrons can be expressed as (Attix, 1986)

$$\frac{1}{\varrho}\frac{\mathrm{d}E}{\mathrm{d}l} = \sigma_0 \frac{N_A Z^2}{A}(E + m_0c^2)B \tag{1.50}$$

where the constant σ_0 is equal to 5.80×10^{-28} cm^2 (per atom), N_A is the Avogadro constant (6.022×10^{27} mol^{-1}), A is the molar mass (g·mol^{-1}), E is the kinetic energy of the incident charged particle (in MeV), and B is a slowly varying function of Z and E having a value of 16/3 for $E \ll 0.5$ MeV, and roughly 6 for $E = 1$ MeV, 12 for 10 MeV, and 15 for 100 MeV.

The comparison of mass radiative and collision stopping powers for electrons (and generally also for protons) in graphite, copper and lead are presented in Figure 1.29.

It can be shown (Attix, 1986) that the mass radiative stopping power is proportional to $N_A Z^2/A$, while the mass collision stopping power is proportional to $N_A Z/A$, i.e., to the electron density. This is why their ratio would

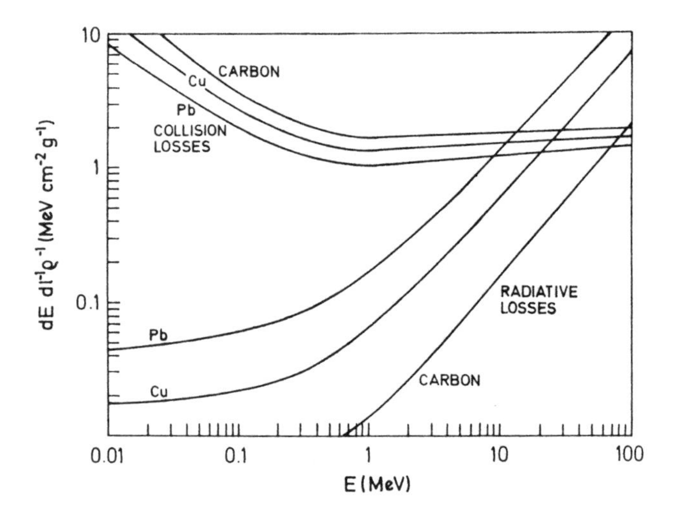

FIGURE 1.29. Mass radiative and collision stopping powers for electrons in various materials as a function of their kinetic energy. Adapted from Attix (1986).

be expected to be proportional to Z. As a matter of fact, the ratio of radiative to collision stopping power is usually given in the form

$$\frac{\left(\frac{dE}{dl}\right)_{rad}}{\left(\frac{dE}{dl}\right)_{col}} = \frac{EZ}{m} \tag{1.51}$$

where E is the kinetic energy of the charged particle, Z is the atomic number of the material, and m is a number variously taken to be 700 or 800 MeV.

In the case of the radioactivity measurement, the electrons of interest, i.e., either beta particles or secondary electrons resulting from gamma radiation interactions, have energies less than a few MeV. This means that radiative energy losses are always only a small fraction of the energy losses due to collisions.

The interaction of charged particles can also be characterized by the *linear energy transfer or restricted linear collision stopping power* (L_Δ) defined as

$$L_\Delta = \left(\frac{dE}{dl}\right)_\Delta \tag{1.52}$$

where dE is the energy lost by a charged particle in traversing a distance dl due to collision with electrons in which the energy loss is less than Δ. Although the SI unit for the linear energy transfer should be J·m^{-1}, usually the unit keV·μm^{-1} is used. As a matter of convenience, the energy Δ (also called *cutoff energy*) is expressed in terms of eV. Thus, L_{100} is understood to be the linear energy transfer for an energy cutoff of 100 eV. From the definition of the linear energy transfer it is obvious that $L_{\infty} = S_{col}$ called usually the *unrestricted linear energy transfer* and marked with a simple symbol L.

The penetration ability of charged particles can be described by their *range* (R), which may be defined by the expression

$$R = \int_0^E \frac{\mathrm{d}E}{(\mathrm{d}E/\mathrm{d}x)} \tag{1.53}$$

where E_0 represents the starting kinetic energy of the particle at the point of incidence in the given material. As long as (dE/dx), which is actually the linear stopping power, is constant along the particle track, then the range equation can be simplified as follows:

$$R = \frac{E}{S_l} \tag{1.54}$$

The range introduced above may be referred to as the CSDA range, i.e., the pathway of the particle of a given type and energy in a given material until it loses all its energy and comes to rest.

The unit for the range is m or quite often, when dealing with low-energy particles, μm or mm.

Instead of the linear stopping power, we may sometimes use the mass stopping power in the range definition [Equation (1.53)], and then the range should be expressed in other units, e.g., in g·cm^{-2} when (S_l/ϱ) is given in MeV·cm^2·g^{-1}. In this case the range is given in terms of *mass thickness* or *areal density* rather than *distance.*

Although for heavy charged particle, such as alpha particles, the range is relatively well specified, there is always a certain *range straggling*. This is the fluctuation in the actual range of individual particles having the same energy. In the case of alpha particles, the straggling may amount to several percent of its mean range.

In designing the measurement arrangement for radionuclides emitting alpha particles, it is usually important to take into consideration their range in air and detector materials. The range of alpha particles in air at 15°C and 101 kPa pressure as a function of their energy is illustrated in Figure 1.30.

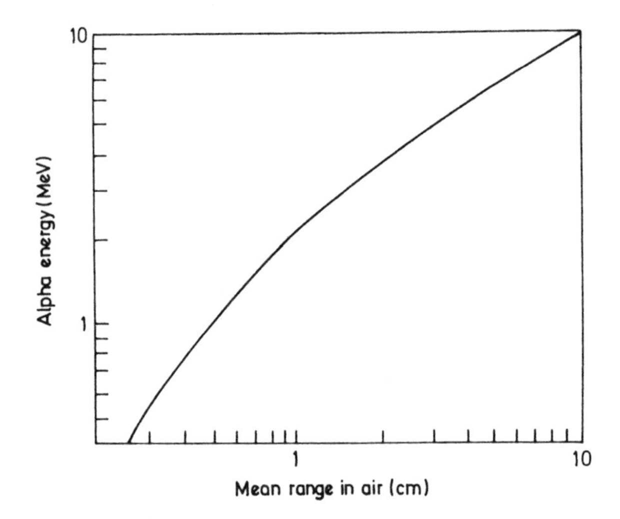

FIGURE 1.30. The range-energy relationship in air for alpha particles with initial kinetic energy from about 0.3 MeV to 10 MeV.

Figure 1.31 shows the range–energy plot for alpha particles in some materials, including silicon and germanium. In this case the range is given in terms of mass thickness.

In addition to the CSDA range, for the light charged particles it is useful to introduce the concept of the *projected range* (R_p), which may be defined as the value of the farthest depth of penetration of the particle in its initial direction. While for heavy charged particles the CSDA range and the projected range are the same, in the case of electrons these ranges may differ substantially. The projected range of electrons due to their tortuous path is equal to approximately half of their CSDA range.

An example of electron projected ranges in silicon and sodium iodide is shown in Figure 1.32. It can be seen that in this representation the values of R_p are similar even for materials with widely different physical properties.

The sensitivity of most detection techniques depends on the number of ion pairs produced by the charged particle in the sensitive volume of a given detector. In this respect the important parameter is the *mean energy expended per ion pair produced* (W_i) defined by

$$W_i = \frac{E_0}{N} \tag{1.55}$$

where E_0 is the initial kinetic energy of a charged particle and N is the mean number of ion pairs formed when the energy of this charged particle is entirely deposited in the material. The mean values of W_i for electrons with

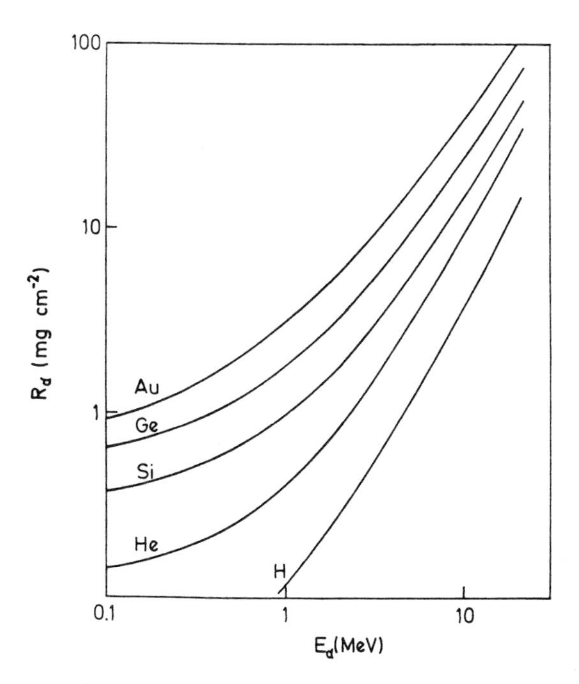

FIGURE 1.31. The alpha particle range given in mass thickness in different materials. Based on Knoll (1989).

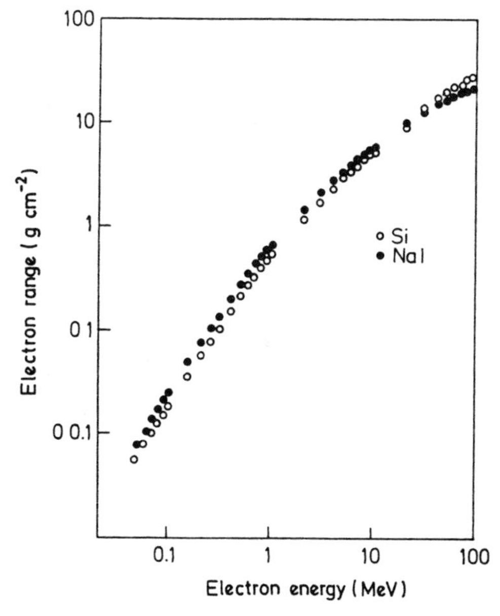

FIGURE 1.32. The range-energy curves for electrons in Si and NaI as a function of their energy. The range is shown in terms of mass-thickness (distance × density). Adapted from Knoll (1989).

TABLE 1.50. The Average Energy Expended per Ion Pair in Some Gases for Electrons and Alpha Particles.

	W_i (eV)	
Gas	Electrons, $E_e > 10$ keV	Alpha Particles, $E_\alpha = 5.3$ MeV
CH_4	27.3	29.1
C_2H_2	25.8	27.4
C_2H_4	25.8	27.9
H_2	36.5	36.4
N_2	34.8	36.8
O_2	30.8	32.2
CO_2	33.0	34.2
Ar	26.4	26.3
Air	33.8	35.1

Based on ICRU (1979).

energy higher than 10 keV and alpha particles of 5.3 MeV in several gases are given in Table 1.50. In the case of electrons, where at higher energies the radiative losses may take place, the ions formed by the bremsstrahlung are to be included in N.

The value W_i depends slightly on the type of charged particles and to a certain extent also on their energy. The mean energy W_i in gases is about 3–5% greater for alpha particles than for electrons. Fortunately, the experimentally found W_i values for electrons in the energy range 1–10 keV varies only by about 2% or less, and above energy 10 keV the mean energy W_i can be considered constant.

In semiconductor detectors instead of ion pairs, one is interested rather in the number of electron-hole pairs (an analog of ion-pairs in a solid). For important detector materials, the mean energy expended per electron-hole pairs in silicon at 300 K for alpha particles is 3.62 eV and 3.68 eV for electrons, and in germanium at 77 K is 2.97 eV for both electrons and alpha particles.

1.2.3 Interaction of Photons

Interaction of photons, as indirectly ionizing radiation, is characterized by processes in which the energy and the direction of the photon is changed. The interaction processes are random, and they occur with certain probability depending on the photon energy and the composition of the medium.

The interaction of indirectly ionizing radiations, including photons, can be described by means of appropriate *interaction coefficients.* The reduction of the number of photons penetrating a material is quantified by the *at-*

tenuation coefficients while the transfer of energy from photons to secondary charged particles may be expressed in terms of the *energy transfer coefficients.* For the deposition of energy the *energy absorption coefficient* is of primary importance.

Similar to stopping power, interaction coefficients for photons can also be defined as linear or related to the density of the material.

The *linear attenuation coefficient* (μ) of a material for photons (and other uncharged particles) is introduced by the expression

$$\mu = \frac{1}{N}\frac{dN}{dl} \tag{1.56}$$

where dN/N is the fraction of photons that experience interactions in traversing a distance dl in a material.

Taking into account the material density ϱ we may define the *mass attenuation coefficient* (μ/ϱ) simply as

$$\frac{\mu}{\varrho} = \frac{1}{N\varrho}\frac{dN}{dl} \tag{1.57}$$

The units of the linear and mass attenuation coefficients are m^{-1} and $m^2 \cdot kg^{-1}$, respectively.

On the other hand, the *linear energy transfer coefficient* (μ_{tr}) and the *mass energy transfer coefficient* (μ_{tr}/ϱ) are defined by the following expressions:

$$\mu_{tr} = \frac{1}{E \cdot N}\frac{dE_{tr}}{dl} \tag{1.58}$$

and

$$\frac{\mu_{tr}}{\varrho} = \frac{1}{\varrho \cdot E \cdot N}\frac{dE_{tr}}{dl} \tag{1.59}$$

where E is the energy of the incident photons, N is their number, and dE_{tr}/dl is the fraction of the photon energy that is transferred to the kinetic energy of charged particles by interactions in traversing a distance dl in the material of density ϱ. The units of the linear energy transfer coefficient and the mass energy transfer coefficient are m^{-1} and $m^2 \cdot kg^{-1}$, respectively.

In the above definitions, the symbol dE_{tr} denotes the sum of the initial kinetic energies of all the charged particles liberated by the interacting photons.

The *linear energy absorption coefficient* (μ_{en}) and the *mass energy absorption coefficient* (μ_{en}/ϱ) are defined in a similar way

$$\mu_{en} = \mu_{tr}(1 - g) \tag{1.60}$$

$$\frac{\mu_{en}}{\varrho} = \frac{\mu_{tr}}{\varrho}(1 - g) \tag{1.61}$$

respectively. The factor g represents the fraction of the energy of secondary charged particles that is lost to bremsstrahlung in the material. The unit of μ_{en} is m^{-1}, while the unit of the mass energy absorption coefficient is $m^2 \cdot kg^{-1}$.

There are three dominant types of interactions of X-ray and gamma photons with matter, which must be considered in the detection of photons or in the measurement of radionuclides emitting photons:

(1) Photoelectric effect (photoelectric absorption)
(2) Compton effect (Compton scattering)
(3) Pair production

Another two interaction processes, Rayleigh (coherent) scattering and photonuclear interactions, are of marginal importance for low-level radioactivity evaluation.

The relative importance of the three major types of photon interactions is illustrated in Figure 1.33, which shows the regions of the atomic number Z and the photon energy E where each interaction predominates. It can be seen that the photoelectric effect plays a major role at the lower photon energies, the Compton effect is dominant at medium energies, while pair production may be significant only for high-energy photons.

In the *photoelectric effect,* the photon interacts with an orbital electron (a bound electron in the inner shell of an atom), transferring all of its energy to the electron. The interacting photon will disappear in the interaction process. The kinetic energy (E_e) of the electron ejected from the atom is given by

$$E_e = E - E_{be} \tag{1.62}$$

where E is the incoming photon energy and E_{be} is the binding energy of the electron (the recoil energy of the atom is neglected).

In the *Compton effect* a photon having an energy E undergoes elastic collision with a loosely bound (essentially regarded as free) orbital electron resulting in a transfer of some energy to the electron and its deflection with a new energy E'. The energy lost by the photon will emerge as the kinetic energy of the electron (also called the recoil or Compton electron) E_e.

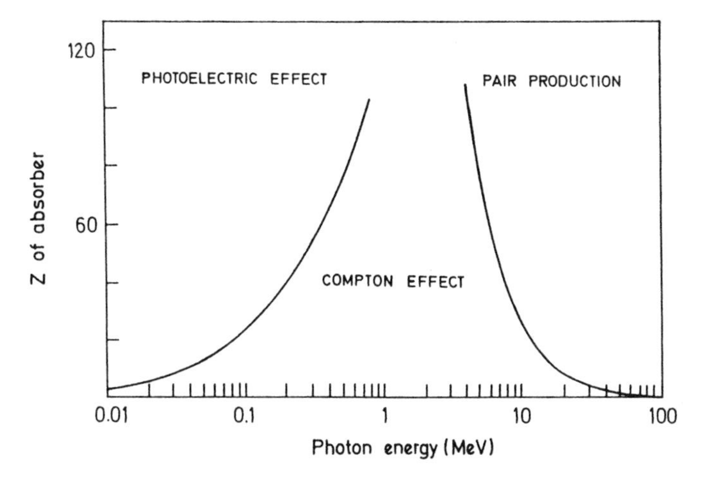

FIGURE 1.33. An illustration of three dominant areas for photoelectric effect, Compton effect and pair production separated by the lines showing the values of Z and photon energy E for which the two neighboring effects are just equal. Adapted from Evans (1955).

Applying the law of momentum and energy conservation, it may be shown that

$$E' = \frac{E}{1 + \frac{E}{m_0c^2}(1 - \cos\theta)} \tag{1.63}$$

and

$$E_e = \frac{E^2(1 - \cos\theta)}{m_0c^2 + E(1 - \cos\theta)} \tag{1.64}$$

where m_0c^2 is the rest-mass energy of the electron and θ is the angle through which the scattered photon is deflected from the direction of the incoming photon.

The maximum energy transfer to the recoil electron occurs when the initial photon is backscattered ($\theta = 180°$). In this case the electron energy is given by

$$E_{emax} = \frac{E}{1 + \frac{m_0c^2}{2E}} \tag{1.65}$$

The Compton electron appears with a continuous energy distribution ranging from virtually zero up to E_{emax}. This is illustrated in Figure 1.34 where the energy distribution of recoil electrons resulting from the primary photons of three different energies undergoing Compton scattering is shown.

When incident photon energies exceed a threshold energy of 1.02 MeV, another interaction called *pair production* becomes possible. In this interaction the interacting photon disappears in the field of a charged particle (usually in the field of an atomic nucleus) and a negatron (electron) together with a positron are produced. These two particles share the kinetic energy that is equal to the difference between the energy of the incoming photon (E) and the threshold energy (corresponding to the two electron rest-mass energies), i.e.,

$$E_{el} + E_{pos} = E - 2m_0c^2 \tag{1.66}$$

where E_{el} and E_{pos} are the kinetic energy of the resulting electron and positron, respectively.

Although the energies E_{el} and E_{pos} are considered, for practical reasons, equal, strictly speaking, the positron usually receives slightly higher kinetic energy than the negatron because of the repulsive force it obtains from the nucleus as opposed to the attractive force that the negatron receives.

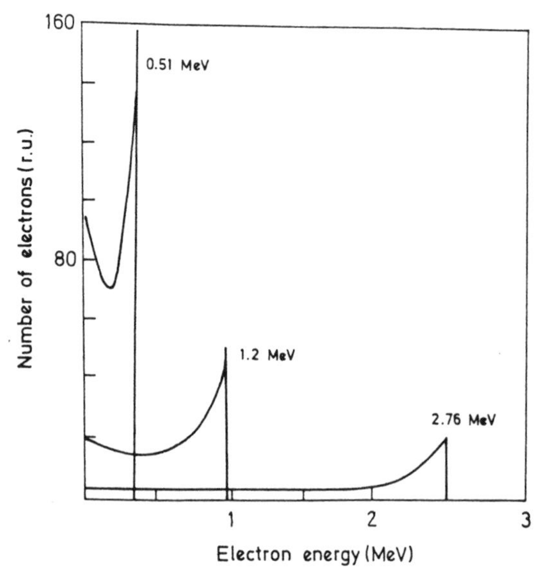

FIGURE 1.34. The energy distribution of scattered electrons from the Compton interaction of photons with given initial energies.

Since the positron is an unstable particle, after slowing down to thermal energies, it annihilates with a nearby electron, with the subsequent emission of two 0.511 MeV photons. These photons move away in opposite directions. The annihilation photons may escape from the detection medium, resulting in the formation of characteristic escape peaks in the measured pulse-height spectra.

The photoelectric effect, Compton effect, and pair production are the principal processes that govern the probability that a photon incident on a given slab of material may be scattered and partially or totally absorbed. If a collimated photon beam of fluence Φ_0 is directed on a slab of material having a thickness l, the fluence Φ that will emerge from such an absorber will be equal to

$$\Phi = \Phi_0 e^{-\mu \cdot l} \tag{1.67}$$

where μ is the total attenuation coefficient. This interaction coefficient can be written in the form

$$\mu = \tau + \sigma_c + \varkappa \tag{1.68}$$

where τ, σ_c, and $\varkappa$ are the components of the total attenuation coefficient due to the photoelectric effect, the Compton effect, and pair production, respectively. The coefficient μ corresponds actually to the total cross section (σ) for the photon interaction with the material atoms. It can be expressed as the sum of the individual cross sections (σ_J)

$$\sigma = \sum_J \sigma_J \tag{1.69}$$

Using this approach, for the total mass attenuation coefficient we may then write

$$\frac{\mu}{\varrho} = \frac{N_A}{M}(\sigma_{ph} + \sigma_c + \sigma_{pp}) \tag{1.70}$$

where N_A is the Avogadro constant, M is the molar mass, and the individual component cross sections refer to those for the photoelectric effect (σ_{ph}), the Compton effect (σ_c), and pair production (σ_{pp}).

Interaction coefficients, such as the mass attenuation coefficient (μ/ϱ), depend on the absorber and also on the energy of interacting particles. In the case of a complex energy spectrum, the mean values of the interaction coefficients are very useful. For example, the mean mass attenuation coefficient for photons with the energy distribution Φ_E can be found using the following expression:

$$\overline{(\mu/\varrho)} = \frac{\int_0^\infty (\mu/\varrho)\Phi_E \mathrm{d}E}{\int_0^\infty \Phi_E \mathrm{d}E} \tag{1.71}$$

A similar equation may be written for the mean mass absorption coefficient. In addition to the fluence particle spectrum Φ_E, both mass coefficients can also be weighted by other radiation field quantities, e.g., the energy fluence.

The obvious advantage of using mass interaction coefficients rather than linear coefficients is that the mass coefficients are independent of the actual density and physical state of the material or absorber under consideration.

In the case of a material that is a mixture of elements, or a chemical compound, the resulting total mass or absorption coefficient of the material will be given by

$$\frac{\mu}{\varrho} = \sum \frac{\mu_i}{\varrho_i} w_i \tag{1.72}$$

where w_i is the weight fraction of the ith element in this material and (μ_i/ϱ_i) is the total mass or attenuation coefficient referring to the ith element.

1.3 REFERENCES

Allisy, A. et al. 1992. "Radiation Quantities and Units, Radiometry," *ICRU News*, December.

Anderson, R. L. et al. 1978. "Institutional Radioactive Wastes," *NUREG/CR-0028*. Washington, DC: U.S. Nuclear Regulatory Commission.

Attix, F. H. 1986. *Introduction to Radiological Physics and Dosimetry*. New York: John Wiley and Sons.

Beckman, R. T. 1988. "Inhalation Hazard to Underground Miners," in *Population Exposure from the Nuclear Fuel Cycle*, E. L. Alpen, R. O. Chester and D. R. Fisher, eds., New York, NY: Gordon and Breach Science Publishers, p. 161.

Bellotti, E. et al. 1992. "Multielement Proportional Chamber – A Study of the Background in the Gran Sasso Underground Laboratory," *Nucl. Instrum. Methods Phys. Res.*, A323:198–202.

Budnitz, R. J. et al. 1983. *Instrumentation for Environmental Monitoring, Vol. 1.* New York: John Wiley and Sons.

Buncel, E. and J. R. Jones, eds. 1987. *Isotopes in the Physical and Biomedical Sciences, Vol. 1, Labelled Compounds (Part A).* Amsterdam: Elsevier Science Publishers.

Chen, C. J., P. S. Weng and T. C. Chu. 1993. "Evaluation of Natural Radiation in Houses Built with Black Schist," *Health Phys.*, 64(1):74–78.

Chung, C. 1989. "Environmental Radioactivity and Dose Evaluation in Taiwan after the Chernobyl Accident," *Health Phys.*, 56(4):465–471.

Cox, W. M., R. L. Blanchard and B. Kahn. 1970. "Relation of Radon Concentration in the Atmosphere to Total Moisture Retention in Soil and Atmospheric Stability," in *Radionuclides in the Environment,* Advances in Chemistry Series 93, pp. 436–444.

Dachev, T. P. et al. 1992. "'Mir' Radiation Dosimetry Results during the Solar Proton Events in September–October 1989," *Adv. Space Res.*, 12:321–324.

Draganic, I. G., Z. D. Draganic and J. P. Adloff. 1990. *Radiation and Radioactivity – On Earth and Beyond.* Boca Raton, FL: CRC Press, Inc., p. 19.

Eisenbud, M. 1987. *Environmental Radioactivity from Natural, Industrial, and Military Sources.* Orlando, FL: Academic Press, Inc.

Evans, R. D. 1955. *The Atomic Nucleus.* New York, NY: McGraw-Hill Book Company, Inc.

Fisenne, I. M. 1982. "Radon-222 Measurements at Chester," in *EML-383,* New York, pp. 192–227.

Fisenne, I. M. 1984. "Radon-222 Measurements at Chester," in *EML-422,* New York, pp. 115–149.

Goldman, M. and R. J. Catlin. 1988. "Health and Environment Consequences of the Chernobyl Nuclear Power Plant Accident," in *Population Exposure from the Nuclear Fuel Cycle,* E. L. Alpen, R. O. Chester, and D. R. Fisher, eds., New York, NY: Gordon and Breach Science Publishers, p. 335.

Granier, R. and D. J. Gambini. 1990. *Applied Radiation Biology and Protection.* New York: Ellis Horwood, p. 269.

Greening, J. R. 1985. "Fundamentals of Radiation Dosimetry," *Medical Physics Handbooks 15 (Second Edition).* Bristol and Boston: Adam Hilger Ltd.

Heusser, G. et al. 1989. "Construction of a Low-Level Ge Detector," *Appl. Radiat. Izot.*, 40:393–395.

Holm, E., C. Samuelson and R. B. R. Peterson. 1981. "Natural Radioactivity around a Prospected Uranium Mining Site in a Subarctic Environment," in *Natural Radiation Environment,* K. G. Vohra et al., eds., New Delhi: Wiley Eastern.

Howles, L. 1993. "1992 Annual Review of Load Factors," *Nuclear Engineering International,* 38(465):19–24.

IAEA. 1981. "Underground Disposal of Radioactive Waste," *Safety Series No. 54.* Vienna: International Atomic Energy Agency.

ICRP. 1987. *Protection of the Patient in Nuclear Medicine,* ICRP Publication 52. Oxford: Pergamon Press.

ICRP. 1991. *1990 Recommendations of the International Commission on Radiological Protection,* ICRP Publication 60. Oxford: Pergamon Press.

ICRU. 1979. *Average Energy Required to Produce an Ion Pair,* ICRU Report 31. Washington, DC: International Commission on Radiation Units and Measurements.

ICRU. 1980. *Radiation Quantities and Units,* ICRU Report 33. Washington, DC: International Commission on Radiation Units and Measurements.

Jacobi, W. 1972. "Activity and Potential Alpha-Energy of Radon-222 and Radon-220 Daughters in Different Air Atmospheres," *Health Phys.*, 22:441.

Jones, J. R., ed. 1988. *Isotopes: Essential Chemistry and Applications II.* London: The Royal Society of Chemistry.

Kathren, R. L. 1984. *Radioactivity in the Environment: Sources, Distribution, and Surveillance.* Chur, London, Paris, New York: Harwood Academic Publishers.

Knittel, J. H. 1989. "Introduction and Background," in *Near-Surface Land Disposal* (Radioactive Waste Management Handbook, Vol. 1). Chur, Switzerland: Harwood Academic Publishers.

Knoll, G. F. 1989. *Radiation Detection and Measurement (Second Edition).* New York: John Wiley and Sons.

Kowalsky, R. J. and J. R. Perry. 1987. *Radiopharmaceuticals in Nuclear Medicine Practice.* Norwalk, CN: Appleton and Lange.

Lal, D. and B. Peters. 1967. "Cosmic-Ray Produced Radioactivity on the Earth," in *Handbook of Physics,* 46(2):551–612.

Liden, K. and E. Holm. 1985. "Measurement and Dosimetry of Radioactivity in the Environment," in *The Dosimetry of Ionizing Radiation, Vol. I,* K. R. Kase, B. E. Bjärngard and F. H. Attix, eds., Orlando: Academic Press.

Mann, W. B., R. L. Ayres and S. B. Garfinkel. 1980. *Radioactivity and Its Measurement.* Oxford: Pergamon Press.

Myrick, T. E., B. A. Berven and F. F. Haywood. 1983. "Determination of Concentrations of Selected Radionuclides in Surface Soil in the USA," *Health Phys.*, 45:631.

NCRP. 1984a. *Radiological Assessment: Predicting the Transport, Bioaccumulation, and Uptake by Man of Radionuclides Released to the Environment,* NCRP Report No. 76. Washington, DC: National Council on Radiation Protection and Measurements.

NCRP. 1984b. *Evaluation of Occupational and Environmental Exposure to Radon and Radon Daughters in the United States,* NCRP Report No. 78. Washington, DC: National Council on Radiation Protection and Measurements.

NCRP. 1985. *Carbon-14 in the Environment,* NCRP Report No. 81. Washington, DC: National Council on Radiation Protection and Measurements.

NCRP. 1987. *Ionizing Radiation Exposure of the Population of the United States,* NCRP Report No. 93. Washington, DC: National Council on Radiation Protection and Measurements.

NCRP. 1989a. *Control of Radon in Houses,* NCRP Report No. 103. Washington, DC: National Council on Radiation Protection and Measurements.

NCRP. 1989b. *Exposure of the U.S. Population from Diagnostic Medical Radiation,* NCRP Report No. 100. Washington, DC: National Council on Radiation Protection and Measurements.

Nelson, C. B. 1988. "Population Exposure from Radon Associated with Uranium Milling," in *Population Exposure from the Nuclear Fuel Cycle,* E. L. Alpen, R. O. Chester and D. R. Fisher, eds., New York, NY: Gordon and Breach Science Publishers, p. 183.

Nero, A. V. 1988. "Radon and Its Decay Products in Indoor Air: An Overview," in *Radon and Its Decay Products in Indoor Air,* W. W. Nazaroff and A. V. Nero, eds., New York, NY: John Wiley and Sons, pp. 1–56.

NRC. 1991. *Comparative Dosimetry of Radon in Mines and Homes.* Washington, DC: National Academy Press.

NRPB. 1991. *Committed Equivalent Organ Doses and Committed Effective Doses from Intakes of Radionuclides,* NRPB Report R245. Chilton, UK: National Radiological Protection Board.

Nunes, P. D. et al. 1973. "U-Th-Pb Systematics of Some Apollo 16 Lunar Samples," in *Four Lunar Sci. Conf. Abstracts.* Massachusetts: M.I.T. Press, p. 1797.

O'Kelley, G. D., J. S. Eldrige and K. J. Northcutt. 1974. "Concentrations of Cosmogenic Radionuclides in Apollo 17 Samples: Effect of Solar Flare of August, 1972," in *Fifth Lunar Sci. Conf. Abstracts.* Massachusetts: M.I.T. Press, p. 577.

O'Kelley, G. D. et al. 1971. "Abundances of the Primordial Radionuclides K, Th and U in Apollo 12 Lunar Samples by Nondestructive Gamma-Ray Spectrometry: Implications for Origin of Lunar Soils," in *Second Lunar Sci. Conf. Abstracts, Vol. 2.* Massachusetts: M.I.T. Press, p. 1159.

Pattenden, N. J. 1991. "A Review of Long-Term Studies of Radioactivity in the Environment from the Chernobyl Accident by AEA Technology," *Nucl. Energy,* 30(6):341–359.

Plaga, T. and T. Kirsten. 1991. "Reduction of Degraded Events in Miniaturized Proportional Counters," *Nucl. Instrum. Methods Phys. Res.*, A309:560–568.

Reedy, R. C., J. R. Arnold and D. Lal. 1984. "Cosmic Ray Record in Solar System," *Science,* 219:127–135.

Roberts, T. R. 1979. *Radio-Gas-Liquid Chromatography.* Amsterdam: Elsevier Science Publishers.

Sabol, J., P. S. Weng and C. H. Mao. 1993. "Water-Borne Radon in Some Hot Springs in Taiwan," *Nucl. Sci. J.*, 30(2):131–136.

Silver, L. T. 1974. "Patterns of U-Th-Pb Distributions and Isotope Relations in Apollo 17 Soils," in *Fifth Lunar Sci. Conf. Abstracts.* Massachusetts: M.I.T. Press, p. 706.

Simpson, J. A. 1983. "Elemental and Isotopic Composition of the Galactic Cosmic Rays," *Ann. Rev. Nucl. Part. Sci.*, 33:323–381.

Skalberg, M. and J. O. Liljenzin. 1993. "Partitioning and Transmutation: The State of Art," *Nuclear Engineering International,* 38:30–33.

Stanislaus, S. et al. 1992. "Results from Beam Tests of MEGA's Low-Mass, High-Rate Cylindrical MWPCs," *Nucl. Instrum. Methods Phys. Res.*, A323:198–202.

Surkov, Yu. A. 1974. "Radioactivity of the Moon, the Planets and Meteorites," in *Soviet-American Conference on Cosmochemistry of the Moon and Planets. Abstracts.* Moscow: Nauka, p. 79.

Tang, Y. S. 1988. "Review of Low-Level Waste Management," in *Population Exposure from the Nuclear Fuel Cycle,* E. L. Alpen, R. O. Chester and D. R. Fisher, eds., New York, NY: Gordon and Breach Science Publishers, p. 229.

Tatsumoto, M. 1972. "U-Th-Pb and Rb-Sr Measurements on Some Apollo 14 Lunar Samples," in *Third Lunar Sci. Conf. Abstracts, Vol. 2.* Massachusetts: M.I.T. Press, p. 1531.

Tykva, R. 1974. "Measurement of Multilabeled Samples" (in German), in *Messung von Radioaktiven und Stabilen Isotopen,* H. Simon, ed., Berlin: Springer-Verlag, pp. 199–224.

Tykva, R. 1975. "Determination of Primordial and Cosmogenic Radionuclides in Lunar Soil Obtained by the Automatic Station Luna 20" (in Russian), *Geokhymia,* 7:1097–1099.

Tykva, R. 1994. "How to Improve Applications of Multilabeling in Radiotracer Methodology," in *3rd Int. Conf. on Methods and Applications of Radioanalytical Chemistry, Abstracts.* Kona, HI: American Nuclear Society, p. 58.

Tykva, R. and B. Bennettová. 1993. "Quantitative Analysis of the Fate of a Pesticide after Its Application to Insects," in *Management of Insect Pests: Nuclear and Related Molecular and Genetic Techniques.* Vienna: International Atomic Energy Agency, pp. 529–536.

Tykva, R. and F. Franěk. 1977. "Nondestructive and Quantitative Evaluation of Radioactive Spots of Two-Dimensional Peptide Maps by an Automated Procedure," *Anal. Biochem.*, 78:572–576.

Tykva, R. and J. Šeda. 1975. "Radiochromatography in Carbon-14-Labelled Compounds Using Continuous Activity Assay by a Semiconductor Detector," *J. Chromatogr.*, 108:37–41.

Tykva, R. and I. Votruba. 1972. "Estimation of Radioactivity in Electrophoresis Gels by a Scanning Semiconductographic Procedure," *Anal. Biochem.*, 50:18–27.

Tykva, R. and I. Votruba. 1974. "Semiconductographic Determination of Labelled Substances in Thin-Layer Chromatography," *J. Chromatogr.*, 93:399–404.

Tykva, R. et al. 1987a. "Heavy Metal Binding to Proteins Extracted from Rats," in *Trace Element Analytical Chemistry in Medicine and Biology, Vol. 4,* P. Brätter and P. Schramel, eds., Berlin: Walter de Gruyter, pp. 521–525.

Tykva, R. et al. 1987b. "High-Sensitivity Semiconductor Spectrometry of Charged Particles in the Intershock Project," in *Proc. of 20. International Cosmic Ray Conference, Vol. 4. International Union of Pure and Applied Physics.* Moscow: Nauka, pp. 418–421.

Tykva, R. et al. 1988. "Radiomicroscopy as a Method of Biological Quantification of New Nontraditional Pesticide's Multicompartment Effect," in *Proc. XVIII Int. Congress of Entomology. Abstracts.* Vancouver: International Entomological Society, p. 478.

Tykva, R. et al. 1992. "A Topographic Method for Studying Uptake, Translocation and Distribution of Inorganic Ions Using Two Radiotracers Simultaneously," *J. Exp. Bot.*, 43:1083–1087.

UNSCEAR. 1977. *Sources and Effects of Ionizing Radiation.* New York, NY: United Nations Scientific Committee on the Effects of Atomic Radiation.

UNSCEAR. 1982. *Ionizing Radiation: Sources and Biological Effects.* New York, NY: United Nations Scientific Committee on the Effects of Atomic Radiation.

UNSCEAR. 1988. *Sources, Effects and Risks of Ionizing Radiation.* New York, NY: United Nations Scientific Committee on the Effects of Atomic Radiation.

USNRC. 1982. "Licencing Requirements for Land Disposal of Radioactive Waste" (10 CFR 61), *Federal Register,* 47:57446–57482.

Valkovič, V. 1980. *Analysis of Biological Material for Trace Elements Using X-Ray Spectroscopy.* Boca-Raton, FL: CRC Press.

Wilkinson, W. L. 1992. "The Nuclear Power Cycle – Achievements and Challenges," *Nuclear Energy,* 31(1):13–19.

Yadav, J. S. and J. R. Arnold. 1990. "Radioactive Material Screening for Gamma Ray Spectrometer Experiment on Mars," *Nucl. Instrum. Methods Phys. Res.*, A295:241–245.

CHAPTER 2

Experimental Arrangements for Low Radioactivities

2.1 FUNDAMENTAL CONDITIONS FOR THE DETERMINATION OF LOW-LEVEL RADIOACTIVITY

Low-level radioactivity represents a unique task, including the identification and evaluation of all radioactive sources of both natural and man-made origin occurring in the environment, as well as radioactive materials used in very small concentrations and amounts in various applications in research, technology, medicine, and many other fields, where these materials are virtually irreplaceable because of their specific characteristics. The measurement and monitoring of radionuclides having a very low-level activity is a special task that requires a comprehensive approach, taking into account such factors as the existing radiation background, the limitations of detection and instrumentation techniques (e.g., efficiency, stability, selectivity, noise, spurious pulses, interference from the surrounding conditions), the statistical nature of radioactive decay, the interference of other than measured radionuclides, and many other effects.

As described in Chapter 1 in detail, radioactivity is characterized by emission of particles. Its level is given by the number of disintegrations of nuclei per second (Bq), i.e., by decay rate. The determination of radioactivity is based on measurement of the number of registered decays per time unit, i.e., the *counting rate* (cps, in low-level measurement, also cpm and cph are used). The counting is carried out by means of detectors in which interactions take place between the input particles emitted from the measured source and the detector mass. The detectors are divided according to the character of these interactions in the appropriate effective volume (for interactions see Section 1.2).

From Chapter 1 it is evident that the level of radioactivity in the low-level sources can vary considerably. Thus, for example, the difference between the ^{14}C activity of two samples (which we shall consider of low activity), on the one hand in radiochronological determination and on the other in tracing work with a labeled compound, can be up to several orders of magnitudes of Bq, and the difference is usually still greater from the point of view of specific activity (Bq/mmol). Therefore in these cases the requirements on the method of determination and on the corresponding equipment will differ considerably. The basic difference consists in the required detection sensitivity: a value that is fully sufficient for the determination of low activity of a ^{14}C-labeled compound will be quite insufficient in radiochronology.

The condition for successful determination of a certain low activity of a radionuclide is thus the attainment of a corresponding detection sensitivity. For its simple expression the *figure of merit*, T, can be used, for which

$$T = E^2 \cdot B^{-1} \tag{2.1}$$

where E means the *detection efficiency* and B the *background*. In order to enable the measurement of low activity and in view of the quadratic dependence, the decisive factor is primarily the detection efficiency of the radiation emitted by the radionuclide. However, a lowering of the background is also desirable.

While the detector for low activities itself and its geometric arrangement relative to the measured sample determine the magnitude of the detection efficiency of the particles emitted by the sample, the background is determined by the localization, equipment, and the working regiment of the laboratory.

The figure of merit is conclusive for assessment of the counting method for low levels. During comparison of two scintillation procedures for radon determination in water, it has been shown (Chereji, 1992) that the counting efficiency is better for a liquid scintillator, but as the background is considerably higher than in a solid scintillator, the figure of merit and, therefore, the lower limit of the measurable activity are favorable for a scintillation detector based on zinc sulfide.

Usually, *relative measurements* of activity are carried out, which represent determination relatively to the defined activity of a standard source (see Section 2.7) of the same radionuclide. In this case the counting efficiency is given by measurement of the counting rate of a standard source with known decay rate. The *absolute determination* of activity is used for low levels quite exceptionally. On the contrary, for samples with identical counting conditions, only *comparison of counting rates* is often performed without any activity measurements.

2.1.1 Detection Efficiency

In the determination of the counting rate we express the detection (counting) efficiency E in a given time interval in percents, using the expression

$$E = \frac{\text{number of recorded counts}}{\text{number of decays in the source}} \cdot 100\% \qquad (2.2)$$

To achieve high detection efficiency it is therefore indispensable to assure first the inlet of the highest possible number of emitted particles into the effective volume of the detector. This can be achieved

- by the choice of the maximum possible *measurement geometry,* i.e., the highest part of the particles entering the effective detection volume from all emitted particles into 4π sr
- by exclusion or limitation of *self-absorption* of emitted particles in the sample itself
- by exclusion or limitation of the *kinetic energy* losses of the detected particles between the sample and the efficient volume of the detector (for example, during their passage through the inlet substructures of the detector which do not contribute to the generation of the response–usually called window)

In the interaction of the detected particle inside effective detector volume, it is indispensable, for the achievement of a high detection efficiency, to

- ensure that the particles generated in interactions of the detected particle with the efficient medium of the detector should take part in the formation of the corresponding response
- ensure, in the active part of the detector, the generation of the response exceeding the noise equivalent of the detector, i.e., the detector noise expressed in the units of magnitude in which the result of the primary interaction of the detected particle is expressed (i.e., the charge equivalent of the noise)
- connect the detector to a suitable electronic device, in order to secure the transmission of the response generated in the detector for its further processing and recording

Ensuring these requirements will differ in dependence on

- the radionuclides, the low activity of which is determined, especially of the type and the energy of the emitted particles
- the character of the sample, for example the aggregate state or the presence of admixtures

- the requirements put on the result of the measurement, for example, the required accuracy of the measurement, or the spectrometric resolution in the analysis of the mixture of radionuclides

In order to present a survey, we present in this section a number of individual viewpoints concerning the increase in detection efficiency. Their concrete application occurs primarily in detectors (Section 2.3) and also partly in electronic instrumentation (Section 2.4).

2.1.2 Background

From Equation (2.1) it follows that sufficiently low background is also important for the achievement of a high detection sensitivity. Measurement of a low activity is expressed in the measured counts per minute (min^{-1} or cpm) or hour (h^{-1}). To reduce it, the following steps are used:

- *screening* of the detector in such a way as to suppress the record of the cosmic radiation particles and the radiation from the medium in the surroundings of the detector
- *limitation of the radiation sources* in the surroundings of the detector, for example, in the screening materials, construction materials of the laboratory, laboratory atmosphere, local geological ground, and further elimination or screening of radiation sources
- *selection of construction materials* of the detector itself, having a high radionuclide purity, and mounting the detector so as to prevent contamination from the assembling medium
- selection of the *optimum counting channel*

The influence of channel selection on natural background for gamma-ray spectrometry is illustrated in Table 2.1. However, the counting channel has to be definitely chosen with respect to the figure of merit.

A very important source of background may also be the *contamination of the detector by a measured sample.* Such contamination should be excluded in low-level counting. An example of this influence by desorption of ^{222}Rn progeny from natural surfaces is given in Figure 2.1.

It follows from the above discussion that the solution of the problems of background is given by the localization and the equipment of the laboratory including the detector itself.

2.1.3 Effect of Specific Activity

The problem of determining the low activity of radionuclides is considerably affected by the level of its specific activity. The requirements or meas-

TABLE 2.1. The Reduction Factor (R_b) of Continuum of the Natural Background.

E_γ (keV)	100–200	150–250	200–300	>300
R_b	4.7 ± 0.9	5.7 ± 1.8	7.4 ± 1.9	8.5 ± 2

Based on Das (1987).

uring arrangements increase with decreasing specific activity. This dependence results primarily from the fact that the effective volume of the detector increases with an increasing amount of the measured sample, and thus its background. This leads to a decrease in the attainable detection sensitivity. Considering, for example, a non-screened ionization detector, its background, expressed by the rate of counts, is approximately 1 min^{-1} cm^{-3} (the dependence of the background on the detection volume is qualitatively equal in detectors of scintillation and semiconductor type). With increasing detector volume the requirements on the screening arrangement increase considerably from the point of view of expenses, volume, and weight. However, the dimensions of the detector proper are limited, e.g., by the increase

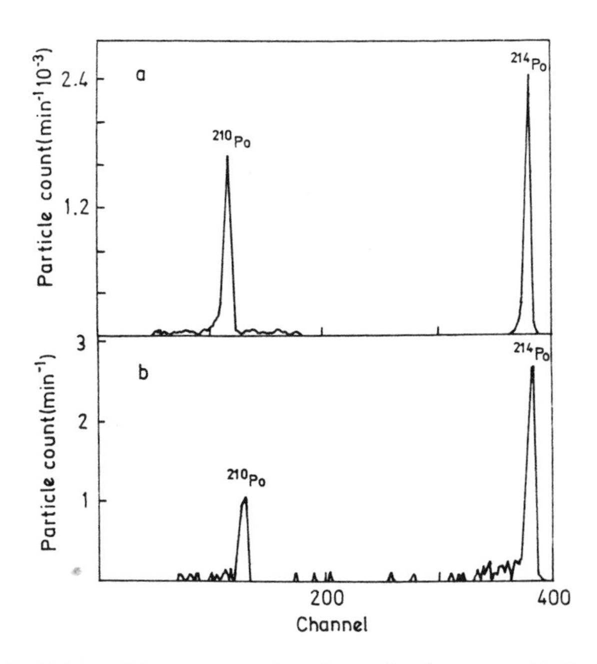

FIGURE 2.1. Alpha-particle energy spectrum from a Cu disc exposed to the ^{222}Rn source (upper graph) and the spectrum measured after removing the disc from the detector. Adapted from Bigu (1992).

of the operation voltage, guaranteeing the collection of the generated charge or the dimensions of the starting semiconductor monocrystal for the semiconductor detector. An increase in the amount of sample is also limited in a number of cases; in the case of solid phase samples it is limited predominantly by self-absorption in the sample, for example at alpha-spectrometry.

The analysis given indicates that the determination of low specific activity brings additional complicating factors, in contrast to the determination of a low total acitivity with the higher specific activity. The difficulties may be circumvented by careful preparation of samples before the measurement, in which the specific activity of the sample is increased, i.e., the concentration of the determined radionuclide is increased (Section 2.8).

2.1.4 Disturbances

In addition to the achievement of a sufficiently high detection sensitivity and a suitable specific activity during sample preparation, a fundamental condition for the determination of low activity of a radionuclide is the elimination of disturbances leading to creation of false counts in the detection arrangement. When measuring low count rates, the elimination of even isolated false impulses is essential. An analysis of these causes of fundamental disturbances and the possibility of their elimination are presented in Section 2.5.

2.1.5 Detection and Spectrometry

Different demands are applied in experimental arrangements for low levels according to the resolved radionuclides in the sample. If we should measure the *activity of only one radionuclide* or a total activity of a mixture of radionuclides, a suitable detector (Section 2.3) with simpler electronic instrumentation (Section 2.4) is sufficient. If low levels of *several radionuclides* should be determined simultaneously, not only is another electronics necessary (Section 2.4) but sometimes another detector may also be required, e.g., for low levels of one gamma emitter a scintillator detector is better than a semiconductor one due to its considerably higher detection efficiency (Figure 2.2). On the contrary, the NaI(Tl) detector does not enable a higher resolution of a complex gamma spectra due to its poor energy resolution given by energy of the detected particle necessary to generate a unit response (Table 2.2).

In low-level measurements the spectra of *alpha or gamma mixtures* are often analyzed using different detectors, with the response proportional to the kinetic energy of the particle and electronic, including multichannel amplitude analyzer.

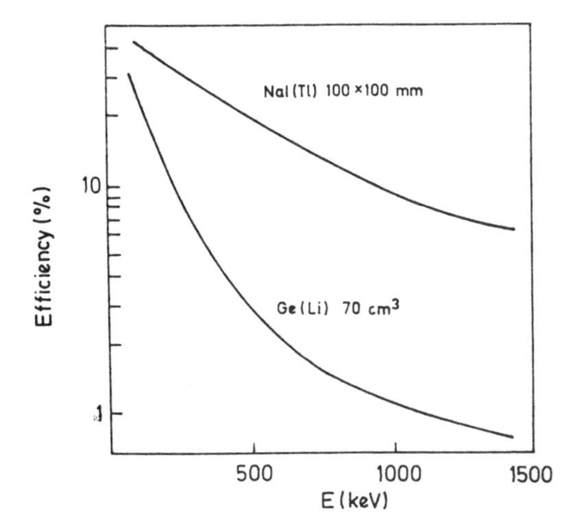

FIGURE 2.2. Comparison of detection efficiency of NaI(Tl) and Ge(Li) detectors for photons of different energy.

The resolving of beta radionuclides is more complicated because all beta spectra are continuous from zero energy and, therefore, cannot be separated into individual channels. Three possibilities exist: to resolve the spectra (1) by means of absorption of the low-energy emitter, (2) by measurement in different wide-channel windows of an amplitude analyzer

TABLE 2.2. A Mean Energy Necessary to Generate a Unit Charge Response $\bar{E}_i$ and the Total Charge Equivalent K_{Eq} for Different Values of the Multiplication Factor M.

Detector Type		$\bar{E}_i$ (eV)	K_{Eq} (pC·MeV^{-1}) $M = 1$	$M = 10^3$	$M = 10^6$
Semiconductor	Si (300 K)	3.62	0.044	–	–
	Ge (100 K)	2.96	0.054	–	–
Gas-filled	Air	32	0.005	5	–
	Argon	26	0.006	6.2	–
	Krypton	23	0.007	7	–
	Xenon	21	0.008	7.6	–
Scintillation	NaI(Tl)	110	–	–	1450
	CsI(Tl)	280	–	–	570
	LiI(Eu)	330	–	–	490
	Anthracen	250	–	–	620
	Stilbene	500	–	–	320
	Plastic and liquid	400–800	–	–	200–400

(Section 2.4), or (3) by chemical separation before measurement. The most often used procedure is the amplitude analysis. Several procedures have been described for such purposes recently (Takiue et al., 1990; Carles et al., 1991; Vapirev and Hristova, 1991). An example of such low-level measurement analyzing the distribution of simultaneously present ^{36}Cl and low-level ^{14}C in plant material (Tykva et al., 1992) is shown in Figure 2.3.

It is evident that spectrometric analysis that always uses only a part of the total sample activity for counting is more complicated than a detection. To make the resolution of several radionuclides easier, various *sample treatments* (Section 2.8) are often applied.

Amplitude analysis in the counting electronics (Section 2.4) can also contribute considerably to the detector background if one selects a counting channel correspondingly to the detected particles. Choice of a counting channel in relation to the ratio of efficiency and background is shown in Figure 2.4. The counts from different particles can also be resolved by means of other electronic circuits (Section 2.4), such as pulse-shape analyzer (Oikari et al., 1987).

Smoothing (Tykva and Jisl, 1986) and/or deconvolution (Jisl and Tykva, 1986) are often used to evaluate either the measured spectra or the planar distribution of radionuclides with higher accuracy.

2.2 LOW-BACKGROUND LABORATORIES

2.2.1 Basic Requirements

From the analysis in Section 2.1 it follows that the concept of "low-level radioactivity" includes a relatively broad spectrum of activity values and specific activities. Although certain general conditions apply for the determination of low-level radioactivities, it is necessary to consider in all cases the concrete interval of values that come into consideration *for a given problem.* This interval defines the requirements for the equipment. This principle applies in a distinct manner mainly to laboratories with high investment requirements. Thus, for example, the location and the equipment of a workplace for the determination of low activities of radiotracers in science and technology will not be suitable for radiochronological determinations. Moreover, the working conditions in radiochronological laboratories would be endangered by the analysis of compounds labeled with radionuclides, for example, from the point of view of a possible contamination. Thus, it is evident that it is essential, for a certain range of low activity, to have a suitable workplace with appropriate equipment and working regimen.

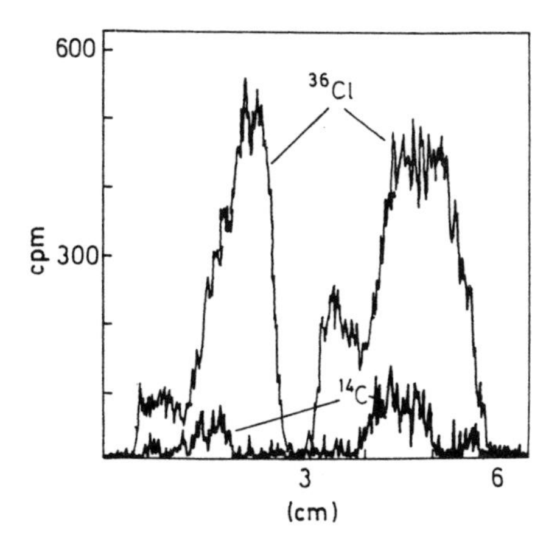

FIGURE 2.3. Comparison of one-dimensional distribution of simultaneously present ^{36}Cl and ^{14}C in the stems of two differently treated plants of *Chenopodium rubrum* (the length of each stem is 3 cm), Tykva et al. (1992).

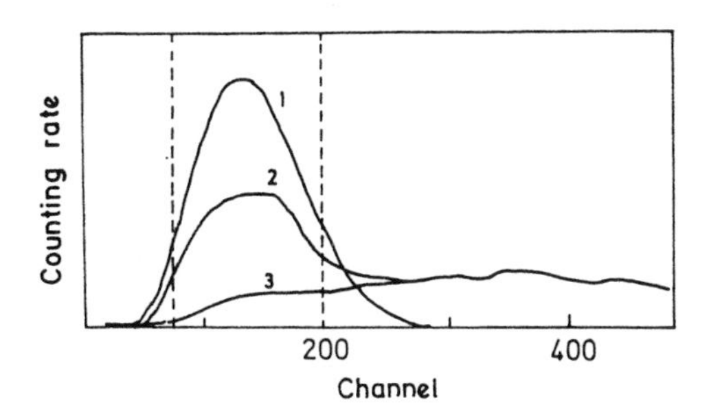

FIGURE 2.4. Measured spectra demonstrating the choice of the discrimination levels (counting channel) to optimize the detection sensitivity (figure of merit) for ^{3}H measurement: 1—^{3}H standard, 2—^{3}H water sample, 3—background. Based on Schoenhofer and Henrich (1987).

In the measurement of low-level radioactivities the *location of the laboratories* is one of the general criteria, i.e., that they should be separated from all areas in which a distinctly higher radioactivity level or sources of disturbances might occur. This separation means not only a sufficient distance, but it must also involve ventilation aeration techniques and electric supply sources and, further, must prevent the influence of the electromagnetic field. Other basic requirements include the exclusion of sparking of connectors and working exclusively with one's own equipment (e.g., glass or auxiliary electronic apparatuses). All new parts of the equipment must be controlled with respect to their radionuclide purity before their introduction into the rooms for work with low radioactivity.

For the *cleanliness of the laboratories,* clothing, and the behavior of the staff, the same rules apply as in modern technological laboratories for the production of semiconductor components. The purity of the laboratory air must be observed not only from the point of view of a possible contamination with radionuclide fallout and radon and its decay products, but also the low concentration of dust particles, which contribute to the formation of radionuclide aerosols (e.g., for this reason, smoking in laboratories or passage rooms is quite unsuitable for low-level radioactivity). The entrance ought to be hermetically closed, and behind it should be a bend, diminishing the danger of contamination from outside. In some cases it is useful to divide the room into several parts, for example, to separate the measurement proper from the preparation of the samples before measurement.

For work with the detectors proper (e.g., assembling or decontamination), it is essential to ensure special conditions. Especially in laboratories where all steps for the measurement of extreme low levels are not realized, it is useful to reserve a closed space with an independent and controlled air circulation for these operations. During the assembling and decontamination operations on detectors, it is necessary to eliminate the possibility of radionuclide contamination by cleaning means by controlling their composition and also by a control measurement before use. It is also indispensable to eliminate the possibility of contamination by the ^{210}Pb radionuclide during soldering.

The constant temperature and humidity of the laboratory air contributes to the stabilization of the measuring conditions, especially during long-term measurements. From the point of view of radionuclide purity, it is suitable to use the air under control in a closed cycle (especially, ^{222}Rn has to be removed).

The general location of the laboratory plays a big role and can have a decisive effect on the reduction of the detector background, which is caused by two basic sources: cosmic radiation on the one hand and radionuclide contamination of construction materials and geological basement on the other.

2.2.2 Sources of the Detector Background

2.2.2.1 COSMIC RADIATION

A permanent source of detector background is cosmic radiation (Janossy, 1950), which has been the subject of continuous intensive physical research since the 1930s. Before the construction of sizable accelerators, it served mainly as the only source of high-energy particles and their interactions; later on it was important in the study of processes in cosmic space. In low above-sea elevations the detector background is caused primarily by the extensive air shower. The mechanism of their formation by interaction of primary cosmic radiation with the earth's atmosphere has been described in detail, for example, by Galbraith (1958).

From the point of view of the effect on detector background, it is desirable to take into consideration the particle and energy composition of cosmic radiation after their passage through the earth's atmosphere at sea level, with predominance of μ-mesons, nucleons, electrons, and photons (May and Marinelli, 1964). These elicit the detector background either by direct response after penetration into the effective detector volume or by secondary particles emitted by interactions of incident cosmic radiation in substances into which the incident cosmic radiation has entered (e.g., in the wall of the ionization detector).

From the point of view of the effect of cosmic radiation on the detector background, the energy of the incident particles plays an important role in achieving high values. In an effort to maximally reduce the background, this fact leads to various laboratories being located underground (Section 2.2.3). For common judgement cosmic radiation incident on the detector can be divided roughly into two parts: a *soft component,* representing about two-thirds, and a *hard one* composed predominantly of high-energy μ-mesons. To decrease their effect on the detector background, a different approach has to be used. A quite special approach is required by the screening for neutrons (for details, see Section 2.2.3).

2.2.2.2 RADIONUCLIDE CONTAMINATION OF MATERIALS

The natural radionuclide contamination in the environment of the detector is determined primarily by the *geochemical composition* of the laboratory surroundings, predominantly by the presence of primordial radionuclides and their decay products (Section 1.1.2.2). For this reason it is necessary primarily to check the ^{238}U, ^{232}Th, and ^{40}K contents in the mountain rock samples along a several km radius of the underground laboratory (Kovalchuk et al., 1982). In addition to the emission of high-energy photons penetrating through the ground or walls of the laboratory, the main danger is produced by radon (Section 3.7) escaping from the soil or rock. Mathematical models have also been elaborated (Revzan et al., 1991) for the description of the penetration of radon of terrestrial origin into buildings. In buildings on the earth's surface, the highest concentration of radon is, naturally, in the cellars, which are, for reasons of screening cosmic radiation and bearing capacity of the floor, most frequently used for the measurements of low activities. As already mentioned, dilution with the outside air by means of ventilation is very restricted or even impossible. It is evident that in the elaboration of a laboratory project designed for the measurement of low activities of radionuclides, it is desirable to choose a place where there is no danger of background increase in this way.

A further source of radionuclide contamination, contributing to an increase of the detector background, consists of the *material of the building* and the *construction material* of the laboratory equipment, mainly that used for the manufacture of the detector shielding, especially of the detector itself (in construction materials of the detector, emission of alpha particles into the effective volume may also take place). Generally, the rule applies that no material can be used without control measurements. The chemical composition itself does not guarantee radionuclide purity. A certain material may be variously contaminated even if produced by the same technology. For example, a case is known where an airplane deck indicator with phosphorescing dial, which contained radionuclides, was processed into aluminum scrap and the whole melt became distinctly contaminated. Pollution by radioactive fallout also plays an important role.

In addition to external contamination in metallurgical or technological processing, a number of materials also contain naturally present radionuclides. Of the uranium decay series, this is mainly ^{214}Bi as a daughter product of ^{226}Ra, which occurs, e.g., in zinc ores and predetermines the inadequacy of zinc or brass for use in equipment for the determination of low activity. Of the thorium decay series, ^{208}Tl is particularly important. A very distinct natural contaminant is radionuclide ^{40}K, the natural presence of which (Table 1.1) elicits in 1 g of natural potassium about 25 electrons per second (spectrum beta with maximum energy 1.33 MeV) and approximately 210 photons of 1.4 MeV per minute. The natural presence of radionuclide ^{14}C (0.34 $Bq \cdot g^{-1}$) affects, for example, the area activity of plastic foils prepared from recent raw materials (average 50 $mBq \cdot 100\ cm^{-2}$). In close proximity to the active volume of the detector for measuring low activities, it is therefore necessary to use plastic materials made of crude oil or coal. The dispersion of the values of contamination for the same kind of material can be considerable. For a preliminary choice it is suitable to select a constructional material with a low natural radionuclide contamination which originates from the time before the first explosions of nuclear weapons in the atmosphere.

In all cases it is essential to control the radionuclide purity of every material before use because the specific activity of building (Wogman and Laul, 1982) or construction (Kamikubota et al., 1989) materials are very variable. A possible control of the radionuclide concentrations in such materials by the use of a special germanium spectrometer system has been described in detail (Arthur et al., 1987).

To illustrate these data, the specific activities of some often used building materials are to be found in Table 2.3. The comparison of radioactivity of

TABLE 2.3. Specific Activities of Natural Radionuclides in Building Materials.

	Specific Activities C ($Bq \cdot kg^{-1}$) (mean; min–max)			
Material	^{40}K	^{226}Ra	^{232}Th	RE[a]
Granites	680 383–1043	58 10–101	59 17–137	191 64–336
Sands	197 65–328	8 3–15	10 4–15	38 25–47
Gravels	152 64–271	10 4–15	12 3–17	38 13–51
Limestones	63 28–137	10 5–14	4 2–8	21 15–32
Marbles	132 103–148	12 11–14	4 3–5	28 24–30
Cements	185 104–271	20 12–28	12 7–16	51 30–61
Bricks	610 477–804	45 28–64	45 22–70	154 123–191
Clinkers	613 435–819	57 20–91	67 34–113	194 116–263
Aerated concretes	368 171–557	40 8–71	33 6–61	114 30–95
Coal fly ashes	443 194–726	86 42–129	72 41–107	215 139–273
Azbestos	191 118–230	9 5–13	14 7–19	43 31–56
Pearlites	1149 1125–1172	53 39–67	76 66–87	248 219–277
Ceramics	543 302–770	29 20–39	39 20–53	125 66–156

[a]RE = $C_{Ra(226)} + 1.26\, C_{Th(232)} + 0.086\, C_{K(40)}$ (radium equivalent).
Based on Wirdzek and Kazimir (1985).

various construction materials is given in Tables 2.4 and 2.5, in which the X-ray fluorescent intensities excited by background radiation are also given. The effect of selection of the starting material of the detector assembly on background is shown in Table 2.6.

After a careful selection of high-radiopurity materials, for ultra-low radioactivity, other sources must be sometimes considered. It has been

TABLE 2.4. Radiation Emitted from Different Materials Measured by Thin-Window Counter.

Sample	Description (purity, origin)	Mass (g)	Area (cm^2)	Alpha (cph/100 cm^2)	Beta (cph/100 cm^2)
Ag	99.969%	171.1	102.4	–[a]	–
Al	Cast plates	154.2	98.1	5.76 ± 1.86	348 ± 8
Au	99.979%	315.1	102.4	–	–
Bi	>99.999%	730	114.7	49.6 ± 2.7	13.6 ± 7.3
C	Madagascar flakes, impure	13.2	101	74.6 ± 1.9	1130 ± 8
Cd	>99.999%	620.8	112.8	–	39.4 ± 3.2
Cu	Powder	56.8	101	35.9 ± 1.6	68.9 ± 4.1
In	>99.999%	541.6	114.3	–	72.6 ± 6.8
Ir	>99.9%	182.3	102.6	134 ± 3	1640 ± 10
Ni	Powder	68.5	101	26 ± 2	114 ± 6
Os	99.92%	162.5	17.4	–	750 ± 24
Pd	99.985%	195	102.3	–	70.8 ± 4.7
Pt	>99.968%	347.9	102.3	–	68.4 ± 4.1
Rh	99.94%	199.4	101.6	20 ± 1.6	415 ± 6
Ru	99.96%	119.5	20.3	144 ± 8	107 ± 21
Sn	Pure, lead-free foil	1.2	62.9	182 ± 5	1260 ± 14
Ta		1.1	27.0	23.3 ± 7.1	146 ± 22
Zn	100%	16.0	88.0	20.9 ± 1.8	12.3 ± 4.7
Pb 1	Mexican	846.5	116.2	59.8 ± 1.7	545 ± 5
Pb 2	USA	832.4	116.2	1170 ± 5	7790 ± 14
Pb 3	Peruvian	845.1	117.5	728 ± 3	7430 ± 8
Pb 4	USA	842.2	116.8	17.8 ± 2.2	238 ± 8
Pb 4B		812.5	113.2	45.4 ± 2.7	296 ± 7
Pb 5	USA	841.8	116.7	58.1 ± 2.8	216 ± 7
Pb 6	Australian	845.1	117.8	99.8 ± 2.9	437 ± 8
Pb 7	USA	850	117.6	737 ± 5	5240 ± 13
Pb 8	Canadian	840.8	117.9	233 ± 6	3257 ± 22
Pb P	Holland	816.1	112.9	–	59.5 ± 7.4
Pb S	U. of C. shield	827.2	114.6	542 ± 4	2980 ± 9
Pb	>99.999%	842.3	114.2	298 ± 4	2910 ± 13
Pb	Powder	99.6	101	263 ± 10	4630 ± 41

[a]Indicates no detectable radioactivity.
Based on Weller (1964).

TABLE 2.5. Count Rates for Intrinsic Radioactivities and X-Ray Fluorescent Intensities of Shielding Materials.

Shielding Material	Mass per Unit Area ($g \cdot cm^{-2}$)	cph 0.1–1 MeV Relative Numbers	$cph \cdot kg^{-1}$			
			^{214}Bi (609 keV)	^{40}K (1460 keV)	^{208}Tl (2610 keV)	Xf-Lines
Yellow lead bricks for radiation protection	8.9	10,100	1.1	1.4	1.1	57,200
Lead electrolytics, 99.99%	12.2	9,700	0.7	0.7	0.4	19,300
Old lead installation	9.7	9,400	1.1	6.1	0.7	16,000
Lead slabs, 40 y old	13.0	2,900	0.4	2.2	0.7	9,320
Bismuth, 99.9%	11.0	−940	0.7	<0.4	2.2	5,470
Lead from Russia	11.3	−2,400	0.7	2.9	0.4	–
Lead from Korea in wire form	10.2	−3,300	0.7	0.7	2.2	1,220
Mercury from China	13.6	−3,700	0.7	2.5	0.4	320

Adapted from Unterricker et al. (1988) and Unterricker et al. (1989).

TABLE 2.6. Comparison of Primordial Radioactivity Levels in the Background of a 132 cm³ Intrinsic Germanium Diode Gamma-Ray Spectrometer before and after Rebuilding with Specially Selected Construction Materials.

Primordial Radionuclide	Energy (keV)	Before (cph)	After (cph)	Improvement Factor
^{235}U	185.72	73	<0.0048	>15,000
^{228}Ac (^{232}Th)	911.07	9.0	<0.0034	>2,700
^{234m}Pa (^{238}U)	1001.03	3.4	<0.0024	>1,400
^{40}K	1460.75	22	0.017	1,300
^{208}Tl (^{228}Th)	2614.47	1.0	<0.0011	>910

Based on Brodzinski et al. (1988).

shown (Brodzinski et al., 1990) that the radioactive background in two 1.1-kg Ge detectors was significantly reduced by selection of materials and careful control of fabrication of both the germanium crystals and copper cryostats. The isotopes ^{57}Co, ^{58}Co, and ^{65}Zn, formed cosmogenically in the crystals, were reduced by minimizing the time between the final zone refinement, crystal growth, installation in the cryostat, and placement underground. An attempt was made to reduce the background from the decay of ^{68}Ge in the detectors by deep mining the ore, rushing it through the refinement, crystal growing, and detector fabrication processes, and storing the germanium underground at all times it was not "in process." Cosmogenically formed ^{54}Mn, ^{59}Fe, and $^{57,60}Co$ in cryostat were minimized by electroforming the cryostat parts. The ubiquitous background from primordial ^{40}K in electronic components was virtually eliminated by selecting low-background components and by hiding the first-stage preamplifier behind 2.5 cm of 450-year-old lead in one unit and special low-background lead in the other. In the same facility (Miley et al., 1991) the underground detectors had copper cryostats completely electroformed from low-background copper. Electroforming is a process analogous to zone refining in its ability to remove chemical impurities.

Another possibility of decreasing background lies in desorption of ^{222}Rn from different metal surfaces by alpha- and beta-recoil processes and other mechanisms (Bigu, 1992).

2.2.3 Shielding

The locality of the workplace itself can make for considerable absorption of kinetic energy from the cosmic radiation particles. Therefore the laboratories for the determination of low activity are often located in cellars of buildings where the *shielding effect of the building* is useful and where the effect of the radiation dispersed on surrounding objects is also lowered, e.g., trees in the vicinity of the building. Moreover, the necessary carrying capacity of the floor is also available for the shielding arrangement, the weight of which is usually several hundreds or even thousands of kilograms. As shown in Section 2.2.2, deciding on the location of the workplace, it is important to carry out measurements of radionuclide contamination of the base of the building and the construction materials used.

When an especially intense suppression of the background is necessary, the workplace is located either in *abandoned mines* or in specially constructed *underground bunkers,* situated, as a rule, several tens of meters under the earth's surface. If the facility is located in former mines, the place must be dry and dustless and must display very low natural radionuclide contamination. Usually, the laboratory is situated in an abandoned limestone or salt mine. In some cases underground laboratories are built as independent bunkers. In such a case the walls are constructed, as a rule, from special concrete with a very low volume activity of natural radionuclides (Section 1.1.2) and possibly with an internal lining with a layer of pure basic rock.

Russian authors (Zdesenko et al., 1985) analyzed the reduction of Ge(Li) background in a salt mine. The measured spectra for various arrangements are shown in Figure 2.5, and the corresponding values are summarized in Table 2.7. From the given values it is evident that the measured spectrum of the gamma radiation originates almost exclusively from the daughter products of ^{238}U or ^{232}Th, and the peak at 1460.6 keV belongs to ^{40}K. From a comparison of measurements on the earth level and in an underground laboratory without shielding, it follows that the gamma background is decreased thirty-five to ninety times. The shielding box is composed of lead, mercury in an envelope of titanium, cadmium, and polyethylene, a part of which contains boron. Under the effect of this shielding, the background caused by natural radionuclides is distinctly decreased, and the majority of the required lines ranges in the counting rate interval from 0.005 to 0.1 h^{-1}. The presence of ^{137}Cs (661.3 keV) and ^{60}Co (1172.5 keV, 1332 keV) may be explained by contamination of the shielding materials in the box, while the traces of ^{144m}In (558.2 keV, 725.2 keV) are assigned by the authors to indium in the eutectic alloy with gallium. These results confirm the principle that, at the same time as the intensive decrease of the effect of cosmic radiation and the detector surroundings, great attention must be paid to radionuclide contamination of all materials used.

The influence of the geochemical surroundings of an underground laboratory has been exposed as a comparison of background selected photopeaks and integral rates measured by a Ge(Li) detector in a mine with a

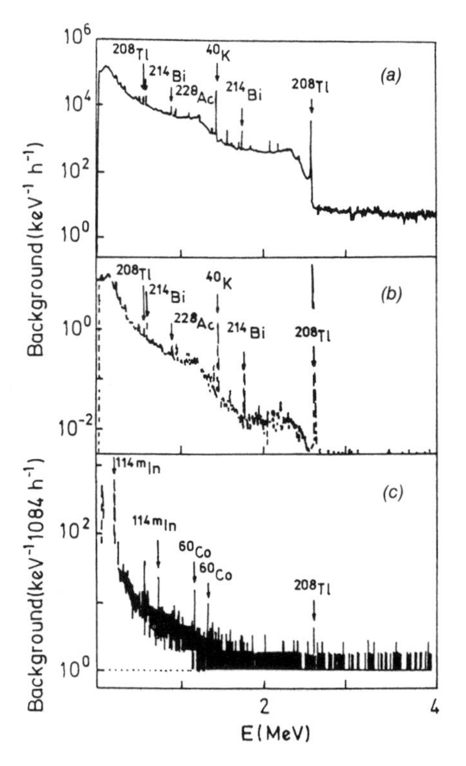

FIGURE 2.5. Background spectra of Ge(Li)-detector of a 35 cm³ volume: (a) ground level without shielding; (b) underground (430 m) without shielding; (c) underground (430 m) with a shielding box. Adapted from Zdesenko et al. (1985).

certain radionuclide concentration and in a salt mine (Table 2.8, where mwe stands for meter of water equivalent).

Winn et al. (1988) described an underground counting facility with a clean room environment, which had been designed and constructed to improve detection of low-level radioisotopes in the environment. The 3.0 m × 4.3 m × 2.4 m counting chamber was placed 14.3 m below ground, had 10.2-cm–thick walls of pre-WWII naval armor plate, and was further shielded by a minimum of 1.2 m of specular hematite. The total overburden of shielding was equivalent to 31.7 m of water. Careful selection of building materials and a special air filtering system maintained a clean room environment with minimum contamination potential. Background improvements were noted relative to an earlier ground-level counting chamber with 30.5-cm–thick walls of pre-WWII naval armor plate. The gamma background continuum was reduced by a factor of three to four in the region of

TABLE 2.7. Peaks of Background (cph) for Different Arrangement of a Ge(Li)-Detector (35 cm^3): A—Ground Level without Shielding; B—Underground (430 m), No Shielding; C—Underground (430 m) + Shielding Box.

Radionuclide	Gamma Energy (keV)	A	B	C	A/B	Ratio B/C · 10^2	A/C · 10^3
^{212}Pb	238.6	877.7 ±7.1	15.3 ±1.6	0.15 ±0.04	57.4 ±4.4	1.0	6
^{214}Pb	295.2	222.7 ±4.7	5.4 ±1.0	0.028 ±0.020	41.2 ±4.7	1.0	8
^{214}Pb	352	387.9 ±0.4	10.2 ±1.0	0.046 ±0.023	38.0 ±4.4	2.2	8
^{208}Ti	511	189.5 ±3.1	2.4 ±0.6	—[a]	79.0 ±5.2	—	—
^{208}Ti	583.1	284.6 ±2.9	3.8 ±0.6	0.026 ±0.02	74.9 ±5	1.5	10
^{214}Bi	609.3	345.3 ±3.0	8.4 ±0.7	0.032 ±0.015	41.1 ±4.3	2.6	10
^{228}Ac	911.2	191.4 ±2.1	2.1 ±0.5	0.02 ±0.01	91.1 ±4.2	1.0	10
^{214}Bi	1120.3	80.8 ±1.9	2.0 ±0.4	—	40.4 ±4	—	—
^{40}K	1460.8	1280 ±3	16.5 ±0.6	0.013 ±0.006	77.6 ±5	12.7	100
^{214}Bi	1764.3	69.4 ±0.9	1.9 ±0.2	0.004 ±0.004	36.5 ±4.5	4.6	17
^{214}Bi	2204.2	18.2 ±0.7	0.6 ±0.1	—	32.5 ±5.4	—	—
^{208}Tl	2614.5	166.1 ±1.0	2.1 ±0.2	0.008 ±0.004	79.1 ±5	2.6	20

[a]Indicates no detectable peak.
Based on Zdesenko et al. (1985).

TABLE 2.8. Selected Photopeak and Integral Rates of Different Gamma Energy Values and Ranges Measured by 10 cm³ Ge(Li) Detector (cph).

Energy (MeV)	1.46	1.76	2.61	≥0.2	≥1.5	2.7–4.0
Surface laboratory	1,940	173	274	428,000	10,700	350
Outside underground laboratory in a mine with radionuclide concentration	4,460	234	380	479,000	12,700	3.2
Salt mine (500 m salt + bundsandstein = 1100 mwe)	650	4	<3	37,800	94	–

Adapted from Unterricker et al. (1988).

0 to 10 MeV. A minimum of 10.2 cm of low-background lead around the detector optimized this factor at four to six. Discrete gammas from airborne natural radon daughters were eliminated by controlling the air near the detector. Detectors (Section 2.3) constructed with low-level materials will bring further improvements in background. A constant background is required for long counts of low-level samples, and the underground detectors are well shielded against surface operations. Performance appraisals of facility detectors include a large dual NaI(Tl) coincidence system, three smaller NaI(Tl) detectors, an HPGe well detector, and gas proportional and Geiger counters. Major electronics for the detectors are located at a ground-level control, and an uninterruptible power is used.

A PC control in such a laboratory is also described (Winn, 1987). A multichannel amplitude analyzer-coupled IBM PC/XT system collects HPGe, dual-coincidence NaI(Tl), and NaI(Tl) background spectra, which are analyzed by computer codes. These codes incorporate numerous efficiencies, integral peak-fitting, and comprehensive I/O features. Spectral files may be transferred to any IBM PC for further analysis. The facility improves detection sensitivity via extended counting times, background reduction, and improved detector efficiency. A constant background extends counting time from one to ten days.

The comparison of beta-gamma coincidence background in two different arrangements in surface and underground laboratories (Niese et al., 1989) is presented in Figure 2.6.

Delibrias and Rapaire (1967) achieved a complete elimination of the nucleonic component and 98.7% elimination of the muons from cosmic radiation under approximately 60 m of calcareous rock (15 $kg \cdot cm^{-2}$). In this way the background of the ionization detector without an additional anti-coincidence shielding decreased from 138.84 ± 0.26 cpm in the laboratory at a shielding of 320 $g \cdot cm^{-2}$ to 2.677 ± 0.033 cpm underground.

In some low-level facilities existing areas characterized by very strong natural shielding are used, for example *transport tunnels* in the Alps. A systematic study (Aglietta et al., 1992) of low-energy background radiation has been performed for different experimental sets under several conditions in the underground experiment at Mt. Blanc laboratory. The information was gathered using three different types of detectors (see Section 2.3): a scintillator (90 tons of liquid scintillator) with an energy threshold higher than 800 keV, NaI ($5' \times 5'$) with 220 keV, and a radon-meter (alpha-spectrometer) running at the same time, has been analyzed during long and continuous time periods. It was found that the variations in counting rate of the low-energy background is due to the presence of ^{222}Rn in the laboratory room. Three types of variations were found: (1) a correlation with the tun-

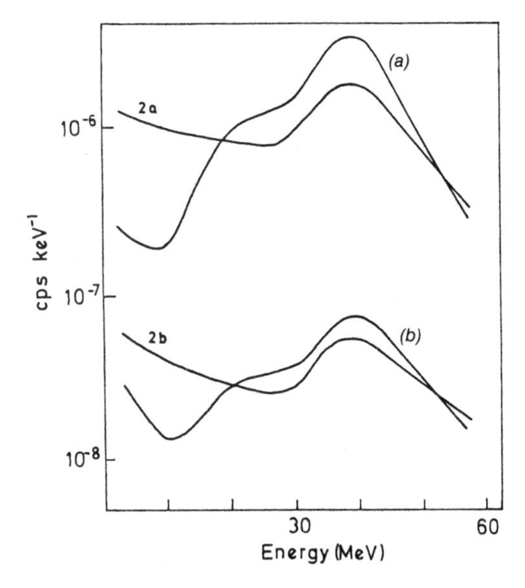

FIGURE 2.6. Beta-gamma coincidence background in two different arrangements (1, 2) in a surface (a) and an underground (b) laboratory. Based on Niese, Helbig and Kleeberg (1989).

nel ventilation system, (2) a daily modulation, and (3) a sporadic increase in the counting rate not correlated to any man-made intervention.

Another analysis in similar experimental conditions has been concentrated on alpha-ray induced background (Hubert et al., 1986). Background spectra of several spectrometers have been recorded in a deep underground laboratory located in the Frejus tunnel. The results show that an alpha-ray induced background during measurements of the ^{210}Pb decay is observed. A possible explanation could be related to the adsorption of the Rn gas on the surfaces of the Ge crystal and/or other parts during the assembly of the spectrometer used.

In addition to the shielding effect of the layers above the measuring arrangement, *mechanical and anti-coincident shielding* are used directly in the workplace (the anti-coincidence circuit and other electronic circuits used to decrease detector background are described in Section 2.4). The basis of the mechanical shielding consists of materials with a high density, a high proton number, a high radionuclide purity, and guaranteeing a lowering of the background caused by cosmic radiation and rationuclide contamination in the vicinity of the measuring arrangement with regard to the mass collision stopping power (Figure 1.28 and the related discussion). This basis material in a very thick layer is supplemented on the inner side for some types of detectors by a thin layer of an additional shielding material absorbing secondary particles set free in the basic shielding material by interaction with cosmic radiation. Another independent shielding layer in the proximity of the detector itself is also used for the shielding of neutrons in order to prevent the activation of the detector material or the emission of a false impulse by the reflected hydrogen nucleus, forming a part of the effective detector volume. For the absorption of neutrons, radionuclide-free material is used inside the outer shielding, displaying a high concentration of hydrogen for neutron scattering (polymers or paraffin) and containing boric acid to absorb thermal neutrons. Such materials are commercially available.

Among the basic shielding materials, lead takes first place. For the determination of low activity, and in view of the proportion of radionuclide ^{210}Pb with a half-life of approximately twenty years, sufficiently old lead should be used, adjusted to the required shapes in a special melt. The low content of antimony in lead is also important: when its content drops from 5 to 0.01 wt%, a decrease of specific activity from 3.4 $Bq \cdot g^{-1}$ to 0.9 $Bq \cdot g^{-1}$ could be measured. In our laboratory we first collected lead for the shielding, which was obtained during an exchange of round glass windows in the Gothic windows of Czech churches and castles. However, in control measurements this lead was found unsuitable, owing to its contamination by radioactive fallout. Therefore we focussed our attention on sufficiently old lead with a low content of antimony, which had not been exposed to atmospheric effects. Such a material was obtained during the demolition of the town of Most in the Czech Republic, from old lead water pipes located in the ground. Its age could be estimated at more than eighty years, based on the technology used for its production. This guaranteed a decrease in the content of ^{210}Pb at a sufficient level (at least sixteen times lower than a "fresh lead"). A radionuclide-free lead is commercially available, predominantly from old wrecked ships.

In laboratories the lead shielding mainly absorbed the soft component of cosmic radiation, while a small loss of kinetic energy of its penetrating component does not prevent penetration into the detector. Comparative measurements have shown that, for this reason, it is not useful to increase the thickness of the lead shielding to over 10–15 cm; moreover, in lead, fluorescent X-radiation is excited by background radiation. This buildup of tertiary background particles was eliminated by internal lining with a material of low proton number such as cadmium. In the presence of neutrons, copper is preferred because it has a lower cross section for radiative neutron capture. But also under copper lining, the background counting rate is increased due to excited levels in relation to the thickness of copper (Table 2.9) and to the energy interval (Table 2.10).

TABLE 2.9. Intensity of 75 keV Lead Fluorescence X-Ray as Function of Thickness of Cu Lining.

Thickness (mm)	0	2	5	9
Rate (cph)	39 ± 3	9 ± 2	2 ± 4	not seen

Based on Verplancke (1992).

A detailed comparison of the influence of a shield arrangement was carried out (Lindstrom and Langland, 1990). Neither cadmium, tin, copper, or plastic (hydrocarbon or fluorocarbon) was desirable as a shield liner, since all these increased the background continuum or introduced characteristic peaks into the background spectrum. Using germanium detectors, two broad peaks in the background result from inelastic scattering of cosmic-ray neutrons in germanium.

The application of inner linings must be carefully considered. A Japanese group (Shizuma et al., 1987) measured background spectra with a Ge detector in a low-background shielding using various inner linings: Lucite, aluminum, iron, copper, and lead. The experimental results show that the background counting rate in the energy region 0–500 keV and backscattered gamma-ray increase in a shielding with low Z linings. This can be qualitatively understood through the gamma-ray absorption and scattering processes in the shielding material.

A recently published list of gamma rays found in a large volume germanium detector under a different lining is summarized in Table 2.11.

For the suppression of the effect of the hard component of the background of the detector itself, a shielding detector is used (or a system of shielding detectors), surrounding the measuring detector proper, and connected with it in anti-coincidence. The anti-coincident arrangement permits the particles, causing the response in the detector for the measurement of low activity to elicit the response in the shielding detector as well. Such a pair of impulses is then not counted. Some results obtained in a relatively simple arrangement in a surface laboratory (Grinberg and Le Gallic, 1961) are presented in Table 2.12. In addition to anti-coincident shielding, the effect of

TABLE 2.10. Background Photons from Excited Copper 13 mm Thick.

Energy (keV)	666.9	770.6	962.1	1115.5	1327.0	1412.0	1481.8
Rate (cph)	6.1	1.4	8.6	2.9	2.2	0.7	1.1

Adapted from Lindstrom et al. (1990).

TABLE 2.11. List of Gamma-Rays Seen in the Background Spectra of a 2000 mm² × 20 mm Thick Low Background Ge Detector in a 10-mm Thick Lead Castle, Lined with 1 mm Cd and 2 mm Cu. Exp.1: Measurement with n-Source on Top of Lead Castle. Exp.2: Like 1, but with Extra 1 cm of Acrylic Lining.

Energy (keV)	Exp.1 (cph)	Exp.2 (cph)	Assignment
23.4		10.9 ± 10.4	^{70}Ge (n,γ) ^{71}Ge
53.4	13.4 ± 5.4	19.1 ± 9.5	^{72}Ge (n,γ) ^{73m}Ge
66.7	4.5 ± 5.0	20.0 ± 8.1	^{73m}Ge (sum: 13.3 ± 53.4 keV)
68.7	16.6 ± 8.0	4.9 ± 4.9	^{73}Ge (n,n',γ) ^{73}Ge[a]
39.6	7.4 ± 4.5	25.5 ± 6.6	^{74}Ge (n,γ) ^{75m}Ge
159.5	7.1 ± 5.6	49.9 ± 8.0	^{76}Ge (n,γ) ^{77m}Ge
186.0		14.0 ± 6.9	^{65}Cu (n,γ) ^{66}Cu
198.3	6.7 ± 4.2	21.8 ± 5.2	^{70}Ge (n,γ) ^{71}Ge
203.1	3.6 ± 2.2	11.9 ± 4.5	Cu (n,γ)
278.3		38.5 ± 6.2	^{63}Cu (n,γ) ^{64}Cu
511.0	39.8 ± 6.0	104.3 ± 7.3	e^+–e^- annihilation
558.2	18.8 ± 3.9	78.1 ± 6.2	^{113}Cd (n,γ) ^{114}Cd
569.6	4.1 ± 3.4		^{207}Pb (n,n',γ) ^{208}Pb
595.6	6.5 ± 5.0	15.8 ± 5.4	^{74}Ge (n,n',γ) ^{74}Ge[a]
651.0	4.9 ± 2.6	10.9 ± 4.0	^{113}Cd (n,γ) ^{114}Cd
669.6	6.7 ± 3.6	7.0 ± 3.5	^{63}Cu (n,n',γ) ^{63}Cu
691.3	10.6 ± 5.3	17.8 ± 7.5	^{72}Ge (n,n',γ) ^{72}Ge[a]
803.3	3.8 ± 3.5	2.8 ± 1.2	^{206}Pb (n,n',γ) ^{206}Pb
805.7		1.7 ± 1.0	^{113}Cd (n,γ) ^{114}Cd
962.1	8.2 ± 2.9	11.6 ± 3.7	^{63}Cu (n,n',γ) ^{63}Cu
1115.5	3.2 ± 1.7		^{65}Cu (n,n',γ) ^{65}Cu

[a]Assymetric and broadened.
Adapted from Verplancke (1992).

TABLE 2.12. The Influence of Different Shielding Arrangements on a Wide-Channel Background of NaI(Tl) Detector (4 × 3 in.).

Measurement Conditions	Background, cpm
Ground level laboratory (A)	36
Underground laboratory (B)	20
B + 10 cm Pb	7
B + anti-coincidence	2.2
B + 10 cm Pb + 5 cm Hg	0.5

Based on Grinberg and Le Gallic (1961).

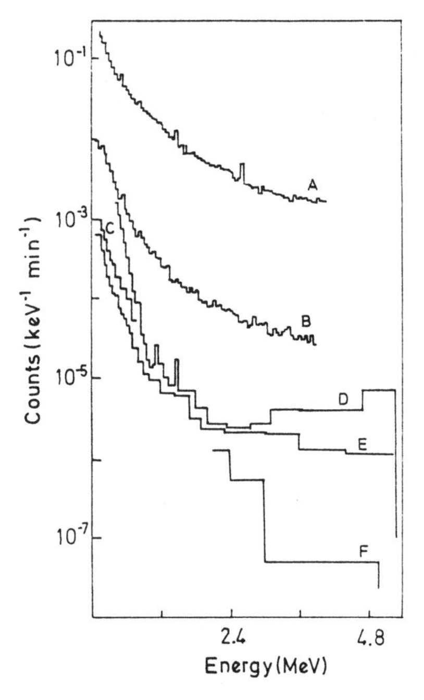

FIGURE 2.7. Background spectra of "radiopure" germanium spectrometer in different cryostats and shielding configurations: A—typical cryostat assembly in 10 cm thick lead shield, B—cryostat assembly rebuilt with radiopure materials in an electronic scintillator anti-cosmic shield above ground, C—7.3 cm thick copper inner shield added, D—indium electrical contact removed, E—copper shield replaced with 10 cm thick shield of 448 year old lead, F—solder electrical connection removed. Adapted from Brodzinski et al. (1988).

discrimination using amplitude shape is also presented (for more details see Section 2.4).

In addition to lead, non-alloyed, preferably old iron of a high radiochemical purity or mercury are also used as basic shielding materials. The high radiochemical purity of the mercury can be achieved by distillation.

Finally, it must be noted that optimizing a low-level detector for decreased background is an iterative process. With every step a new generation of background activities shows up (Verplancke, 1992). A complex evaluation of various steps can be considered from results obtained 1438 m underground in a gold mine (Figure 2.7).

2.3 LOW-LEVEL DETECTORS

In accordance with the general conception of this work we shall include in the section on detectors a discussion of problems important for the determination of low activity. The topics concerning radioactivity detectors in general are included only to the extent necessary to understand this specialized theme, and they can be found described in detail in the appropriate monographs (Knoll, 1979; Tsoulfanidis, 1983).

2.3.1 Starting Viewpoints for Selection of Detectors

The choice of a detector for certain measurements forms one of the basic conditions for successful work. As shown in Section 2.2, when optimizing the conditions for the measurement of low activity of radionuclides, several basic factors come into consideration: localization of the workplace, selection of radionuclide-free surroundings and materials, choice of suitable electronic circuits, stability of the equipment, and suitably combined shieldings. The application of these viewpoints creates the necessary conditions for measurements. However, the attainable efficiency of detection, which affects the desirable increase in detection sensitivity [Equation (2.1)], is determined predominantly by the detector itself, which also predetermines the processing and the arrangement of the sample; therefore, great attention should be devoted to the detector itself.

In evaluating the suitability of the detectors for a given use, it is always important to consider individual alternatives critically, with respect to their advantages and drawbacks for actual needs. If the conditions of the workplace permit it, the development or at least specific adjustments of the detectors directly at the workplace frequently represent the most suitable approach. For many applications the commercially available detectors are unsuitable because they afford only limited possibilities.

The selection of detectors for certain uses is determined by four fundamental variables:

- emitted radiation
- detection or spectrometry
- nature of the sample
- requirements on the precision of the measurements

These criteria should be considered in a complex manner when choosing a detector for measurements leading to the fulfillment of a set goal.

2.3.2 Detector Noise

In the field of semiconductor technology, there is a principle somewhat simplified, mentioned, that a good semiconductor element is one with a *low noise.* This principle can be transferred to detectors for low activity and utilized as a common general requirement. In physical systems great attention is paid to noise (D'Amico and Mazzetti, 1986). The signal-to-noise ratio is crucial under the working conditions of the detector, both for the quality of the detection and for spectrometry.

The magnitude of noise can be expressed in several different manners. Usually, it is determined by the use of mono-energy radiation (for example, conversion electrons, alpha particles, gamma rays, characteristic X-radiation) on the basis of the determination of the full width of the corresponding peak at its half maximum (FWHM). The noise expressed in this way is usually determined in eV, or in percent of energy corresponding to the peak. The equivalent noise charge ENC may then be determined for, e.g., silicon detectors from the relation (Nicholson, 1974)

$$\text{FWHM(eV)} = 5.28 \times 10^{19} \cdot \text{ENC(C)} \tag{2.3}$$

The noise of the detector itself, FHHM_d, is determined from the relation

$$\text{FWHM}_m^2 = \text{FWHM}_d^2 + \text{FWHM}_e^2 \tag{2.4}$$

where FWHM_m is the measured value and FWHM_e means the noise of the sensing electronic devices recording the detector response (Section 2.4). This value is usually measured by means of signals from the pulse generator introduced onto the input of recording electronics. At the same time there is introduced a condenser of a capacity equivalent to the sum of the detector capacity at the working bias and the parasitic capacity at the input member of the recording electronics. It should be stressed that the detector noise usually assumes a higher value than the noise of the electronics, so that it is decisive for the total noise.

Of the *detector noise components* the intrinsic component elicited by fluctuations in the number of particles generated during the interactions of each, the detected mono-energetic particles with the medium of the effective detector volume cannot be eliminated. The *Fano factor* has been introduced in an attempt to quantify the departure of the observed statistical fluctuations in the number of the charge carriers generated in the interactions of the detected particle from Poisson statistics (Section 2.6.2). It is evident that the magnitude of this component of the noise depends on the type and the energy of the detected particle and the detector used (Yamaya, 1981).

Other components of the noise can be restricted by constructional arrangements, selection of the starting materials for the detector, partial technological operations during its construction, and the use of certain working conditions made possible by the experiment itself.

The detector noise should be considered as the sum of several components. According to the character of each of them, their contributions can be lowered. Thus, the frequency-dependent component can be eliminated outside the frequency channel desirable for transmitting the real counts

TABLE 2.13. Dependence of the Noise of the Silicon Detector on Temperature during the Detection of 62.2 keV Electrons.

T (K)	233	253	273	293	313
FWHM (keV)	3.67	3.93	4.70	5.77	10.25

Not published, own results.

generated by the measured sample (for details see Section 2.4). For some types of detectors, the most important effect among these conditions is the temperature change of the detector (Table 2.13), which, from a certain value, is threatened by water vapor condensation from the surrounding air and therefore requires additional measures. However, for some types of samples these measures are not possible (e.g., evaporation of the sample during evacuation of the measured site to prevent condensation of water vapor and, consequently, the background increase due to internal contamination of the detector).

The importance of decreasing the detector noise to improve the energy resolution in spectrometry is evident. With decreasing noise the counting rate in the channel (or keV) increases, thus increasing the signal-to-noise ratio. When recording a continuous energy spectrum of beta particles, the noise is also important for a low activity, especially in the case of a low-energy emitter. The value of the noise determines the lower discrimination level of the recording channel, i.e., the lowest level of the detector response that can still be recorded. In detectors with a linear relationship between the amplitude of the voltage response and the kinetic energy loss in the effective detector volume, the energy interval of the recorded particles thus increases as does the efficiency of the detector [Equation (2.2)].

When measuring low activities of beta emitters, measurement above the discrimination level is used as a rule, guaranteeing the exclusion of noise from the record. However, in all instances the spectral composition of the background must be taken into consideration simultaneously, since the counting efficiency alone is not decisive. The aim is to set the counting channel in such a way as to achieve a maximum value of the figure of merit [Equation (2.2)]. For the sake of clarity, Figure 2.4 shows the spectra of the background and the counts evolved by ^{3}H in the liquid scintillation spectrometer.

2.3.3 Gas-Filled Detectors

2.3.3.1 DETECTOR TYPES

Gas-filled detectors are, with few exceptions (gas scintillators, without practical application in low levels), based on the *ionization of molecules of the filling,* after their interactions with a detected particle. They are among the older types of detectors for low radioactivity and their present development is – for example, in comparison with semiconductor detectors – only relatively slow. Their use can be considered for measurements of low activity of the *emitters of alpha particles, electrons, and low-energy photons.* They are unsuitable for the low activity of gamma emitters, owing to low detection efficiency. Ionization detectors are only utilizable to a limited extent for spectrometry of low-activity sources, because the average energy required for the formation of the positive ion-electron pair by ionization in a common gas filling (approximately 30 eV) causes bad energy resolution of the detector, and thus it decreases the counting rate belonging to the channel (or keV). For this reason ionization detectors and chambers are used predominantly for the measurement of low activity of a single radionuclide.

Nowadays, plastic scintillators are often used as *shielding detectors* instead of ionization detectors in anti-coincidence arrangements (Section 2.4). The basic reasons are their simpler long-term maintenance and easy formation of the plastic scintillator according to experimental conditions. Therefore scintillators are also preferred in the measurement of low activity in beta-gamma coincident arrangements.

In principle all gas-filled detectors represent condensers in which the space between electrodes is filled with a suitable gas. When putting voltage onto the detector, an electric field is created between its electrodes. The magnitude and the distribution of the electric field has a decisive influence on the detector properties. On impact of a particle emitted by a radionuclide into the gas filling, ionization of gas molecules takes place: interaction of the incident particle results in a part of its kinetic energy being lost to release the electron from the gas molecule, so that a positive ion is formed (the charge of each of the generated parts of the molecule is $1.602 \cdot 10^{-19}$C).

At the relatively small intensity of the electric field the current of the detector practically does not depend on its working voltage. Its value corresponds to the number of ionic pairs formed in the detector volume in one time unit. This operational regimen is characteristic of *ionization chambers.* Most applications of ionization chambers involve their use in current mode in which the rate of ion formation within the chamber is measured. However, for low-level radioactivities pulse mode operation is preferable (Figure 2.8) because of higher sensitivity and also the ability to measure the energy of the incident radiation.

By increasing the voltage, the situation on the detector changes and the gas amplification takes place, caused by the fact that the molecules of the gas filling are ionized by the impacts of the ions from the primary ionization, which are sufficiently accelerated in the electric field of the detector.

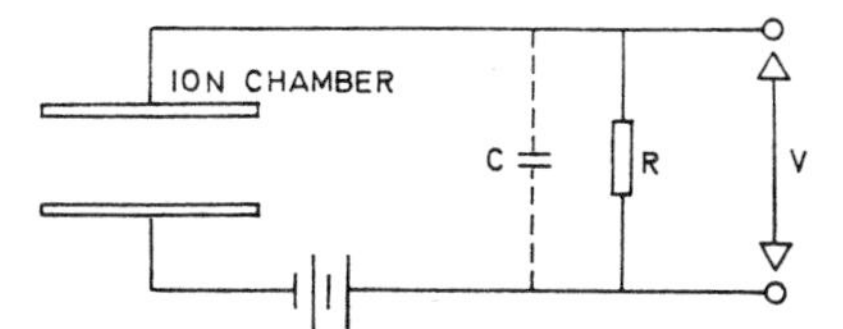

FIGURE 2.8. Equivalent circuit of ionization chamber operated in pulse mode: C—capacitance of the chamber and any parallel capacitance, V—output pulse with amplitude proportional to the generated charge response.

Therefore, the gas amplification increases with increasing field intensity. As a consequence, the current increase is higher than would be expected from the formation of ionic pairs as a consequence of primary ionization. If the charge response resulting in this way is proportional to the charge generated in the primary ionization, the detector is called a *proportional counter* (the gas amplification ranges from 10^3 to 10^4). The voltage response, i.e., the amplitude of the output signal, is thus proportional to the energy of the detected particle, which was absorbed during its interaction with the gas filling. When all the kinetic energy of the detected particle is absorbed in the effective detector volume and the collection of the generated charge is perfect, a proportional detector may be used for spectrometric determinations. It is evident that the probability of secondary ionization depends on the intensity of the electric field. Therefore, only a very limited area around the central anode with high intensity exists in the cylindrical proportional detector, where the ionization by accelerated electrons generated in the primary ionization is considerable, while in the rest of the volume this phenomenon practically plays no role. Only the area in very close proximity to the central electrode contributes significantly to the magnitude of the gas amplification. Thus, it is independent of the place of the inlet of the detected particle and also of the localization of the primary ionization.

In contrast to proportional counters, in another operational field of the gas-filled detectors called Geiger-Müller (GM) counters, the value of the gas amplification (approximately 10^8) causes such an increase of the total charge that all output voltage signals have the same amplitude.

Similar to ionization chambers, the formation of the voltage response takes place in proportional or GM counters. The charge response elicits a potential response $U(t)$ to the resistance inserted in series with the source of the direct current potential for the detector U_d (for details see Section 2.4). If U_d is gradually changed under otherwise constant conditions of the measurement and if measurement is taken at each value of the counting rate N elicited by the constant disintegration rate, we obtain a dependence called

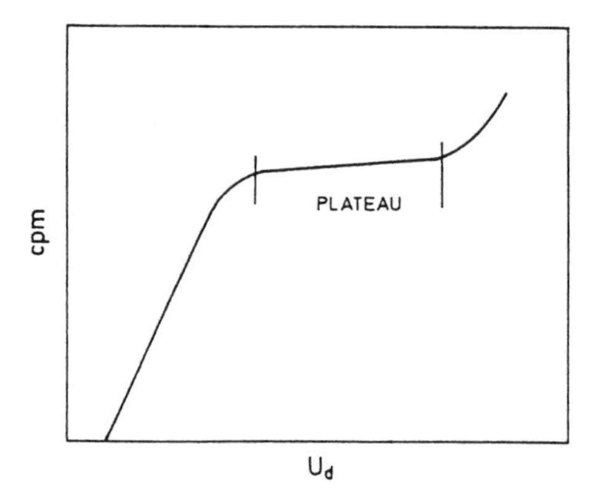

FIGURE 2.9. Counting curve of gas-filled counters (U_d—counter bias).

counting curve (Figure 2.9). The region of minimum slope is called the counting plateau, and it represents the area of operation. Usually, the working voltage of the counter is set at approximately one-third of the plateau. A good counter is characterized by a long plateau (usually 200–400 V) and a small slope (e.g., less than 1% per 100 V), which contributes distinctly to the stability of the conditions of the measurement.

Other types of gas-filled detectors are not used for low-level measurements. The essence of gas-filled detectors is described in detail elsewhere (Rossi and Straub, 1949; Wilkinson, 1950).

2.3.3.2 GRIDDED IONIZATION CHAMBERS

Ionization chambers are rarely used for the determination of low-level activities of radionuclides. Their application is limited practically only to the form of large-area gridded chambers (Figure 2.10) for very low alpha activities in spectrometric measurements. They are used especially for the environmental samples or mixtures of transuranium elements because of their large active detector area (parallel-plate type up to 0.05 m^2 and cylindrical

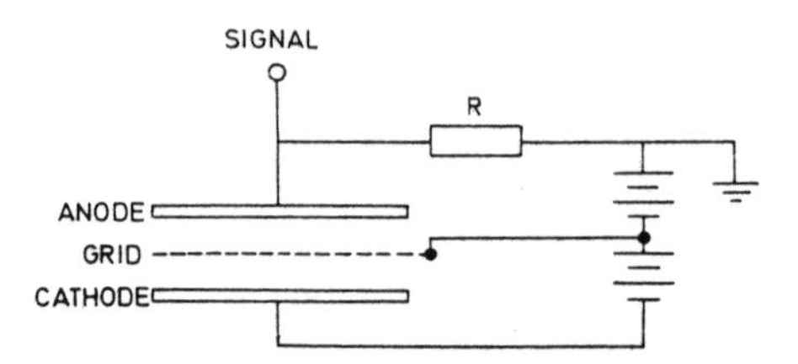

FIGURE 2.10. Principle configuration of the grided ionization chamber.

type up to 2 m^2). This large area represents an undoubtedly advantageous feature in comparison with silicon detectors, the size of which is limited by the diameter of the starting single crystal, which does not exceed six inches. Thus, when combining such detectors closely to each other in a mosaic, usually only an area of tens of cm^2 is achieved. The sample for a large-area chamber is usually prepared by spraying a suspension of finely ground material in a fixing agent onto metal-coated foils. In such a way it is possible to measure without any previous sample treatment, although removal of organic matter by low-temperature ashing or chemical separation before sample preparation is also applied.

After deposition of aerosols on a circular tin plate dish 20 cm in diameter by electrostatic precipitation, energy resolution of 35 to 70 keV for the ^{210}Pb line of a layer thickness of between 30 and 100 $\mu g \cdot cm^{-2}$ is obtained (Hötzl and Winkler, 1984). Using this uncomplicated procedure for detection of artificially produced alpha emitters (^{241}Am, ^{239}Pu, ^{242}Cm) in environmental air, levels of units of $\mu Bq \cdot m^{-3}$ can be estimated. Measurements of actinides in water samples from the primary circuit of a nuclear power plant have also been carried out by Rösner (1981).

If lower specific activities have to be analyzed, recourse must be made to chemical separation or extraction techniques (Section 2.8), with subsequent electrodeposition on small area planchets. A detection limit is comparable for such samples, applying either grid ionization chamber or silicon detectors, although the chamber background is somewhat higher. Moreover, some silicon detectors for low-level alpha spectrometry are commercially available while homemade ionization chambers have to be used. Shielding of smaller volumes of silicon detectors is considerably easier.

Other types of chambers for low levels, such as for example drift chambers (Povinec, 1980), are used quite exceptionally.

2.3.3.3 PROPORTIONAL COUNTERS FOR SAMPLES IN SOLID OR LIQUID PHASE

The schematic arrangement of a flow counter for solid or liquid samples is shown schematically in Figure 2.11. The sample on a small dish is inserted under the detector and measured in 2π sr geometry, usually in proportional region.

The following is required of the *dish material* itself:

- high radionuclide purity
- a higher proton number, guaranteeing an increased back scattering of emitted particles from the base and thus an increase in detection efficiency

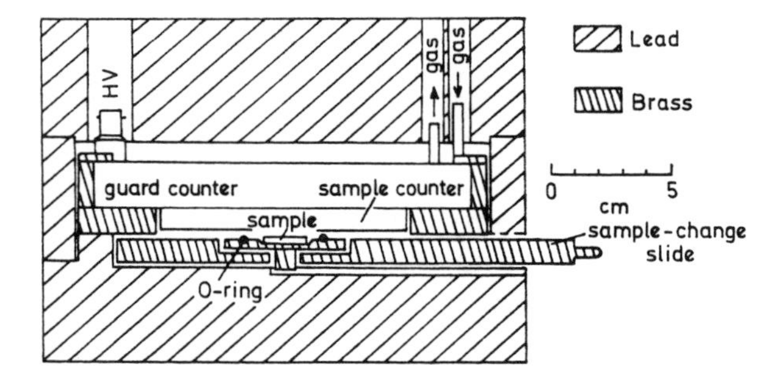

FIGURE 2.11. Schematic view on 2π proportional flow counter. Adapted from Winkler and Rösner (1989).

- stability in contact with solvents of solid samples or with liquid samples
- easy technological preparation, attaining a smooth bottom surface

These criteria are well satisfied by selecting appropriate sorts of steel. The background of every dish is controlled before use. Their repeated use is undesirable.

The *sample* introduced onto the dish is usually solid. Most frequent samples are

- a dry residue prepared from the sample directly on the dish
- membrane or glass filter put into the dish with a dry cell suspension
- chromatographic or filter paper containing the sample
- a sample worked up by radiochemical procedure

The detector itself is used almost exclusively in its flow-through version (flow-rate is approximately 0.3 $cm^3\ s^{-1}$), both under the proportional regimen and in the GM. Depending on the working field, either a flow-through gas or a mixture of gases is used; moreover, the proportional detector has a preamplifier, located, as a rule, next to the detector, in order to restrict the parasitic capacity. The detector usually has a thin window below 1 $mg \cdot cm^{-2}$ (e.g., Mylar). When measuring low activity of low-energy beta emitters, a windowless arrangement is more suitable, which sometimes deteriorates the counting characteristics and, consequently, the stability of working conditions. In order to eliminate the effect of the electrostatic charge a fine conducting net is sometimes used in place of the window. A mechanical cover is used for shielding, with an anti-coincidently connected detector. This detector is either an ionization detector, or it is made of a plastic scintillator.

Liquid samples are measured much less often. The basic danger consists of the possibility of evaporation and subsequent contamination of the detector arrangement.

At present a gas-flow type proportional counter is used only exceptionally. It has been applied, e.g., for low levels of alpha emitters such as uranium and thorium contained in the base material of LSI or VLSI (Minobe et al., 1987), determination of ^{241}Pu in environmental or biological material (Winkler and Rösner, 1989) or the total alpha-beta activity in measurement of environmental contamination in the surroundings of a uranium ore mill (Mondspiegel et al., 1990).

If a small amount of sample is measured, a detector may be used to advantage, with a small effective volume (so-called needle), which permits a less demanding achievement of a low background (Figure 2.12). In the working arrangement given and using an aluminum base for the sample, the measured count rate of background was 0.117 ± 0.007 min^{-1}.

The application of measurement geometry 4π sr is limited by the character of the samples, so that it is generally not used for low levels.

In all instances the use of this type of detector is limited, owing to self-absorption in the sample to measurements of higher specific activities, especially for alpha emitters and low-energy beta emitters.

2.3.3.4 INTERNAL GAS COUNTER

The internal gas counter is filled with gas, the activity of which is determined. In most cases the detector is *evacuated and filled* with a measured amount of gas; its flow-through version is used quite exceptionally. In view of the measurement geometry 4π sr and the negligible self-absorption, its

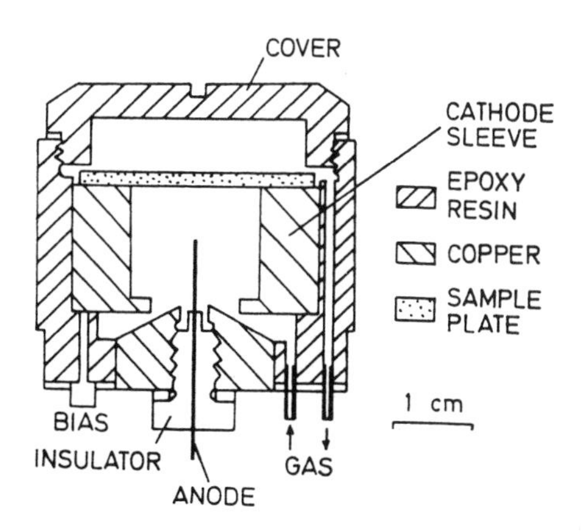

FIGURE 2.12. Schematic view on proportional flow counter for small samples ("needle counter"). Adapted from Zastawny and Rabsztyn (1986).

advantage consists mainly in the high detection efficiency for low-energy beta emitters or also X-rays. The 100% counting efficiency in the effective volume is diminished only by a relatively small number of particles emitted near the walls or ends of the effective volume and escaping without generation of the charge response exceeding the equivalent noise charge (so-called wall or end effect). Usually, the internal detector has a relatively large volume (up to several dm^3), and it is filled at high pressure (up to about 1 MPa), so that it enables – in view of its high counting efficiency – the determination of low activity with a sufficiently lowered background. A specific feature of the lowering of the background is that, when using a higher filling pressure and a filling consisting of a hydrocarbon with a larger number of hydrogen atoms in the molecule (e.g., propane), an effective shielding of neutrons is required in view of the increase of the probability of a false response by reflected protons.

The sample is obtained either directly in the form of a gas (e.g., separated krypton), or, more commonly, it is converted to a form of suitable gaseous filling for the detector from the solid phase (Section 2.8). When measuring low activity, the sample in the gaseous phase usually forms an independent filling of the detector; in some instances it is mixed with another, radionuclide-free gas to obtain the necessary properties of the detector filling. Internal detectors are used both in the GM and – more frequently – in the proportional region. Preference is given to the latter for the following three reasons:

(1) In view of the lower operating voltage on the detector in the proportional region, it is possible to use a higher filling pressure at the same voltage.
(2) In view of the lower gas amplification, a higher concentration of electronegative admixtures is admissible, facilitating an easier preparation of the sample.
(3) By simple amplitude discrimination some components of the background may be decreased (for example contamination of construction materials of the detector itself by alpha emitters).

The simplest type of internal gas counter is formed from a cylindrical cathode with a wire anode in the axis, which is surrounded both by an independent shielding detector and by mechanical shielding. The shielding detector, connected in anti-coincidence, is formed either by a plastic scintillator or a simple or double wreath of parallel-placed GM detectors. An example of the arrangement is shown in Figure 2.13 (Šilar and Tykva, 1977). The outer part of the detector was made from a copper pipe from the time before the Second World War, which was obtained during the exchange of the piping in a brewery. The cathode is made of the vacuum gilded inner

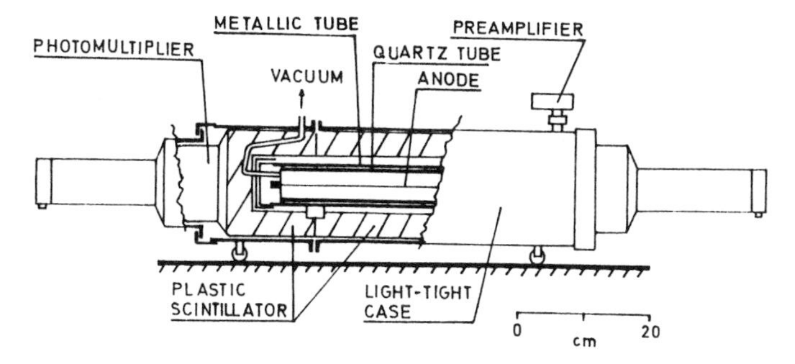

FIGURE 2.13. Scheme of an internal gas counter with plastic scintillator in anti-coincidence at the Charles University, Šilar and Tykva (1977).

surface of a pure quartz tube, which is inserted into the copper tube. The molybdenum wire for the anode was repeatedly annealed before assembly in a radionuclide-free medium. All further constructional materials were thoroughly controlled for radionuclide purity before use. The surfaces of the insulating materials were defatted in a non-contaminating medium, and the assembling was done in a radionuclide-free atmosphere. The photomultipliers in anti-coincident connection were made of glass with a very low content of potassium. All materials of further parts were also controlled carefully for radionuclide purity. The solder used did not contain lead or any other active contaminants.

Such a type of internal proportional gas counter also applied for resolving ^{3}H- and ^{14}C-activity by a method using a filling mixture of $C_2^3H_2$ and $^{14}CO_2$ (Tykva, 1967).

Other types of large-volume detectors are less widespread, primarily because of their difficult production requirements and demanding maintenance.

The Oeschger type (Houtermans and Oeschger, 1958) has its own detector separated from the shielding system of detectors by a thin plastic foil (Mylar), metal-coated on both sides (Figure 2.14). The shielding detectors between the outer metallic wrapping and this foil are made using parallel stressed wires. The foil prevents the penetration of low-energy beta particles, emitted in the filling of the internal detector, into the multiwire shielding detector connected in anti-coincidence. At the same time the separating foil lowers the background via the Compton electrons set free from the detector cathode, and it also decreases the danger of radionuclide contamination in the detector wall.

In the Curran type (Moljk et al., 1957) the foil is substituted by a grid system of parallel wires at the same voltage as the outer counter body (Figure 2.15). In this way the partial background produced in the Oeschger

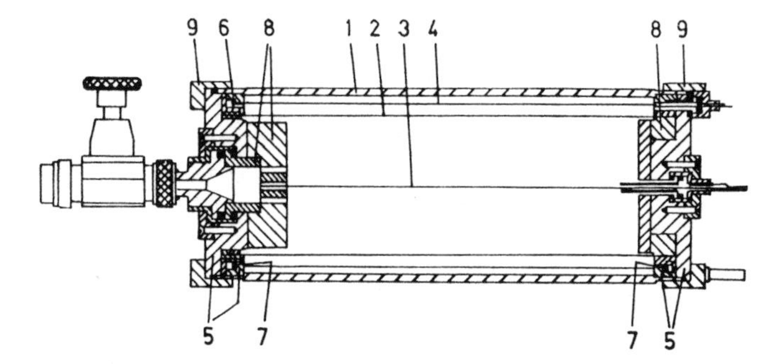

FIGURE 2.14. Scheme of the Oeschger type of internal gas counter (IGC): 1—detector wall, 2—metallized polymer foil as IGC wall, 3—IGC anode, 4—anodes of anti-coincidence ionization detectors, 5—insulating ring, 6—contacting ring of the metallized foil, 7—fastening ring of the anode wires in anti-coincidence, 8—insulating of the IGC anode, 9—frontal closures. Adapted from Gfeller and Oeschger (1962).

type by Compton electrons is decreased. Another advantage is the ability to choose the thickness of the inner cathode according to the requirements of the experiment. In the multi-element counter (Povinec, 1980) the inner counter has been replaced by seven or more element counters of the same dimensions arranged in a hexagonal form separated by cathode wires from each other (Figure 2.15). This makes it possible to use a higher gas pressure at the same working voltage and to decrease the counter background using a system of internal anti-coincidences between element counters.

To special types of internal gas counters belong small counters (with volume of the order of magnitude of units of cm^3) that are used mainly for radiocarbon dating (Section 3.3) of milligram-sized archeological and environmental samples (Sayre et al., 1981). However, for measurements of such very low specific ^{14}C-activity in small samples, even more accelerator-based mass spectrometry has been used (Section 2.3.6.5).

Internal gas counter detectors, in their flow-through version, are much less widespread because flow-through systems do not afford the sufficiently long time intervals necessary for measurements of low activities with

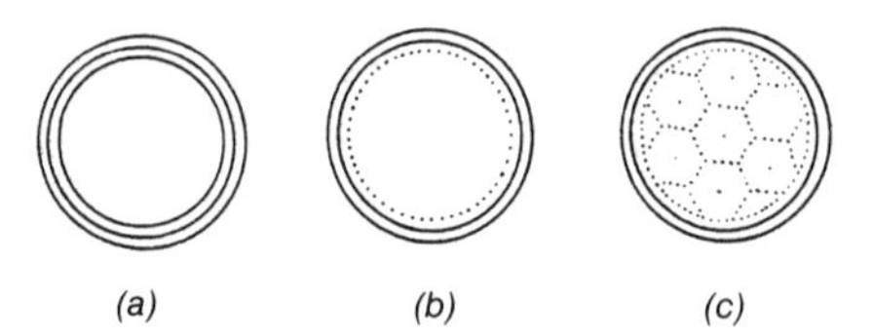

FIGURE 2.15. Types of special internal gas counters: (a) Oeschger type, (b) Curran type, (c) multielement type.

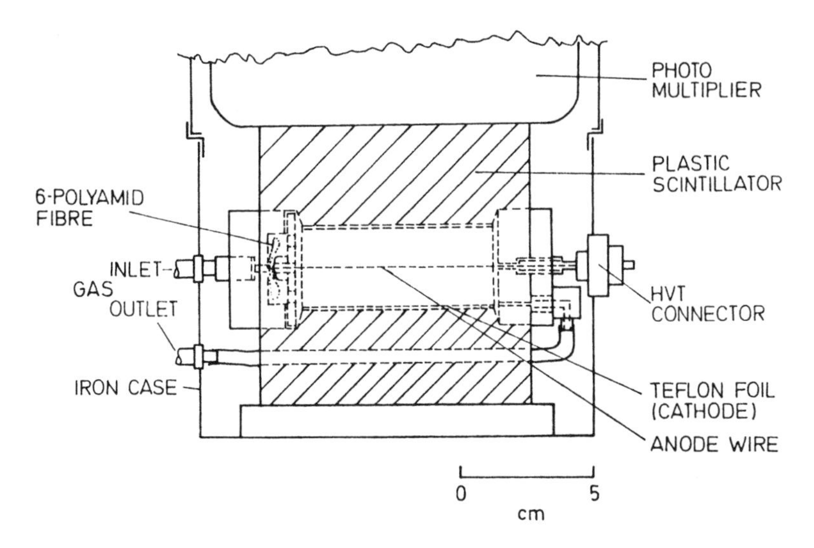

FIGURE 2.16. Scheme of gas-flow internal counter.

greater accuracy (Section 2.6). In our laboratory we have made a detector (Figure 2.16) in which the cathode is made of a gilt side of a teflon foil, which is inserted into the cylindrical hole in plastic scintillator, connected in anti-coincidence (Tykva and Kokta, 1967).

2.3.4 Scintillation Detectors

2.3.4.1 DETECTOR PRINCIPLES

Detectors based on the emission of photons produced in the scintillator after its interaction with a detected radiation is—after autoradiography—the oldest detection technique. If the ionizing radiation transmits a part of its entire energy (or its whole energy) to the scintillator during its passage through it, it emits scintillation photons. This emission of photons is an intramolecular property that has its origin in the electronic structure of the scintillating molecules. Through excitation of their electrons and their subsequent return to the ground electronic state, emission of photons with a characteristic spectrum takes place, i.e., *luminescence* (for a more detailed description see Section 2.3.4.4). Among the extensive developments we select information and operation procedures that are useful for the present-day measurements of low levels of radioactivity. General findings on scintillation measurements and the physical explanation of individual phenomena are surveyed in detail in the literature (Birks, 1964).

The operation of a scintillation detector may be divided into the following four stages:

(1) Absorption of the incident radiation by the scintillator
(2) The scintillation process, i.e., the transfer of the energy lost by a detected particle in the scintillator to the emission of scintillation photons
(3) Transfer of the photons emitted in the scintillator onto the cathode of the photosensitive radiation (i.e., photomultipliers or photodiodes)
(4) Absorption of the scintillation photons by the cathode and emission of photoelectrons

When photomultipliers are used, stage (4) can be subdivided into (4a) collection of photoelectrons on the first dynode and (4b) multiplying process in the photomultiplier.

The light response in the scintillator generated during the interactions is thus registered by a photomultiplier from the outlet of which the voltage response is led via an amplifier into a multichannel amplitude analyzer (Section 2.4). The linearity of the transfer between the energy lost by the registered particle in the scintillator and the amplitude of the voltage response at the photomultiplier outlet is jeopardized (and the spectrometric utilization as well) by self-absorption of scintillation photons in the scintillator, in consequence their reflection at interfaces and by scattering. For the measurement of low activity, the effect of self-absorption should be considered most important. From the point of view of mere detection, linearity is not necessary, but self-absorption can be so high that it significantly decreases the efficiency and thus prevents the measurement of low activity [Equation (2.1)]. In addition to non-activated organic and inorganic scintillators (for example, anthracene or NaI), the use of which is limited to small thickness only, self-absorption manifests itself mainly in liquid scintillators, where it is called quenching.

In all the stages mentioned, many factors that affect the characteristics of the scintillation detectors are operative. The course of the scintillation process depends on the type of scintillator. Generally, two basic types of scintillating material are known: inorganic and organic scintillators.

2.3.4.2 INORGANIC SCINTILLATORS

The majority of inorganic scintillators is created by activation with low concentrations of certain admixtures. Owing to the presence of this admixture, luminescence takes place. Activated alkali metal halides are typical (e.g., NaI activated by Tl), as well as sulfides activated with silver or copper, or also an excess of the ions of the basic material (e.g., ZnS with an excess of Zn ions). On the basis of their different structures, they are classified into three groups: the most common are scintillators activated with admixtures; further possibilities are pure crystals and self-activated scintillators.

The most widely used inorganic scintillator nowadays is NaI(Tl) with about 0.1 mol% of activator per mol of NaI. In view of its high density, this scintillator is mainly used in the detection and spectrometry of X and gamma radiation. The interaction mechanisms of photons in matter have been analyzed in Section 1.2.3. Photoelectric absorption makes it possible to measure the energy of the incident gamma radiation by means of the corresponding photopeak. On the contrary, Compton scattering interactions create a continuous Compton background, which considerably influences the measured spectrum (Figure 2.17). It evident that for low-level measurements the Compton continuum must be suppressed. This is made possible by placing another detector around the primary detector and selecting those events that are most likely corresponding to full-energy absorption by means of coincidence (Section 2.4). This arrangement is called *anticompton shielding.* Such a combination of two separated gamma-ray detectors operated in coincidence under mechanical shielding is shown in Figure 2.18. The measuring germanium detector is placed at the center of a coincidence shield, which is composed of two large NaI(Tl) detectors (30 cm diameter) and three smaller movable NaI(Tl) detectors to ensure the maximum anticompton shielding for various sample sizes and/or shapes.

From these facts it follows that inorganic scintillators arranged in a certain manner are suitable for the spectrometry of simple mixtures of gamma emitters. Therefore, they are used primarily as shielding detectors for germanium gamma-ray spectrometers (Section 2.3.5). Such suppression of the Compton continuum can improve sensitivity in most low-level measurements by a factor of two or three (Klucke and Beetz, 1987). Nevertheless,

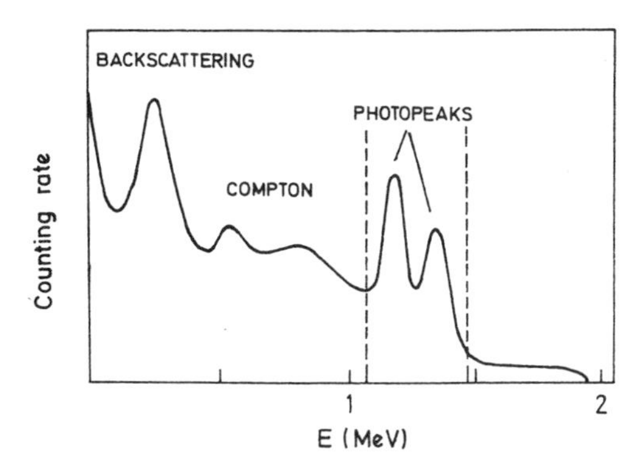

FIGURE 2.17. Amplitude spectrum of ^{60}Co measured by NaI(Tl).

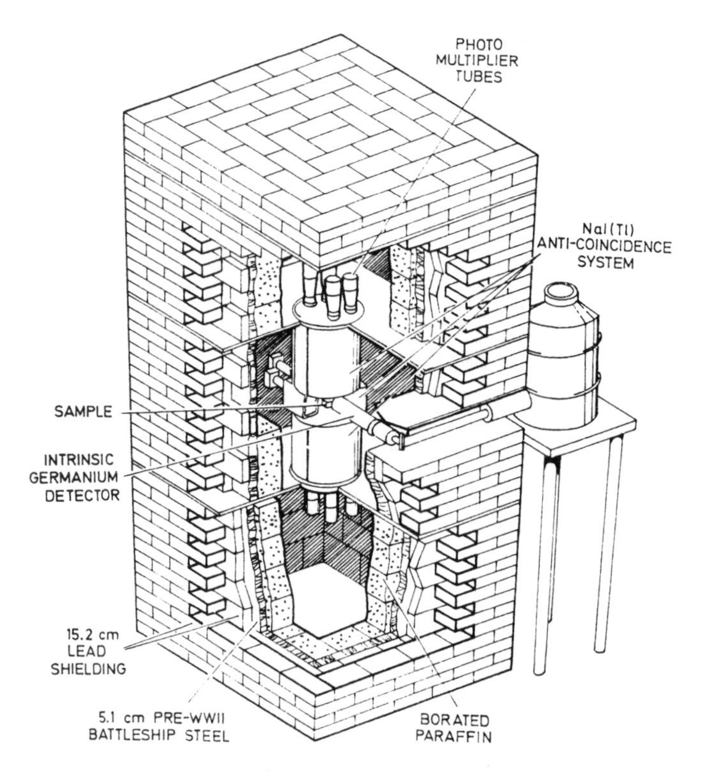

FIGURE 2.18. Scheme of anticompton schielding of a gamma-ray spectrometer in a shielding box. Adapted from Wogman (1981).

NaI(Tl) detectors for low-level gamma detection are also applied (Huimin et al., 1989). Other inorganic scintillators such as CsI(Tl) photodiode (Kilgus et al., 1990), are applied only rarely.

2.3.4.3 ORGANIC SCINTILLATORS

There are three systems of organic scintillators: *unitary* (primarily pure crystals, e.g., anthracene), *binary* (two-component solutions in liquid or solid form, e.g., *p*-terphenyl in polystyrene), and *tertiary* (primarily liquid scintillators in which the third component ensures the shift of the emission spectrum of photons into the region of maximum spectral sensitivity of the photomultiplier, for example 1,4-di[2-(5-phenyloxazolyl)] benzene (POPOP) in a solution of 2,5-diphenyloxazol in toluene (Figure 2.19). When biological material is introduced into a liquid scintillator, still more complex systems are used, generally called scintillation cocktails. The emission of photons from a scintillator takes place in consequence of one or several transfers of the excitation energy of the molecules. These transfers can proceed with varying efficiency in different types of scintillators, and they also have different emission spectra. In order to be able to compare the scintillators, it is suitable to express the total effectiveness of the scintillator, which will summarily characterize the transfer of the energy withdrawn from the ionizing particle onto the energy of the emitted scintillation photons. The process is described in the next section about liquid scintillation counting.

From the point of view of the measurements of low activity and in contrast to inorganic scintillators, organic scintillators are characterized by low density and a low proton number, a linear response to the electrons, and the possibility of their existence in a liquid or plastic form. Organic scintillators are suitable for detection of low levels of alpha or beta radiation. They are also utilizable for spectrometry, but in view of their low energy resolution, they do not permit analysis of the low activities of individual radionuclides in more complex mixtures.

2.3.4.4 LIQUID SCINTILLATION COUNTING

Among low-level measuring systems, liquid scintillation counting (LSC) is – together with germanium spectrometry – most widespread, and it is treated from various aspects in a number of specialized publications (Bransome, 1970; Noujaim, 1976; Crook and Johnson, 1978). The extended use of this detection procedure – especially in measurements of the samples of biological material – is caused by the high detection efficiency for low-energy beta emitters (elimination of self-absorption and the counting geometry of

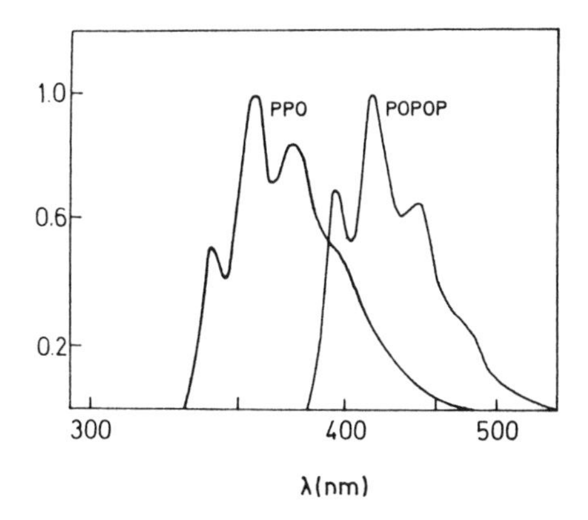

FIGURE 2.19. The emission spectra of two components of a tertiary scintillator.

practically 4π sr), the highly sophisticated spectrometers with PC-controlled measurements and evaluations, and the development of scintillation cocktails, which permit the introduction of samples with different characters.

2.3.4.4.1 *The Influence of Processes in LSC on Low-Level Measurements*

The simplest scintillation medium for liquid scintillation must contain at least a *primary organic solvent* and a *scintillation component.* The role of the primary solvent is to dissolve the scintillation component, accept the energy of the detected particles, and transmit it to the scintillation component. In view of the nature of the samples measured, two types of solvents are used: mainly dioxane for polar (aqueous) solutions and toluene for nonpolar samples. In order to achieve a better solubility of the sample in the scintillation solution, a secondary solvent is sometimes added, such as methanol, ethanol, or dimethoxymethane, but more frequently various commercially available cocktails, developed for some types of biological samples or inorganic salts, which are insoluble in common organic solvents, are used. Since the point of solidification of some primary solvents (dioxane 285 K, *p*-xylene 286 K) is in the range of temperatures at which the samples are sometimes measured in the liquid scintillation spectrometer, the so-called defroster, for example, ethylene glycol, is frequently used in scintillation solutions. In the molecules of the primary solvent (similar to scintillation components), π-electrons exist, which form the outer interatomic bonds, and σ-electrons, which form the bonds in the atom. π-Electrons that are less strongly bound and more mobile are responsible for the fluorescing properties of these molecules.

Of the energy loss by the ionizing particle in the solvent, a part produces ionization and excitation of the molecules, which corresponds to the relative proportion of π-electrons among all the electrons of the solvent molecules (e.g., in benzene this part is about 14%). The residual part of the energy is consumed for σ-electron excitation and ionization, and it is dispersed by non-radiation processes. The molecules with excited π-electrons are formed directly by excitation or more frequently by recombination of the ionized molecules with slow electrons. π-Electrons at higher energy levels are rapidly transformed to the first excited singlet level S_{1x} by non-radiation conversion processes (100 fs). The number of molecules of the primary solvent existing in the π-excited state S_{1x} after absorption of the ionizing particle with kinetic energy W will be equal to

$$A = sW \tag{2.5}$$

where s is the conversion factor of the primary solvent.

The molecules of the primary solvent X do not possess the properties of a scintillator. They have a small quantum yield and there is little probability that they will get rid of the excess energy ($S_{1x} - S_{0x}$) by emission of photons even though the lifetime of the excited state S_{1x} is relatively long (30 ns). Before the irradiation can take place, the excess energy is transmitted between the primary solvent molecules by thermal diffusion (Brown's molecular movement of the solvent and the scintillation component) and excitation migration (gradual rapid formation and extinction of excited dimers-excimers between the adjacent excited and non-excited molecule), until it comes to the molecule next to the molecule of the primary scintillation component Y. The dipole-dipole resonance causes the transfer of the energy from molecule X to molecule Y. The difference of energies $S_{1x} - S_{1y}$ is converted to heat. B molecules of the primary scintillation component will be in the π-excited singlet state S_{1y}

$$B = fA = fsW \qquad (2.6)$$

where f is the quantum efficiency of the energy transfer between the solvent and the primary scintillation component, which depends on the type of the solvent used and the substance concentration of the primary scintillation component.

During the fluorescence persistence ($\tau_y \sim 1.5$ ns) the molecules of the primary scintillation component come to the ground level S_{0y}, emitting P photons:

$$P = aB = asfW \qquad (2.7)$$

where a is the quantum fluorescence efficiency of the primary scintillation components, which depends on the substance concentration of the primary scintillation component and is practically independent of the solvent properties. The scintillation process in a binary scintillation solution is shown in Figure 2.20.

In *tertiary systems* it is also necessary to consider the energy transfer between the molecules of the primary and the secondary scintillation component (characterized by quantum efficiency of the energy transfer between their molecules) and the emission of fluorescence photons by a secondary scintillation component (characterized by quantum fluorescence efficiency a').

The liquid scintillation solution as a whole is characterized mainly by the *scintillation spectrum*, the *absolute scintillation efficiency* S, and the *scintillation response* L. In binary systems the scintillation spectrum of the primary scintillation component is used (or, in tertiary systems, the emission spectrum of the secondary scintillation component). The absolute scintilla-

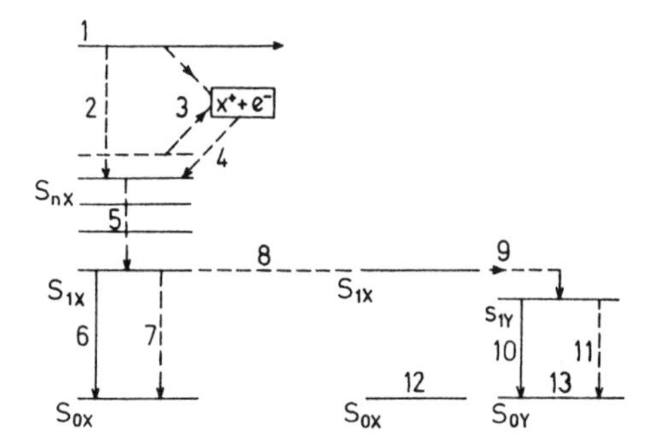

FIGURE 2.20. The scintillation process in a binary scintillation solution: 1—ionizing particle, 2—excitation of the solvent, 3—ionization of the solvent, 4—recombination, 5—internal conversion, 6—fluorescence of the solvent, 8—migration of the excitation energy, 9—excitation of the molecules of the scintillation component, 10—fluorescence of the scintillation component, 11—internal quenching of the scintillation component, 12—solvent molecules, 13—molecules of the scintillation component.

tion efficiency S is defined as the fraction of all incident particle energy that is converted into the energy of the generated photons. In a binary system the following relationship applies for S:

$$S = sfa\overline{E}_j \tag{2.8}$$

where $\overline{E}_j$ is the average energy of photons emitted by the primary scintillation component. In a tertiary system the following relationship applies for S:

$$S = sff'a'\overline{E}_j \tag{2.9}$$

where $\overline{E}_j$ is the average energy of photons emitted by the secondary scintillation component. The scintillation response L is the energy irradiated by fluorescence photons after absorption of an ionizing particle of energy W in the scintillator.

As has already been mentioned, the main advantages of the liquid scintillation measurement are the 4π geometry of the measurement and the exclusion of losses of kinetic energy of detected particles by absorptions in the sample itself and on the way between the sample and the detection medium. This, however, has its negative side, because it decreases the efficiency of the excitation energy transfer and also, the efficiency of the transfer of fluorescence photons through the scintillation medium. Such phenomena are together called *quenching,* and they appear as a decrease in the measured counting rate in quenched samples in comparison with non-quenched. Quenching plays a very important role in low-level measurements, in which it always has to be considered. In individual samples it differs both in quality and quantity. The following types of quenching are known:

- ionization quenching
- quenching by dilution and concentration
- chemical quenching
- color quenching
- phase quenching
- photon quenching

The first five quenchings take place in the scintillation solution. The photon quenching includes events that decrease the number of photons on the way between the scintillation solution and the photocathode of the photomultiplier.

In ionization quenching for electrons with a kinetic energy of $W > 150$ keV, a linear relationship applies between the scintillation response L and the electron energy

$$L = SW \tag{2.10}$$

but at lower energies, this relationship ceases to be linear. This fact is explained by ionization quenching, i.e., by the interaction of excited molecules and the ions of the primary solvent with other excited and ionized molecules, leading to a decrease of the number of excited molecules of the primary solvent. The ionization quenching depends on the density of ionized and excited molecules of the solvent, which is expressed by the relation:

$$\frac{\mathrm{d}L}{\mathrm{d}r} = S\frac{\frac{\mathrm{d}W}{\mathrm{d}r}}{1 + kB\frac{\mathrm{d}W}{\mathrm{d}r}} \tag{2.11}$$

where S is the absolute scintillation efficiency of the solvent for high-energy electrons, $B \cdot \mathrm{d}W/\mathrm{d}r$ means the density of ionized and excited molecules, and k is the quenching parameter. It follows that the ionization quenching is important mainly for particles with a higher linear stopping power (alpha particles, protons, or low-energy electrons). For example, for electrons with an energy of 5 keV the scintillation efficiency can be decreased by about 40%.

Quenching by dilution and concentration results when the scintillation solution is diluted with a sample or a secondary solvent and a part of the energy of the ionizing particles may be absorbed by the diluent, which does not contribute to the scintillation formation. Dilution also decreases the scintillation migration in the primary solvent and the probability of the transfer of the excitation energy from the molecule of the primary solvent to the molecule of the primary scintillation component. With increasing substance concentration of the primary scintillation component in the scintillation solution, the scintillation efficiency increases at the beginning up to a maximum value. Further concentration increase does not cause further changes of the scintillation efficiency in the case of some primary scintillation components, for example butyl-PBD, where the abbreviation means 2-phenyl-5-(4-biphenyl)-1,3,4-oxidazole, but in some other cases, for example PPO (2,5-diphenyl-oxazole), its decrease does take place. This phenomenon is called self-quenching or quenching by concentration. Larger amounts of secondary scintillation components may also cause concentration quenching.

In chemical quenching many measured organic samples compete with the molecules of the primary scintillator and absorb the excess energy of the molecules of the primary solvent. This phenomenon is called chemical quenching, and it is characterized by the relative factor of quenching g. When indicating—for the sake of simplicity

Energy absorption:

$$\begin{gathered} M + E \rightarrow M^* \\ \text{rate} = 1 \end{gathered} \tag{2.12}$$

Photon emission:

$$\begin{gathered} M^* \rightarrow M + h \cdot \nu \\ \text{rate} = k_1[M^*] \end{gathered} \tag{2.13}$$

Quenching:

$$\begin{gathered} M^* + Q \rightarrow \text{loss} \\ \text{rate} = k_2[M^*][Q] \end{gathered} \tag{2.14}$$

where M is the molecule of the primary scintillation component in the ground electronic state, M^+ in the excited state, Q is the molecule of the quencher, $[M^+]$, $[Q]$ are molecular concentrations, k_1, k_2 are relative rates of the reaction. Then the total number of possible fluorescences is

$$S_0 = k_1[M^*] + k_2[M^*][Q] \tag{2.15}$$

and the number of real fluorescences from the quenched solution will be

$$S = k_1[M^*] \tag{2.16}$$

Then the relative quenching factor is

$$g = \frac{S}{S_0} = \frac{1}{1 + \frac{k_1}{k_2}[Q]} \tag{2.17}$$

The relative quenching factor g is independent of the energy of the ionizing particle. For example, oxygen is a strong chemical quencher. In the case of some organic materials, it is useful to oxidize the sample in order to eliminate this effect and bring the oxidation product obtained ($^{14}CO_2$, or ^3H_2O) into the scintillation medium (Section 2.8).

Color quenching decreases the number of scintillation photons formed during their passage through the absorption material in a mixture of scintillating solution–sample, the quenching effect can be determined on the basis of Beer's law (Horrock, 1980):

$$I = I_0 e^{-alc} \tag{2.18}$$

where I_0 is the number of photons generated by the scintillation event, I is the number of photons after the passage through the scintillator, a is the coefficient of absorption, l the distance covered by the photons, and c is the concentration of the colored quencher. It follows that, in comparison with chemical quenching, color quenching depends on the site where the scintillation photons were formed. In consequence the impulse spectrum of two samples with an equal degree of quenching is higher and narrower in chemical quenching than is the spectrum in the case of color quenching. In all cases possible, it is suitable to eliminate the color quenching reproducibly, for example, by decolorizing the sample with hydrogen peroxide.

Phase quenching is used to indicate that a mixture of sample-solvent-scintillation component is not homogeneous and that part of the energy of the ionizing particles is not transmitted into the scintillation solution. In vials with a homogeneous distribution of radionuclide, a part of the radionuclide is always close to the vial wall, and therefore a percentage of the particles emitted by them falls on the wall before losing its entire energy in the scintillation solution. This is the so-called wall effect already described for internal gas counter (Section 2.3.3.4).

The two further forms of phase quenching may occur in samples with inhomogeneous distribution in the vial. The first one is formed when the sample is a part of the second phase (for example, of the membrane filter, a part of the gel, precipitate, etc.) and a part of the energy of the particles emitted is absorbed by this phase without a scintillation response. The second type is the inhomogeneity of the sample distribution or the distribution of the scintillation components in the scintillation solution in consequence of their absorption, or also diffusion through the walls of the measuring vial. In this case there is an instability of the measurement in time because the value of these changes also depends on the time between the preparation of the sample and its measurement.

Photon quenching is the loss of scintillation photons between the scintillation solution and the photocathode of the photomultiplier. The phenomenon is affected by the difference of the wall thickness and the transmittance of the light through the walls, the formation of optical destructions, impurity of the vials or condensation on their surface, changes within the scintillator volume with respect to the spectral sensitivity of the entire photocathode, changes of optical coupling by aging and changes in the refractive index.

Phosphorescence is the emission of photons owing to the excitation of molecules of some scintillation media (mainly on the basis of dioxane) or some types of measuring vials by solar or some other type of light. It is mainly observed during the detection of low-energy parts of the spectrum. The fluorescence is followed by the emission of photons at approximately 1 ns after excitation of the molecules, while in phosphorescence this time is longer—from 10 ms to several days. The occurrence of phosphorescence can be decreased to a minimum by preventing the access of light to the sample before its measurement. We observed a very unfavorable effect on the increase of the background by exposing, for example, glass vials with ^{3}H samples to direct sunlight, even for a short time.

Chemiluminescence is the emission of photons elicited by a chemical reaction taking place in a mixture of sample and the scintillation medium. It is frequently observed, e.g., in cases when a basis solvent or peroxide is used for the suppression of the color quenching. The duration of chemiluminescence depends on the rate of the chemical reaction and the lifetime of the molecules in the excited state. Bioluminescence is a special case of chemiluminescence. In this case the emission of photons is usually caused by bacteria, fungi, or an exothermic reaction catalyzed by proteins, enzymes, etc. In modern spectrometry this phenomenon is automatically controllable.

Some types of measuring vials (for example plastic ones) may be charged with *static electricity* during transport to measuring. The light flashes created during the electric discharge may thus increase the background.

In view of the fact that a single photon is formed during the above-described effects (phosphorescence, chemiluminescence, bioluminescence, electrostatic effects), they are frequently called *one-photon events*. Elimination of such effects for low-level LSC is very important.

2.3.4.4.2 The Measurement by LSC

Scintillation photons formed in the liquid scintillation solution are recorded by a photomultiplier from the output of which the appropriate voltage response is amplified and analyzed in a multichannel amplitude analyzer (Figure 2.21). The number of photons that come to the photocathode FN from the scintillation solution depends on the optical quenching. The number of photoelectrons emitted from the photocathode depends on the course of its spectral sensitivity (the bialkalic photomultipliers have their spectral sensitivity peak at about 380 nm). Since the background noise of photomultipliers may be decreased by their cooling (Section 2.2), in the past a cooling system was a component of the majority of spectrometers

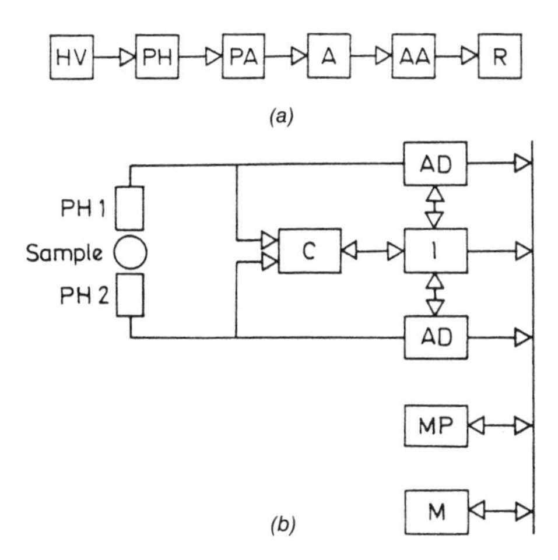

FIGURE 2.21. Basic block schemes for LSC. (a) The functional scheme: HV—high voltage supply, PH—photomultiplier, A—amplifier, AA—amplitude analyzer, R—recording device. (b) Coincidence arrangement controlled by a microprocessor: C—coincidence circuit, AD—convertor, I—interface, MP—microprocessor, M—memory.

produced. When using low-noise photomultipliers, the importance of cooling is no longer crucial, but maintaining a constant temperature during the measurement is still important for the stability of the apparatus.

The background caused by the photomultiplier noise and the one-photon events are considerably decreased by using two photomultipliers in coincidence (Figure 2.21, for details see Section 2.4). Previously, only the signal from one multiplier was analyzed, while the second one served as a monitor for coincident impulses. Thus the signal amplitude depended on the position of the formation of the scintillation event in the scintillation solution. This position-dependence resulted in a deterioration of the energy resolution ability and thus of the spectrometric possibilities of the apparatus. Addition of the signals from both photomultipliers in the summation circuit improved the energy resolution.

For current analyses a suitably set interval, the so-called counting window, is used for a given radionuclide, the lower discrimination level of which corresponds either to the level that cuts off the noise impulses, or for more than one beta nuclide to the level excluding the impulses elicited by a radionuclide with a lower energy. The width of interval, i.e., the position of the upper discrimination level, is determined by the interval of the energies of the detected particles. Sometimes multichannel analysis is also carried out, especially to resolve a simple alpha spectrum (analogously to NaI(Tl)- and Ge-detectors, for a complex alpha spectrum, silicon detectors with higher energy resolution have to be used).

The detection efficiency, which is changed according to the quenching of individual samples, can be determined in several ways.

The method of *internal standard* (IS) is the oldest procedure for the determination of the detection efficiency: first, the sample itself is measured and after addition of the known activity of a radioactive standard (the most common internal standard for ^{3}H and ^{14}C is toluene or *n*-hexadecane), it is measured a second time. The condition must be observed, i.e., that the addition of the standard does not change the experimental conditions (e.g., quenching). The detection efficiency is determined on the basis of the increase of the counting rate caused by internal standard and according to its activity. The great disadvantage of this method is that during the pipetting and the determination of the volume of the internal standard,a certain error may be made. Furthermore, a longer time is necessary for the preparation and measurement of the samples, and measurement of the samples after addition of the internal standard is no longer possible. This is the only method in which the detection efficiency is not determined from the calibration curve, but directly by computation.

The method using *the ratio of two channels* (SRC) is a procedure during which two measuring windows are set on the apparatus, comprising certain parts of the spectrum, elicited by the measured beta radionuclide. The ratio of the counting rate recorded in these two channels depends on the degree of quenching. This method is suitable only for a single radionuclide.

The *external standard* (ESC) method is a procedure in which a gamma emitter (most frequently ^{226}Ra, ^{137}Cs, or ^{133}Ba), which was previously located in a sufficiently shielded container, is placed near the measuring vial with the sample after the measurement of the sample. The Compton electrons elicited by the irradiation of the sample by this external standard increase the counting rate. On this basis the detection efficiency is determined from the calibration curve. The disadvantage of this method is that the number of counts from the external source also depends on the volume and the method for determining the ratio of the color and chemical quenching, and it is not suitable for heterogeneous samples, but it is very suitable for samples containing more then one radionuclide. The method of *ratio of channels of external standard* (ESR) represents an improvement on the ESC method by the determination of the ratio of the counting rate elicited by the external standard in two counting channels. In this method the effect of changing the volume of the sample is reduced.

The method uses *the shift of the average amplitude* of the impulses (ESP) from the external standard of radionuclide ^{133}Ba. ESP is defined by the relationship:

$$\text{ESP} = \frac{P_r - P}{P_s - P} \tag{2.19}$$

where P_r is the average amplitude of impulses in the reference, P_s in the measured sample, and P in the sample with an infinite quenching. The method is also suitable for strongly quenched samples.

The *H-number* method makes use of the shift of the inflection point of the Compton edge of the external standard of radionuclide ^{137}Cs. H is defined by the relationship:

$$H = P_0 - P_q \tag{2.20}$$

where P_0 is the amplitude of the impulse of the inflection point in the reference vial and P_q in the measured sample.

From the comparison of the last two methods it follows (McQuarrie et al., 1980) that the H-number method is less precise because it uses local parts of the external standard spectrum while the ESP method uses the whole spectrum. Between ESP and H the following relation exists:

$$H = 226.5 \log \text{ESP} - 5.1 \tag{2.21}$$

The method of *spectral index of the external standard,* SIE, is another method for the determination of the degree of quenching. It is derived from the first moment $M(1)$ of the amplitude distribution of the external standard:

$$M(1) = \frac{\sum_{L}^{U} XY(X)}{\sum_{L}^{U} Y(X)} \quad (2.22)$$

where $Y(X)$ represents the number of impulses with an amplitude between X and $X + \Delta X$ and L and U are the lower and upper limits on the axis of impulse amplitude. For SIE L is chosen above the region of the wall effect and U at the end of the spectrum. The value of SIE does not depend on the sample volume, its activity, or the form of the measured spectrum. The calibration curve between SIE and the detection efficiency is smooth, and for strongly quenched samples (chemically and by color), the values of SIE differ at an equal efficiency. In order to eliminate this difference, a point on the axis of impulse amplitude is used, under which an accurately determined percentage of all impulses is located. On the base of the value of the shift of this point (CQI), the differences in the detection efficiency during color and chemical quenching may be corrected on the basis of the following relationship:

$$E_{col} = F(\mathrm{CQI,SIE}) \cdot E(\mathrm{SIE}) \quad (2.23)$$

where F(CQI,SIE) is the compensation factor between the color and chemical quenching and E(SIE) is the detection efficiency at chemical quenching, characterized by the value SIE.

The method of *spectral index of the sample* (SIS) is analogous to the preceding one. The value SIS is deduced from the first moment of the distribution of the impulse amplitude of the sample measured, and in this case external standard is not used. The method is suitable for samples with one radionuclide only.

With exception of the IS method, the detection efficiency of a sample is determined in all other cases from the *calibration curve,* on the basis of the value of the indication parameter of the quenching. For the construction of this curve a set of reference samples may be used, which have equal activity but different quenching and which are usually supplied by the producer of the apparatus.

LSC is widely used, not only for low levels. Therefore, some of the commercially available spectrometers are produced with relatively high background, small vial volume and poor energy resolution. Nevertheless, the development of these spectrometers has been in progress. Thus, e.g., with background decreased to 0.60–0.65 cpm in ^{3}H-channel (4-mL water sample + 16-mL scintillator in a polyethylene vial), the measurement of 1.0 ± 0.1 TU is possible after electrolytic enrichment using a commercially available spectrometer (the background of 0.35–0.40 cpm in ^{14}C-channel is reached).

Due to the development of spectrometers including low-noise ^{40}K-free photomultipliers, the previously developed, rather complicated synthesis of primary (usually $^{14}C_6H_6$ or $C_6{}^3H_6$) or secondary (e.g., $^{14}CH_3OH$) solvent is not often applied (Shibata et al., 1987). With respect to the lower noise of modern spectrometers positively influencing their energy resolution, the multichannel amplitude analysis (1024 channels) can often be used for direct measurements of various alpha spectra (Figure 2.22), while time-consuming radiochemical treatments are used only for special cases (Yu-Fu et al., 1992).

Some authors have analyzed the influence of anti-coincidence and/or lead shielding on the decrease of background (Arnold, 1983; Schoenhofer, 1991). The concentration of radon in the gaseous space above the scintillator during ^{3}H measurements has to be taken into consideration (Murase et al., 1989). However, the decisive influence is the radionuclide purity of the vial material. Because "low-level" quartz vials with very good transparency for scintillation photons are rather expensive, usually various commercially available vials from borosilicate glass (^{40}K and ^{226}Ra free), teflon, or polyethylene are applied. Also, special material combinations have been developed such as an "ideal vial" of teflon and copper (Polach et al., 1983).

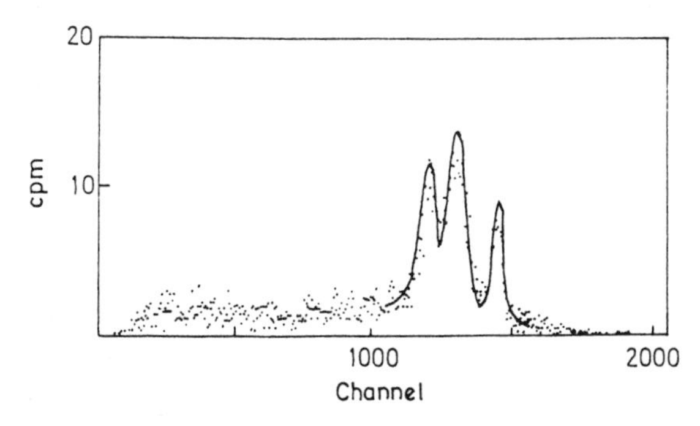

FIGURE 2.22. Alpha spectrum of mineral water (8 ml water, 12 ml Instagel, teflon vial): left peak—^{226}Ra (4.60 + 4.78 MeV), middle peak—^{222}Rn (5.49 MeV) + ^{218}Po (6.00 MeV), right peak—^{214}Po (7.69 MeV). Adapted from Schoenhofer and Henrich (1987).

The vial volume of 20 cm^3 is frequently used. For very small samples, smaller vials (e.g., 2 cm^3) inserted into the standard vials are applied, while for low specific activities larger volumes with an appropriate counting arrangement are used. Also, the vial holder should be radionuclide free (Butterfield and Polach, 1983). The counting efficiency is mainly determined by the counting geometry and the scintillation medium. The choice of the medium can considerably influence the obtainable value (Takiue et al., 1989).

The comparison of the measurement of weak beta emitters using either LSC or internal gas counting is given by several authors in detail (Tschurlovits et al., 1982; Polach, 1987). While the LSC has been developed, there has been almost no progress in gas counting during the last two decades.

2.3.5 Semiconductor Detectors

2.3.5.1 DETECTOR PRINCIPLES

The results achieved in research and applications, both in technological and diagnostic processes for semiconductor structures – in the first place in integrated circuits of various types and degrees of integration – create starting conditions for the development of semiconductor structures for analysis of low activity of radionuclides. Therefore the present-day development of detection procedures and their application is substantially broader in comparison to scintillation and ionization detectors.

We consider it more precise to use the term *semiconductor detection structures* instead of semiconductor detectors because, in the detection of ionizing radiation, the monolithic sensor is frequently replaced by an element that is integrated in the effective volume such as silicon strip detectors used in high-energy physics (Rancoita and Seidman, 1982). Moreover, the surface parts may include, in addition to the effective part, further regions with different structures, for example, a passivating structure decreasing the leakage current. For the sake of brevity, we shall use the shorter term *detectors*.

In view of the high-energy resolution, semiconductor detectors are irreplaceable for analysis of complex energy spectra of emitted particles, but their use is frequently advantageous for measurements of the low activity of a single radionuclide. While silicon detectors are used with a high detection efficiency for charged particles, germanium detectors are designated for photons.

For low levels the use of semiconductor material is of great advantage because the detector dimensions can be kept much smaller than the equivalent gas-filled or scintillation detector, and the energy resolution is considerably higher even in comparison with scintillation detectors, due to considerably lower energy needed to generate charged particles in interactions of incident radiation (Table 2.2). These advantages are utilized especially in measurements of *very small samples with resolving of more radionuclides* present simultaneously at the same measured site (Tykva, 1987). The advantages of semiconductor detectors for measurements of low-level activity has been, well known for a long time (Chwaszczewska and Przyborski, 1967; Kuhn, 1968).

Review articles are devoted to semiconductor detectors, where a detailed analysis of their properties and preparation may be found (Dearnaley and Northop, 1966; Bertolini and Coche, 1968; Walton and Haller, 1982). The basic charge response of a semiconductor detector is formed by electron-hole pairs created along the path of the detected particle through the detector. Analogously to ionization chambers in pulse mode operation, the motion of the generated charge carriers in an applied electric field enables charge collection and the following voltage signal. The basic materials for detectors are silicon and germanium; other materials (e.g., CdTe, HgI_2, InSb or GaAs) have rarely been used in low-level measurements until recently. The basic properties of intrinsic (i.e., without any admixtures) silicon and germanium are summarized in Table 2.14.

In practical technology it is virtually impossible to achieve such high purity that the material may be obtained in intrinsic form. Moreover, some admixtures are added during preparation of single crystals to tailor the

TABLE 2.14. Properties of Intrinsic Silicon and Germanium.

	Si	Ge
Atomic number	14	32
Atomic mass	28.09	72.60
Stable isotope nucleon number	28-29-30	70-72-73-74-76
Density (300 K); $g \cdot cm^{-3}$	2.33	5.33
$Atoms \cdot cm^{-3}$	$4.96 \cdot 10^{22}$	$4.41 \cdot 10^{22}$
Dielectric constant	12	16
Forbidden energy gap (300 K); eV	1.115	0.665
Intrinsic carrier density (300 K); cm^{-3}	$1.5 \cdot 10^{10}$	$2.4 \cdot 10^{13}$
Intrinsic resistivity (300 K); Ω cm	$2.3 \cdot 10^{5}$	47
Electron mobility (300 K); $cm^2/V \cdot s$	1350	3900
Hole mobility (300 K); $cm^2/V \cdot s$	480	1900
Electron mobility (77 K); $cm^2/V \cdot s$	$2.1 \cdot 10^{4}$	$3.6 \cdot 10^{4}$
Hole mobility (77 K); $cm^2/V \cdot s$	$1.1 \cdot 10^{4}$	$4.2 \cdot 10^{4}$
Energy per hole-electron pair (300 keV); eV	3.62	
Energy per hole-electron pair (77 eV); eV	3.76	2.96

properties of detectors prepared from these crystals. To illustrate the effect of this doping, we use silicon as an example, although other materials behave in a similar way.

Tetravalent silicon in the normal crystalline structure forms covalent bonds with the four nearest silicon neighbor atoms. When a pentavalent admixture is present in small concentration (an order of magnitude of 10^{-6} or less), the admixture atoms will occupy a substitutional site within the silicon lattice, taking the place of a normal silicon atom. After all corresponding covalent bonds have been formed, there is one valence electron left over. Admixtures of this type are referred to as donors and such material as *n*-type silicon, with electrons called the majority carriers and holes called the minority carriers. Correspondingly, after the addition of a trivalent admixture (acceptor) *p*-type silicon is created, with holes as the majority carriers. If donors and acceptors are present in equal concentration, the semiconductor is said to be compensated.

The function of the semiconductor detector is made possible by forming a junction. At present, the introduction of doping admixtures at the semiconductor surface is carried out mainly by ion implantation using a beam of appropriately accelerated ions: either donor (e.g., phosphorus) on *p*-type or acceptors (e.g., boron) on *n*-type. To a smaller extent *surface-barrier* technology is applied based on a contact metal-semiconductor (e.g., gold on *n*-type silicon) prepared by special technological operations. The main advantage of implanted detectors is the simple preparation of complicated detector structures applied, for example, in high-energy physics. For low levels of radionuclides, only monolithic semiconductor are desirable so that this advantage is negligible.

The electron-hole pairs generated during the interaction of the detected particle can be collected only from the depleted layer of the detector where all free carriers are removed by working voltage. According to the ratio of this effective detection layer to the whole detector thickness, the detectors are divided into *partially or totally depleted detectors.*

2.3.5.2 *SILICON DETECTORS*

The basic characteristics of silicon ($Z = 14$, density 2.328 g·cm^{-3} at 300 K and the mean energy per electron-hole pair 3.62 eV at 300 K) afford very good conditions for the use of silicon detectors for detection and spectrometry of low activities of alpha and beta emitters. Moreover, the necessary properties are not affected by the technological method used to form the junction. Both the implanted type and that with a surface barrier are suitable. The advantage of the latter is a relatively simple production (Tykva, 1989) and uncomplicated structures (Kopeštanský et al., 1991).

In addition to the high efficiency of detection of alpha particles and electrons, and the high resolution at room temperature, silicon detectors achieve a low background, mainly due to their low efficiency of the background particles with a low specific ionization and because of the thin detector layer (only planar geometry is used). This layer is determined in spectrometry by the range of the detected particles with maximum energy (e.g., for alpha particles of 6 MeV energy, a silicon layer of 34 μm is sufficient). For the detection, where the total absorption of the kinetic energy of the incident particle is not necessary, it is only desirable that the layer should permit a maximum number of detected particles to generate such a number of electron-hole pairs, so that the generated charge would be larger than the equivalent noise charge [Equation (2.3)]. The great advantage of silicon detectors is the possibility of decreasing the detection volume to a size corresponding only to the detected radiation. For low levels it is thus possible to tailor the thickness of the depleted layer and, in this way, to minimize the background for attainable counting efficiency, especially for totally depleted detectors (Tykva et al., 1992). The attainable figure of merit of beta emitters, including low-energy radionuclides (e.g., ^{14}C), is then higher than those of the gas counter or liquid scintillation spectrometer (Tykva, 1977).

Taking into consideration the low efficiency of the detection of photons and the high-energy component of the background, it is frequently possible to measure the low activity without any shielding if the radionuclide purity of the immediate vicinity of the detection element is high. For measurement of extremely low levels, it is useful to avoid contamination of silicon with ^{32}Si (Plaga, 1991). At the same time the background can be limited by choosing a corresponding counting channel. So the background of a 200 mm^2 planar surface-barrier detector in ^{14}C-counting channel can be decreased to 0.19 cpm with ^{14}C-counting efficiency of 25% (Tykva, 1977).

In comparison to ionization and scintillation detectors, silicon detectors permit not only a substantially *higher energy resolution,* but also a simpler arrangement for measurement and a lowering of investments and demands on operation. These advantages are limited for samples with a low specific activity by self-absorption in the sample and a corresponding increase in the effective detector area (i.e., by an increase of the background and also by the noise increase). The characteristics of a surface-barrier type with effective detection area of 10 cm^2 (Tykva et al., 1990) are shown in Figure 2.23.

Apart from low-level beta counting, the silicon detector is used predominantly for detection and spectroscopy of low-level alpha emitters in solid planar samples at 2π counting geometry, i.e., similar to a gas-flow counter (Matyjek et al., 1988). The counting geometry 4π sr is quite exceptionally

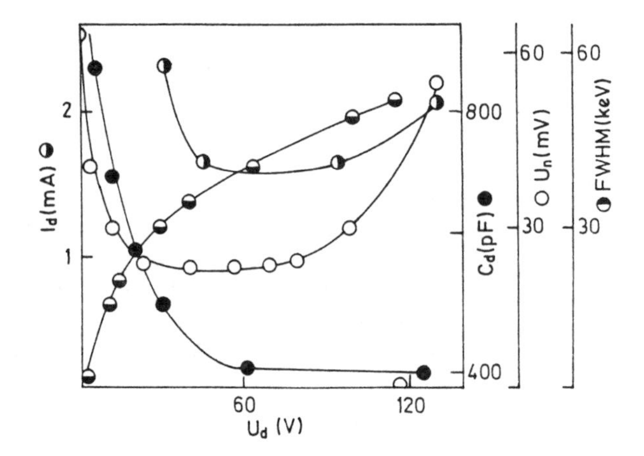

FIGURE 2.23. Characteristics of a surface-barrier silicon detector for determination of actinides at 293 K (effective area of 10 cm²): U_d—detector bias, I_d—reverse current, U_n—wide-band noise, FWHM—resolution of ^{241}Am, Tykva et al. (1990).

used, which is achieved by two detectors (Johnson, 1967). In this case the sample is placed between the foils of an area density less than 1 mg cm^{-2} (Mylar).

The thickness of the depleted layers of implanted or surface-barrier Si-detectors is usually up to about 2 mm. This is sufficient for total energy absorption of alpha particles from all radionuclides or for low-energy beta emitters. For total absorption of X-rays (e.g., ^{55}Fe) or high-energy beta particles, deeper depleted layers are necessary. For such purposes Si(Li)-detectors are sometimes used with a depleted layer up to about 5 mm. These detectors are prepared by compensation of acceptors in *p*-type silicon by lithium ions. For measurement of low levels of radioactivity at the laboratory temperature, Si(Li)-detectors are not suitable because high long-term stability is desirable.

2.3.5.3 GERMANIUM DETECTORS

Germanium detectors today afford the best conditions of all detection systems for analysis of *complex gamma spectra*. On the other hand, in comparison with scintillation crystals, it shows—for the measurements of low activities—a disadvantage existing in the lower detection efficiency (Figure 2.2). The detection efficiency with a NaI(Tl) scintillator is higher than 4:1 under 200 keV, and this part increases distinctly with increasing energy (e.g., it is 8:1 for 500 keV). From the point of view of achieving a sufficiently high sensitivity for the measurement of a low activity [Equation (2.1)] it means that it is important to pay great attention to the suppression of the background of a germanium detector. For this reason and with respect to the influence of Compton scattering (Section 2.3.4.1), shielding of germanium detectors has been investigated very extensively.

A certain disadvantage of germanium detectors also exists in the relatively low energy per hole-electron pair (2.96 eV at 77 K). Thus at room temperature, due to thermal excitation of electrons, an unacceptable increase of the reverse-current component generated in the detector volume occurs. Therefore germanium detectors have to be cooled with liquid nitrogen.

Earlier production technologies afforded germanium single crystals with high concentrations of impurities. Therefore Ge(Li)-detectors were used exclusively, in which the acceptors in *p*-type germanium were compensated by drifted lithium ions. These detectors are still used, but the present production technology of the starting high-purity single crystals (HP crystals) already permits the use of germanium alone for the preparation of detectors (HP Ge-detectors).

As has been already mentioned, the counting efficiency for a certain combination of incident radiation–detector type is primarily influenced by the counting geometry. For low levels mostly the well-type coaxial detectors are used, introducing the sample into a hole in the detector (Kemmer, 1968), and planar types (Hedvall et al., 1987) are used only exceptionally. For the configuration sample-detector, the Marinelli beaker around the detector offers lower efficiency than a disk sample (Walford et al., 1976). The detector volume should correspond to the energy of the incident gamma rays.

The residual background peaks in Table 2.15 were measured under simple mechanical shielding (Figure 2.24). A combination of more complex mechanical shielding with anti-coincidence of multiwire gas counters is shown in Figure 2.25. Another possibility of anti-coincidence shielding is offered by a plastic scintillator.

A basic scheme of anticompton shielding is shown in Figure 2.26. The reduction factor of the interfering peak obtained by anticompton shielding is presented in Table 2.16.

Finally, it has to be noted that the background can be significantly reduced by material selection and careful control of fabrication of both the germanium crystals and cryostat (Brodzinski et al., 1990). The fluctuation of background spectrum measured by germanium detector can be caused by the concentration variations of ^{222}Rn and ^{220}Rn with their daughters in the air. In order to diminish these fluctuations, non-active gas should be directed into the vessel for shielding.

TABLE 2.15. Background Peak Count Rates for n-Type HP Ge-Detector Measured Sixty Hours.

Parent	Nuclide	Energy (keV)	Count Rate (cph)
^{238}U	^{234}Th	63	13.3
^{230}Th	^{230}Th	67	0.7
^{226}Ra	^{226}Ra	186	8.3
^{220}Rn	^{214}Pb	295	3.6
		352	4.3
	^{214}Bi	609	3.0
^{210}Pb	^{210}Pb	46	1.1
^{228}Ra	^{228}Ac	338	0.7
		911	1.0
		969	0.7
^{220}Rn	^{212}Pb	239	1.4
	^{208}Tl	583	1.6
^{40}K	^{40}K	1461	1.1
^{137}Cs	^{137}Ba	662	0.4

Adapted from Murray et al. (1987).
Detector characteristics: diameter 43.9 mm; length 55 mm; detector to window 5 mm approx; window 500 μm beryllium; resolution 1.83 keV FWHM at 1.33 MeV; peak to Compton 50.5:1; efficiency 19.2% relative to 3″ by 3″ NaI crystal; 25 cm source to detector distance.

2.3.6 Miscellaneous Detection Procedures

The ionization, scintillation, and semiconductor detectors described in the preceding parts of this chapter are basically used in low-level measurements. Other types of ionizing radiation detectors (Knoll, 1979; Tsoulfanidis, 1983) are used for low levels only exceptionally or not all. Therefore let

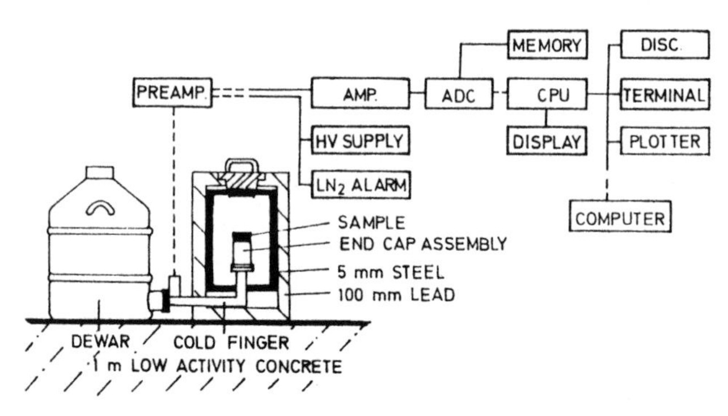

FIGURE 2.24. Simple shielded germanium detector with counting arrangement. Adapted from Murray et al. (1987).

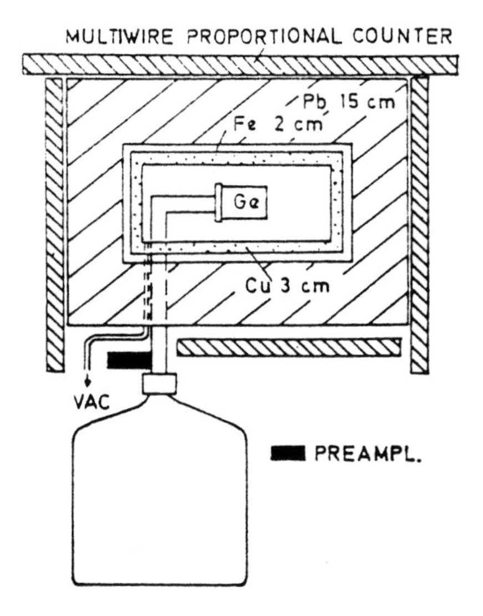

FIGURE 2.25. Scheme of Ge(Li) detector in shielding box and in anti-coincidence with multiwire proportional counters. Adapted from Heusser et al. (1989).

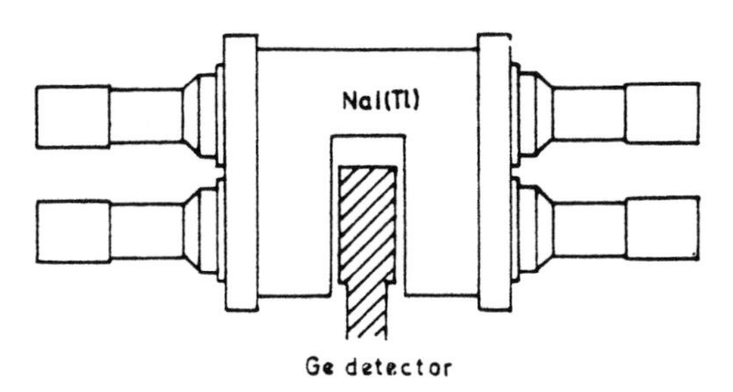

FIGURE 2.26. Scheme of the anticompton shielding of Ge detector with several photomultipliers. Based on Das and Comans (1990).

TABLE 2.16. The Experimental Values of the Reduction Factor of the Interfering Peak R_p *Using Anti-Compton Arrangement.*

Radionuclide	Half-Life	E_γ (keV)	R_p	Radionuclide	Half-Life	E_γ (keV)	R_p
^{46}Sc	84 d	889	7.3 ± 0.5	^{134}Cs	2.1 y	563	17 ± 2
		1121	9.1 ± 0.9			605	13 ± 2
^{59}Fe	45 d	143	2.4 ± 0.2			796	14 ± 2
		192	7.7 ± 0.4	^{152}Eu	13.3 y	122	$\approx 10^2$
		1095	1.1 ± 0.1			245	$\approx 10^2$
		1292	1.1 ± 0.1			344	3.5 ± 1.5
^{75}Se	120 d	121	48 ± 2			779	20 ± 4
		136	55 ± 5			965	3.5 ± 1
		265	17 ± 1			1087	1.6 ± 0.2
		401	1.0 ± 0.1	^{154}Eu	8.8 y	1004	3.0 ± 0.5
^{124}Sb	60 d	603	11 ± 1			1278	2.8 ± 0.3
		1692	$>10^2$			1408	5 ± 1
^{181}Hf	42 d	133	10 ± 2	^{182}Ta	144 d	1189	2.2 ± 0.4
		482	2.5 ± 0.5			1222	2.5 ± 0.5
						1231	2.5 ± 0.5

Based on Das (1987).

us briefly mention only those which are utilizable for certain determinations of low levels.

2.3.6.1 AUTORADIOGRAPHY

In the first place it is desirable to introduce the detection method based on *activation of the silver bromide grains* in the emulsion (Fisher and Werner, 1971; Rogers, 1979). This non-destructive method is meant primarily for the localization of a beta nuclide in a planar sample. Its advantage results from the high topographic resolution enabling the precise localization of the emitter and thus the visualization of the active site in the sample with great accuracy. This advantage applies mainly for ^{3}H, while the range of the electrons emitted with higher energy spectra from ^{14}C or ^{35}S already produces diffuse illustrations. According to the size and the character of the imaged experimental system, we usually separate the *micro-* from the *macroautoradiography.* The difference is in the fact that, in the case of macroautoradiography, the blackening of a certain area of the film is evaluated directly, while in the case of microautoradiography we evaluate, after magnification, either using a common optical or an electron microscope (Hutchison, 1984). Autoradiography is usually applied for detection of one radionuclide because the resolution of two simultaneously present beta emitters is limited (Veselý and Tykva, 1968).

Although the exposure of a preparation can be prolonged up to several weeks without an important increase of the background, the utilizability of autoradiography for low levels is very restricted owing to the low detection efficiency. Illustrative data are afforded by the exposure of a ^{3}H-sample, the activity of which has been determined in an internal gas counter (Tykva and Veselý, 1965): for a visual blackening on a perfectly mounted stripping film with a low background, it is necessary that about 10^{6} emitted beta particles should impinge on 1 cm^{2} of a macroautoradiograph.

An increase in detection efficiency may be achieved by decreasing the energy losses of the emitted particles in the sample itself (self-absorption) by introducing the scintillation medium directly into the sample. In this way the emitted low-energy beta particles are converted into scintillation photons with a substantially lower self-absorption. Sometimes this method is called *fluorography,* and today various scintillation substances are commercially available (for example, in the form of sprays), for which the preparations can be saturated. Thus considerable increase of the ^{3}H-detection efficiency up to one order of magnitude may be achieved by a simple adjustment (Tykva and Pavlu, 1973). Also the detection efficiency of ^{14}C (or ^{35}S) may be increased several times in this way.

2.3.6.2 TRACK DETECTORS

Track detectors are also used for a non-destructive imaging of the distribution of radionuclides emitting *alpha particles* in a planar sample. They can also partly contribute to the estimation of the kinetic energy of the detected particles. The particle transfer through a dielectric foil results in a trail of damaged molecules along the track, which can be made visible by *etching*. Track detectors require individual track counting so that they are utilizable for relatively low activities of alpha emitters. However, evaluation of the particle energy from the form of the etched track is limited because many of the details of the original track are lost in the etching. Details concerning track detectors are to be found in the appropriate monograph (Fleischer et al., 1975).

2.3.6.3 CHERENKOV COUNTERS

If a fast, charged particle is passing through an optically transparent medium with a refraction index n greater than one, Cherenkov radiation takes place. A detailed analysis and possible applications are summarized in the appropriate monograph (Jelley, 1958). If a good quality liquid scintillation spectrometer is available, low levels of high-energy beta radiotracers can be measured using Cherenkov radiation in a ^{3}H-counting channel, such as ^{32}P or ^{42}K. According to our experience, the ^{32}P-counting efficiency in approximately 1 ml of non-quenched aqueous samples and low potassium glass vials represents approximately 40% of the value achieved after the introduction of the sample into the scintillator based on dioxane (Bray, 1960). In view of the comparable value of the background for an equal volume of the measuring medium and at the same type of vials, a decrease of the figure of merit to approximately 16% takes place.

For low levels this value can be somewhat increased by adding a liquid with a higher n to or instead of water ($n = 1.33$) to the sample. A corresponding threshold energy for emission of Cherenkov radiation by beta particles of water $E_t = 263$ keV is decreased, e.g., for glycerol ($n = 1.4622$, $E_t = 190$ keV) (Swailem and Moussa, 1985).

In comparison to liquid scintillation counting, the possibility of measuring somewhat higher activity should also be taken into consideration and carried out in a sample volume increased to the whole vial (without addition of the scintillation medium). It can also be used for the detection of Cherenkov radiation emitted in the glass of the vial during the location of the dry sample, for example, low-level radioactive suspension on a dry membrane or a glass filter put into a counting vial (Haviland and Bieber, 1970).

2.3.6.4 THERMOLUMINESCENCE

Exposure of the thermoluminescent material to a continuous source of radiation leads to a progressive buildup of trapped charge carriers. The thermoluminescent dosimeters (TLD) function as an *integrating detector* in which the number of trapped electrons and holes is a measure of the number of pairs formed by the radiation exposure. After the exposure period, the TLD sample is warmed and the recombinations of the released charge carriers result in emission of photons in the visible region. The total number of these photons is registered by a photomultiplier. The whole detection process is discussed in the appropriate monograph (Cameron et al., 1968). In low-level measurements this procedure is applied to the dating of heated light-bleached mineral material (Aitken, 1985). The reliability of such an integral technique is limited if the decay series is not in secular equilibrium (e.g., the ^{226}Ra to ^{222}Rn decay). Therefore, the analysis of low levels of natural radionuclides in small mineral samples for use in thermoluminescence dating is desirable (Murray and Aitken, 1988).

2.3.6.5 ACCELERATOR MASS SPECTROMETRY

All methods of measurement of low levels of radionuclides described so far in this chapter were based on decay counting. The counting of ^{14}C atoms instead of their decays enables smaller sample sizes and shorter measuring times for radiocarbon dating. Further applications of this promising approach can be expected in low-level measurements in the future (Arnold, 1987) although the decay counting will continue to exist due to its relative simplicity and some other advantages. The principal arrangement of such a device for accelerator mass spectrometry (AMS) consists of several main components (Figure 2.27). From the ion source accelerated Cs^+ ions are focused on the sample surface and the negative ions produced are accelerated by an extraction voltage of 40–60 kV. The inflection magnet is used to select the mass of the isotope to be injected into the accelerator. The accelerated particles then pass three analyzing systems. Firstly, the particles are separated according to their energy per charge in the electrostatic deflector and then in the analyzing magnet according to their momentum per charge. The beams of ^{12}C or ^{13}C, respectively, are stopped in Faraday

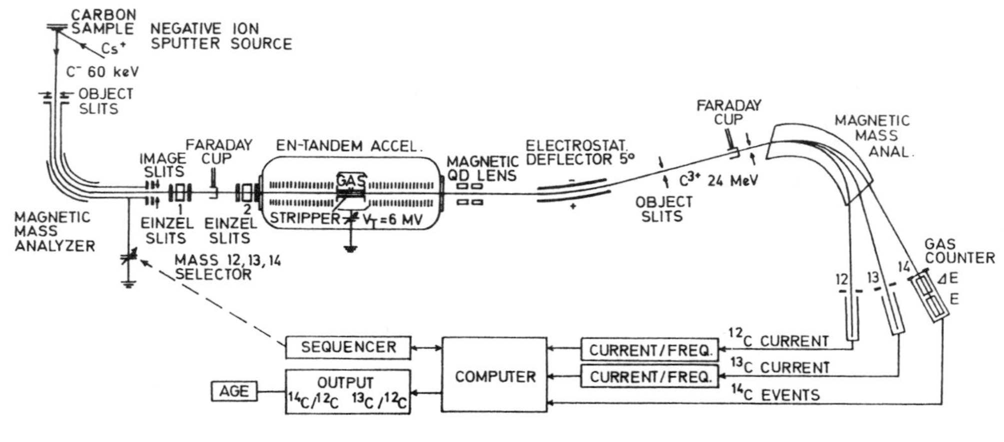

FIGURE 2.27. Scheme of the high-energy mass spectrometer for ^{14}C-dating. Based on Beer et al. (1981).

cage. The ^{14}C atoms are counted in an ionization chamber permitting distinction between isobars ^{14}C and ^{14}N.

The results obtained by AMS technique in radiochronology in a different arrangement have confirmed the possibility of ^{14}C-dating of less than 1 mg sample in a relatively short time interval (Bertsche et al., 1991).

From the short description of AMS it is evident that it is a rather expensive apparatus that requires considerable investment.

However, it is also possible to take into consideration the determination of mass spectrometry without using an accelerator, for example for enriched or depleted uranium samples (Sánchez et al., 1992) or for the estimation of the buildup of ^{241}Am in a plutonium fraction previously separated from environmental samples or biological material (Winkler and Rösner, 1989). Some mass spectrometric techniques have proven their abilities as versatile tools for determinations of low levels of radionuclides (Baxter and Scott, 1987; Fairbank, 1987).

2.4 INSTRUMENTATION FOR THE PROCESSING AND EVALUATION OF DETECTOR SIGNALS

Low-level radioactivities require highly sensitive and stable instruments and systems, which are usually based on the detection of individual particles or photons emitted by radioactive sources. The radiation detector is actually a transducer that converts the information about the incident particle into a measurable electrical signal. The character of this signal reflects interaction processes in the detector material and its evaluation can reveal some useful parameters concerning the measured radiation, namely:

(1) Number of particles
(2) Energy of particles
(3) Type of particles
(4) Time correlations – the time when a particle interacted with a detector
(5) Position and place where a particle hit the detector

Using appropriate processing and evaluation of the electrical signal produced at the output of a suitable detector, one can obtain information which may be interpreted in terms of desired quantities characterizing radiation sources, radiation fields, the transfer or absorption of the radiation energy in matter (dosimetry and radiation protection related quantities), and other radiation parameters or the outcome of their interaction with different materials. Our interest is primarily in obtaining the final information derived from the detector response in the form of the activity of radioactive sources.

The magnitude and properties of signals generated by various detectors vary substantially according to the type of detector, detection medium, its size and working conditions, including for example applied voltage, pressure of the filling gas, temperature, etc. In order to extract the required information from a detector response in an optimal manner, it is necessary to take into account all relevant parameters of the signal available at the detector output and to match associated electronics accordingly. Furthermore, a brief but systematic appraisal of electronic techniques for the processing and evaluation of signals from the detectors used in measurements of low-level activity of radioactive substances will be presented and discussed.

2.4.1 Basic Measuring Arrangements

Any system for measuring radiation or radionuclides, based on the use of active detectors (i.e., gauges or sensors converting the interaction of radiation with an appropriate material into a promptly available electrical response), usually consists of four well-defined parts (Figure 2.28).

(1) *Detection unit,* this comprises one or more detectors along with necessary electronic circuitry such as working resistors, coupling capacitors, voltage dividers, etc., and sometimes also preamplifiers or other circuits to match the high output impedance of the detector to the low input impedance of coaxial cables or input characteristics of the following electronic blocks.

(2) *Voltage power supply,* this provides the detectors with an external (high) voltage needed for their proper operation.

(3) *Processing unit,* this includes all necessary blocks taking part in the treatment of the detector signal prior to its evaluation, especially the amplification and shaping of the signal, and in some applications also a special treatment (e.g., filtration) in order to enhance the property of the processed signal so that the required information can be extracted with higher accuracy or selectivity.

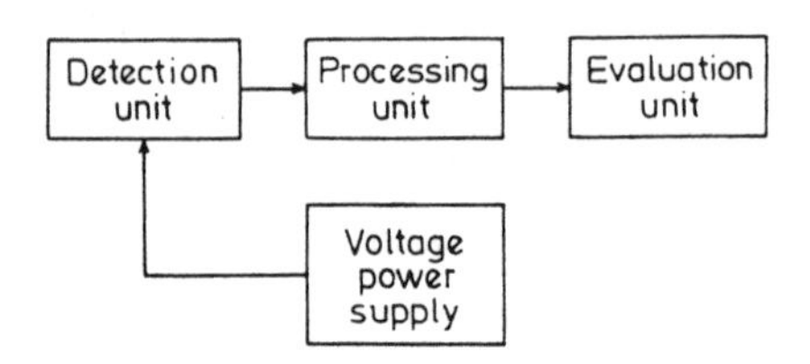

FIGURE 2.28. Basic configuration of a system for the measurement of radiation and radioactivity.

(4) *Evaluation unit,* this is where the signal is sorted out, and after an appropriate selection, converted into unified pulses that can be counted; may include such blocks as an analog-digital converter and digital devices for the storage, handling, and evaluation of collected data.

The design and choice of individual parts of the electronic instrumentation depend mainly on the type and performance of the radiation detector as well as on the purpose of the measurement, the character of the experiment, and other requirements, which in the case of low-level radioactivities may include low background, high detection efficiency, long-term stability, selectivity, etc.

The measurement of radioactive sources having very low activities requires techniques based on the careful selection of suitable detectors and associated electronic instrumentation which usually have to be modified to meet some special demands. The requirements on electronics depend primarily on the characteristics of the detector used and its operation mode.

2.4.2 Signal Processing and Shaping

The first electronic circuitry or block to which a signal from the detector is passed is usually a *preamplifier.* This unit picks up the detector output signal in order to ensure some preliminary shaping and amplification. The other important role of the preamplifier is the conversion of a relatively high detector impedance to the low impedance of the output stage (buffer) of the preamplifier, which then can transfer the signal through a low-impedance transmission line or cable to other electronic blocks. In this way the information carried by the signal can be preserved, and its distortion by noise and other negative effects can be kept to a minimum.

From the electronic point of view a detector can be considered as a source of a current whose parameters reflect processes taking place in the detector after a particle within it has lost a sufficient amount of its energy. Such processes may include the movement and amplification of the primary charge carriers (the ionization-based detectors), or the emission of light photons and their subsequent conversion into photoelectrons, which are then multiplied and collected to give an electrical signal (the scintillation and Cherenkov detectors).

The current signal from the detector may be integrated directly at its output and then amplified as a voltage signal, or it can be integrated using an active integrator based on an operational amplifier. The first option uses a simple *voltage-sensitive preamplifier* while in the second case we refer to a *charge-sensitive preamplifier.* The obvious requirement for both types of preamplifiers is that they possess a high input and a low output impedance so that they can drive succeeding components.

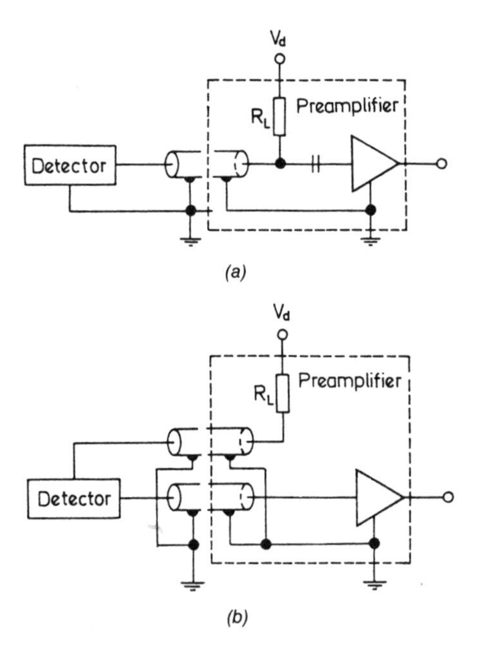

FIGURE 2.29. Two basic configurations for supplying the detector bias through a preamplifier: one cable (a) arrangement and two cable (b) arrangement.

Preamplifiers and amplifiers now usually use integrated operational amplifiers as their building blocks. The operational amplifier is characterized by a very high open-loop gain, very high input, and very small output impedance. The response of such an amplifier can be modified as desired by means of an appropriately selected feedback.

The preamplifiers may be connected to a detector either by a single cable, which serves as both the detector supply voltage and the detector output signal, or by two separate cables: through one a detector supply (bias) is provided while the other is used to transfer the output detector signal to the amplifier. These basic configurations are shown in Figure 2.29.

2.4.2.1 VOLTAGE-SENSITIVE PREAMPLIFIERS

The function of a voltage-sensitive preamplifier is simple: it provides at its output a voltage pulse which is proportional to the voltage pulse applied to its input. The maximum pulse-height of the input pulse is usually related to the ratio of the charge Q collected by the detector and the integrating capacitance C at its output. When the detector capacitance (which is an important pare of C) changes, this can directly affect the output voltage of the preamplifier, which leads to the distortion of the energy spectra from detectors with varying capacitances. Consequently, such types of preamplifiers do not provide the optimum solution, for example, for semiconductor detectors where the capacitance depends on the detector bias.

A simplified arrangement of a voltage-sensitive preamplifier and its modified version coupled to a detector are shown in Figures 2.30(a) and 2.30(b). The first preamplifier [Figure 2.30(a)] consists of an inverting operational amplifier with a high open-loop gain A and a feedback accomplished by two resistors R_1 and R_2. Since $A \gg R_2/R_1$, for the preamplifier output voltage one can write a simple expression

$$v_0(t) = -\frac{R_2}{R_1} v_i(t) \tag{2.24}$$

where its input voltage $v_i(t)$ is actually an output detector signal representing the integrated charge divided by the relevant capacitance. A further figure, Figure 2.30(b), presents an arrangement of a modified version of a voltage-sensitive preamplifier with a detector.

To the category of voltage-sensitive preamplifiers belong also emitter or cathode followers, which were popular in the past and which performed the role of an impedance transformer rather that of than a preamplifier. Their voltage gain was always slightly smaller than one.

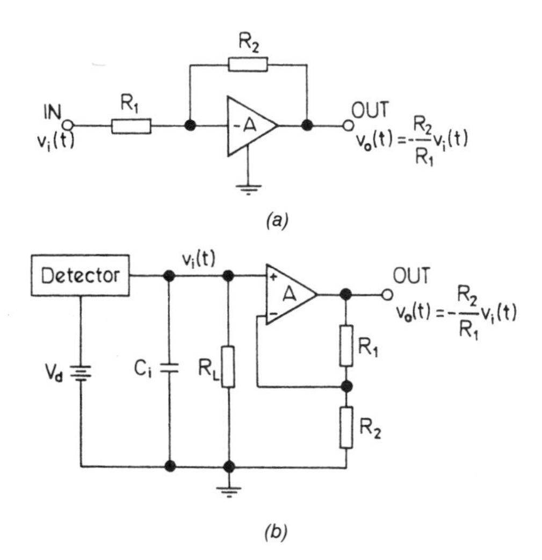

FIGURE 2.30. Typical basic configurations of a voltage-sensitive preamplifier, (a) based on an inverting operational amplifier, (b) using a universal operational amplifier, together with a detector.

2.4.2.2 CHARGE-SENSITIVE PREAMPLIFIERS

The output voltage signal of charge-sensitive preamplifiers is directly proportional to the charge at their input. This charge actually represents a time integral of the detector output current. Changes in the detector capacitance have no effect on the preamplifier output voltage.

The basic diagram of a charge-sensitive preamplifier and its coupling to a detector are illustrated in Figure 2.31.

Taking into account properties of the operational amplifier, the output voltage of a charge-sensitive preamplifier with integrating feedback resistor R_f and capacitance C_f [Figure 2.31(a)] can be expressed in the form

$$v_0(t) = -A\frac{\int_0^t i_i(t')dt'}{C_i + (A + 1)C_f} \tag{2.25}$$

where C_i is the parallel combination of the detector capacitance and the

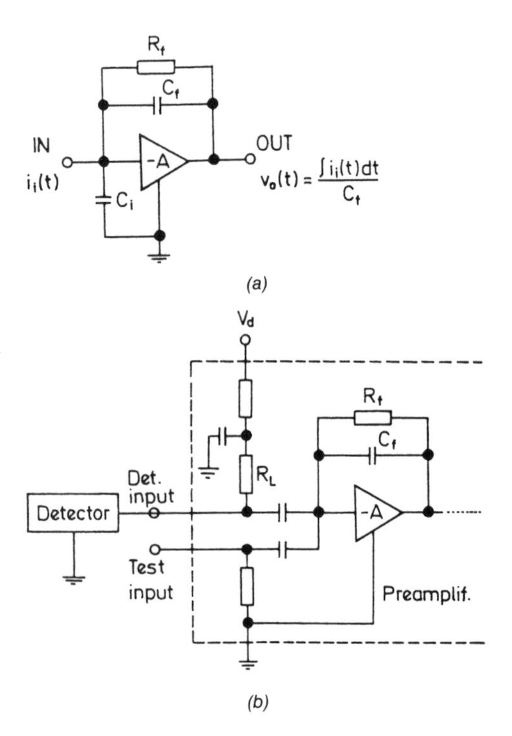

FIGURE 2.31. Charge-sensitive preamplifier: (a) basic diagram, (b) configuration of a detector coupled to a preamplifier input.

capacitance of the cable between the detector and the preamplifier. Again, since the gain A is sufficiently high so that

$$A \gg \frac{C_i + C_f}{C_f} \tag{2.26}$$

for $v_0(t)$ we can write

$$v_0(t) = -\frac{\int_0^t i_i(t')dt'}{C_f} \tag{2.27}$$

This means that with a negative feedback, the voltage gain of a charge-sensitive preamplifier is virtually independent of the actual gain of the operational amplifier, and in addition, it does not depend on the detector capacitance and stray capacitances. As a matter of fact, the resulting gain depends only on the value of the capacitance C_f, which can be very stable with respect to time and temperature.

For energy spectrometry an important parameter of a charge-sensitive preamplifier is its noise, which depends on the noise of the input FET, the input capacitance C_i, and the load resistor R_L. In general, the noise increases as the input capacitance increases. The noise of the preamplifier can be measured using a test input through which a pulse from a high-precision generator can be applied, and a pulse-height distribution is evaluated as a response to the injection of a known charge. The FWHM of this voltage distribution (2.35 V_n) can then be used as a measure of the noise.

As shown already in Section 2.3.2, in the case when a particle losing an energy E forms in a detector a charge Q and the response of the detector-preamplifier system is the Gaussian distribution in terms of voltages around the peak at V (with a standard deviation V_n), one can write the following expression

$$\frac{\text{ENC}}{Q} = \frac{V_n}{V} = \frac{E_{\text{FWHM}}}{2.35E} \tag{2.28}$$

where ENC is the so-called *equivalent noise charge* corresponding to a charge which when applied at the output of a detector would generate an output pulse V_n. The parameter ENC is usually given in C or as a number of ion pairs. The noise can also be expressed in terms of equivalent energy E_{FWHM} in eV or keV. The relationship between ENC (in C) and E_{FWHM} (in eV) is for silicon detectors as follows:

$$E_{\text{FWHM}} = 5.28 \times 10^{19}\text{ENC} \tag{2.29}$$

while for germanium detectors, taking into account a different value of E/Q (energy per one electron-hole), this relationship is given by

$$E_{FWHM} = 4.27 \times 10^{19}\text{ENC} \qquad (2.30)$$

The output pulse of a preamplifier has a fast rise time (5 to 40 ns) followed by a relatively slow exponential decay (50 to 100 μs). The long decay (integration) constant is necessary for the full collection of charge carriers from the detector. While the maximum pulse-height of the output signal represents the energy, its rise time is used for timing and sometimes also for identification of different types of particles based on a pulse-shape discrimination. The preamplifier for use in timing applications should be faster than the charge collection time of a detector. The extra speed, however, is usually at the expense of its stability, gain, and noise.

The *gain* of a charge-sensitive preamplifier coupled with a detector is conveniently expressed in terms of mV/MeV, mV/pC, or mV/ion pair.

Typical parameters of the detector-preamplifier systems are given in Table 2.17.

The complete diagram of a typical low-noise charge-sensitive preamplifier is presented in Figure 2.32 (Mann et al., 1991). The feedback loop consists of the capacitance C_f and the resistor R_f. The test input is provided in order to apply an external reference pulse for testing the performance of the preamplifier or the entire electronic system, without the detector itself.

The preamplifier is usually located very close to the detector and is connected to it by a short, electrically shielded capable in order to minimize the capacitance C_i and the distortion of the output detector signal by a long interconnecting cable.

In some applications it may be desirable to shorten the preamplifier output pulses using the so-called *active reset technique* based on a fast dis-

TABLE 2.17. Characteristics of the Commercially Available Preamplifiers Coupled to Some of the Most Commonly Used Detectors (C_i Is the Sum of the Detector and Stray Capacitance, ip = Ion Pair, 1 fc = 10^{-15} C).

Type of Detector	Noise (at C_i = 0)	Gain (mV vs. Q or E)
Proportional counters	350–500 ip	50–200 mV/10^6 ip
Scintillation detectors	<fc	5 mV/pC
Silicon detectors	<2–3 keV	10–40 mV/MeV
Germanium detectors	<600 keV	100–500 mV/MeV

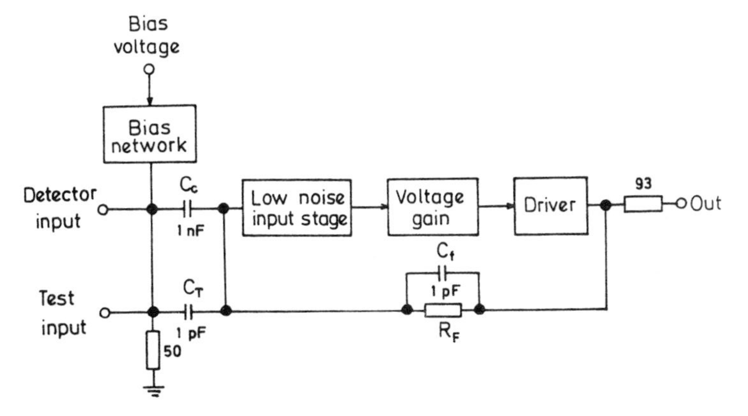

FIGURE 2.32. Block diagram of an AC-coupled charge-sensitive preamplifier. From Mann et al. (1991).

charge of the feedback capacitor. One such method uses a *pulse optical feedback* with a light emitting diode (Knoll, 1989).

2.4.2.3 PULSE SHAPING

The pulse from the detector and preamplifier have to be modified further in order to

- allow an accurate measurement of the energy distribution
- improve the signal-to-noise ratio and thus achieve the highest energy resolution
- make pulses shorter to avoid their possible overlapping
- assure a precise timing for time correlation related applications (e.g., coincidence and anti-coincidence measurements)

Pulse shaping is usually performed using passive or active *RC shaping networks* and sometimes also *delay lines.* Single and combined differentiation and integration shaping networks, along with the illustration of their response to a step input voltage, are shown in Figure 2.33 (Mann et al., 1980).

Figure 2.33(a) represents a simple *RC differentiation circuit,* which is also known as a *high-pass filter* because it attenuates low-frequency signals more than high-frequency signals. As long as the time constant of such a circuit is short enough, its output approximates the differential of the input

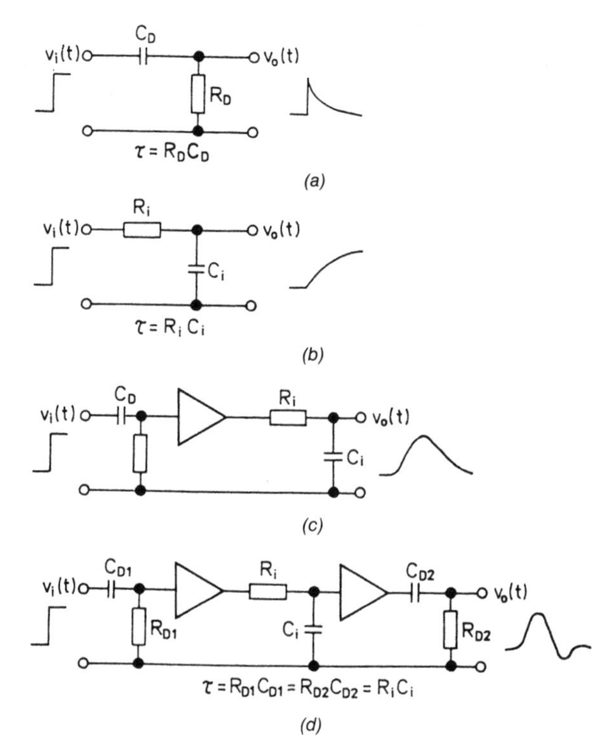

FIGURE 2.33. Basic RC shaping networks: (a) single RC differentiation, (b) single RC integration, (c) combined differentiation-integration, (d) combined double differentiation-integration (v_i—input pulse, v_o—output pulse, V_m—maximum output pulse). The triangles shown represent buffers or amplifier-gain stages providing isolation between the individual shaping circuits. After Mann et al. (1980).

signal. The differentiation circuit is often used to reduce or clip the pulse duration.

The output of an RC integration circuit [Figure 2.33(b)] is approximately proportional to the time integral of the input signal. The integrated property of this circuit is effective for any shape of the input pulse provided the time constant RC is sufficiently large. The circuit can also be used as a *low-pass filter,* because it suppresses high-frequency signals much more than low-frequency signals.

The response of a CR-RC shaping network, i.e., a shaping circuit consisting of sequential differentiating and integrating stages [Figure 2.33(c)], for a step input voltage V can be expressed as

$$v_0(t) = \frac{V R_1 C_1}{R_1 C_1 - R_2 C_2}\left(e^{-t/R_1C_1} - e^{-t/R_2C_2}\right) \tag{2.31}$$

In processing the detector signal, CR-RC shaping is usually carried out using equal differentiation and integration time constants. In such a case, i.e., $R_1C_1 = R_2C_2 = RC$, the output signal $v_0(t)$ is given as (Knoll, 1989)

$$v_0(t) = V\frac{t}{RC}e^{-t/RC} \qquad (2.32)$$

The output pulse of the network shown in Figure 2.33(c) is referred to as a *unipolar pulse,* while the output of the configuration from Figure 2.33(d) is commonly called a *bipolar pulse.*

Another electronic component that is also often used for pulse shaping is a *delay line* based on its delay time and reflection.

2.4.2.4 LINEAR AMPLIFIERS

An amplifier is the most important signal processing block, which should preserve the information of interest and fulfill two main requirements, namely:

(1) To amplify the signal from the preamplifier or sometimes directly from the detector
(2) To shape this signal into a form most suitable for the particular measurement

Some amplifiers are designed to be universal, and they are sufficiently flexible to handle a variety of measurement conditions, while others are specially designed to optimize specific requirements or functions.

In nuclear spectrometry *linear amplifiers* are supposed to produce output pulses, which are directly proportional to the energy lost by the particles in the detector. Such an amplifier accepts relatively long pulses (with a tail time of about 50 to 100 μs) with a rather short rise time (in the range of 5 to 100 ns). In order not to affect the energy resolution, its gain should be very stable and its contribution to the noise kept to a minimum.

The gain of linear amplifiers can be changed manually or electronically (by a digital control) continuously in the range from about 5 or 10 to 1500 to 2000. The temperature gain stability of current amplifiers is better than 0.005%/°C for a typical range of operating temperature from 0 to 50°C and their noise contributions in terms of the RMS voltage level referred to the amplifier input is lower than about 4 μV.

An absolutely linear amplifier would exhibit an ideally constant gain which would not depend on the pulse height of the input signal. The gain of a real amplifier fluctuates and this results in its non-linear performance. The properties of linear amplifiers regarding their linearity can be characterized by two parameters: a *differential* and an *integral non-linearity.*

The *differential non-linearity* (DN) is a measure of the change in amplifier gain as a function of amplifier output signal. This parameter is usually expressed in percentage and may be defined as

$$DN = \frac{(\Delta V_0/\Delta V_i)_A - (\Delta V_0/\Delta V_i)_A}{(\Delta V_0/\Delta V_i)_B} 100\% \tag{2.33}$$

where the symbols used refer to Figure 2.34 (Tsoulfanidis, 1983). The ratio $(\Delta V_0/\Delta V_i)_A$ is the slope of the amplifier gain curve at the point A (where the non-linearity is measured), while $(\Delta V_0/\Delta V_i)_B$ represents the slope of the straight line as shown in Figure 2.34.

On the other hand, an *integral non-linearity* (IN) is defined as the maximum vertical deviation between the straight line shown in Figure 2.34 and the actual amplifier gain curve, divided by the maximum rated output of the linear amplifier. Consequently, the integral non-linearity can be expressed in percentage as

$$IN = \frac{V_m - V_L}{V_{max}} 100\% \tag{2.34}$$

The integral non-linearity of current commercial amplifiers is usually better than about 0.05% over the total output range.

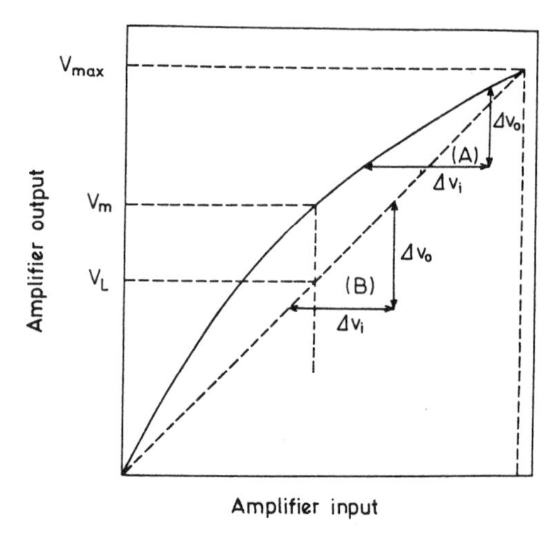

FIGURE 2.34. Illustration of the performance of an ideal and real amplifier with explanation of the symbols used in the definitions of a differential and integral non-linearity. Modified from Tsoulfanidis (1983).

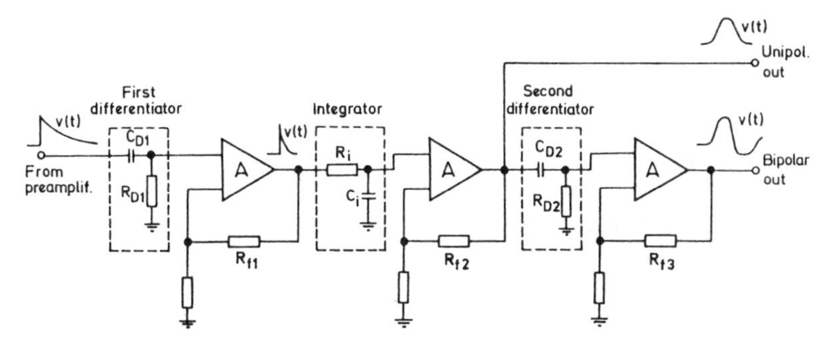

FIGURE 2.35. Simplified configuration of a typical linear amplifier with RC shaping producing both unipolar and bipolar output pulses. From Mann et al. (1980).

In addition to the amplification, the other essential function of a linear amplifier consists in pulse shaping in order to achieve the best signal-to-noise ratio in the evaluation of pulse amplitudes of individual pulses. A typical linear amplifier (Figure 2.35) usually has one or more independent integrators and differentiators. Both time constants can be selected in such a way as to satisfy the requirements given by the detector parameters as well as by experimental conditions.

Most linear amplifiers are equipped with so-called *pole-zero cancellation circuits* which can be adjusted to eliminate undesired undershoots in the shape of output pulses which, if uncorrected, may cause a baseline shift and by this affect the quality of the measuring results.

In some energy-related measurements, it is often useful to obtain more detailed information about a small part of the total pulse-height spectra. A special *biased amplifier* is used for this purpose in order to expand the region of interest into the full input range of the available pulse-height analyzer. Such an amplifier produces proportional output pulses only when corresponding input pulses are within the selected biased levels.

Timing reference pulses may be derived from the normal output pulses of the linear amplifier, but for special applications where fast response is required, fast *timing amplifiers* give better results. These amplifiers are characterized by their wide bandwidth and short integrating and differentiating time constants which are in the nanosecond range compared with the spectrometry amplifiers with time constants in the microsecond range. Most timing amplifiers usually give negative pulses because the majority of timing electronic blocks require negative input signals.

Some of the latest types of high-performance current amplifiers are designed in such a way that they can be programmed and controlled by a computer. The amplifier parameters, which can be remotely programmable, include coarse gain, fine gain, time constant, and some others. Computer-controlled amplifiers are more and more often used in automated or remotely controlled systems involving radiation and radionuclide measurements including long-time low-level radioactivities.

In some measurements the linear amplifier is used together with a *linear gate,* which functions as a switch for linear signals. When the gate is open, linear pulses can pass through it with a minimum of distortion or change. The linear gate is used whenever it is necessary to gate a signal prior to its pulse-height analysis. Some linear gates can operate in either coincidence or anti-coincidence mode, i.e., the input of the linear gate can be disabled or enabled for an adjustable gate period. Specially designed linear gates include also a variable linear delay and a stretcher generating an output pulse with an amplitude equal to the input pulse and shaped to a uniform rise time and width.

2.4.3 Electronic Blocks for Evaluation of Detector Signals

2.4.3.1 GENERAL CONSIDERATIONS

Detector signals are usually evaluated after appropriate processing and treatment in amplifiers or other analog circuits. Only in a few cases can a signal be taken directly from the detector and used for the extraction of specific information. This may include some special applications such as, for example, fast timing or pulse-shape analysis where a signal is picked up from the anode or the dynode of the photomultiplier.

The output pulse from a detector contains information which can be obtained using appropriate electronic blocks. In the field of radioactivity measurements, we are especially interested in the following four types of measurements.

(1) The number of *pulse counts* can give us the information about the *number of particles* striking the detector sensitive volume. The final results may then be interpreted in terms of field quantities, such as fluence or, most important for the evaluation of radiation sources, directly in terms of activity, activity concentration and other parameters related to the number of disintegrations. Since every radionuclide has a unique decay scheme, and particles emitted differ as to their type and especially as to their energy, there is no universal counting system that can be used for any radioactive source or radiation. A system for particle counting usually relies on pulse-height analysis and sometimes also on pulse-shape discrimination and time correlation in order to identify and select for further registration only those pulses which really represent the desired type of disintegration.

(2) The *pulse-height distribution* of the signal taken from the output of a

detector or rather a linear amplifier represents the distribution of the energy lost by particles in the detector. This distribution can be converted (using an appropriate unfolding or deconvolution method, which depends primarily on the detector) into the *energy distribution* of measured radiation. Most techniques in low-level radioactivity evaluation are based on *radiation energy spectrometry.*

(3) In the case of a mixed radiation, i.e., particles of a different nature and mass or charge (for example gamma-beta or beta-alpha radiation), the pulse-height analysis cannot distinguish between different types of particles. The *identification* of such particles (producing pulses of the same amplitude) is in some cases possible using a *pulse-shape discrimination* method based on the characteristic shape and form of the pulses corresponding to different types of particles.

(4) The *time* a detector or amplifier output pulse appears can be *correlated* with the time a particle (initiating such a pulse) interacted with the detector. Although there is always some delay between these two time marks and the delay is subject to some fluctuations, this correlation is very useful in many measurements and experiments. One can study, for example, the distribution of time intervals between output pulses of a single detector and use the information obtained for the evaluation of some parameters such as dead time, the presence of spurious pulses, etc. The *time correlation* is widely used especially in *coincidence* and *anti-coincidence* counting systems based on the evaluation of time information from two or more detectors.

2.4.3.2 COUNTING

The block diagram of a sample counting system is shown in Figure 2.36. In this arrangement, all the detector events that generate an output response having a pulse-height greater than the selected threshold of the pulse-height discriminator are counted by a counter (scaler). The discriminator level is

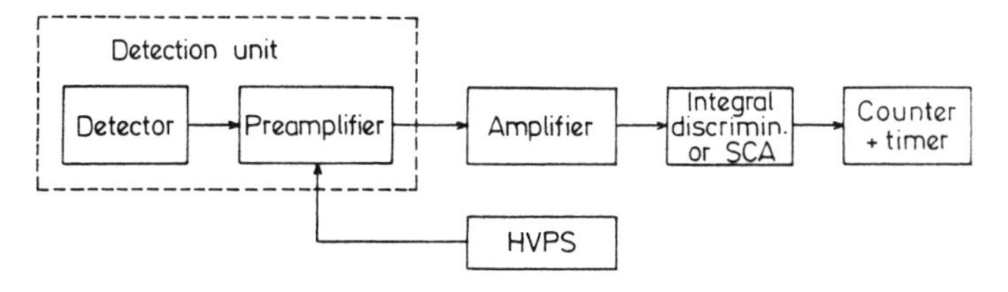

FIGURE 2.36. Configuration of a simple counting system (HVPS—high-voltage power supply, SCA—single-channel analyzer).

usually set just above the unwanted noise signal, which is rejected in this way.

In more powerful counting systems single-channel analyzers (SCA) instead of simple integral discriminators are used. In this case, only pulses having an amplitude higher than the lower discrimination level and at the same time lower than the upper discrimination level can be counted.

The properties of the output signal from different detectors and also the operation characteristics of both preamplifiers and amplifiers have already been discussed. Now we may describe the function and parameters of other electronic blocks and units used in counting systems.

For their proper operation, all active radiation detectors need an *external voltage* (detector voltage, detector bias) provided by a *high-voltage power supply* (HVPS). The HVPS unit can deliver a positive or negative voltage required for the operation of various types of detectors. Most detectors work with the positive bias, which is usually applied through a load resistor.

The most important parameters of the HVPS are as follows:

(1) The *voltage range* (including maximum and minimum voltage available) and *polarity*. Different types of detectors need different voltages: for ionization chambers this voltage is usually below 1000 V; proportional counters work with typical voltages from the range of 500–1500 V; GM counters, depending on their form and size, require a voltage between about 800 to 2500 V; for the operation of silicon surface-barrier detectors a voltage below 100 V is enough; other types of semiconductor detectors may need voltages from as low as 100 V to as high as 3000 V.

(2) The *stability* of the supplied voltage with respect to possible changes or fluctuations of a power line voltage, temperature, and long-term effects and other drifts. In general, detectors such as scintillation detectors and proportional counters require much higher stability than, for example, ionization chambers or semiconductor detectors. In the case of semiconductor detectors, hum and noise may play a more important role than stability. Typical high-voltage variations due to line voltage fluctuations are usually less than 0.005% per 10% of line voltage changes. The temperature instability of most HVPS units is better than 0.005%/°C, while their long-term drift at an ambient temperature and constant line and load is usually below 0.01%/h, 0.02%/8 h and 0.03%/24 h.

(3) Another parameter of the HVPS is the *maximum current* available. While photomultipliers in scintillation detectors draw continuous current from the HVPS, most semiconductor detectors require very little

current. The HVPS for scintillation detectors must be able to supply current up to 1–2 mA and sometimes even higher (10 mA is usually the upper current rating of most HVPS units). For semiconductor detectors, which usually require very low current, bias supplies with high-impedance outputs giving current up to 100 μA are used most.

(4) The fluctuations of the voltage delivered by the HVPS are also affected by the hum and noise, which are usually given in terms of *peak-to-peak voltage* ripples. Good HVPS sources have noise and ripple below about 2–10 mV.

The voltage of the HVPS can be adjusted manually in steps or continuously using a helipot. Some sophisticated power supplies can be controlled electronically and used in computer-based measuring systems. Such an HVPS unit may also be used for compensation of total gain drifts of the scintillation spectrometers.

The basic device for the pulse-height selection is an *integral (amplitude) discriminator.* This circuit produces a unified or standard-sized output pulse as a response to an input pulse whose height exceeds the preset discrimination level (threshold). A schematic of a simple integral discriminator and the time diagram illustrating its operation are shown in Figure 2.37. Here, as a threshold element, a voltage comparator is used. These types of circuits are now available as monolithic integrated elements with very good parameters as far as stability is concerned. The integrated comparator is characterized by fast response, high gain, and two electrically matched input terminals, which show very good temperature stability. The comparator

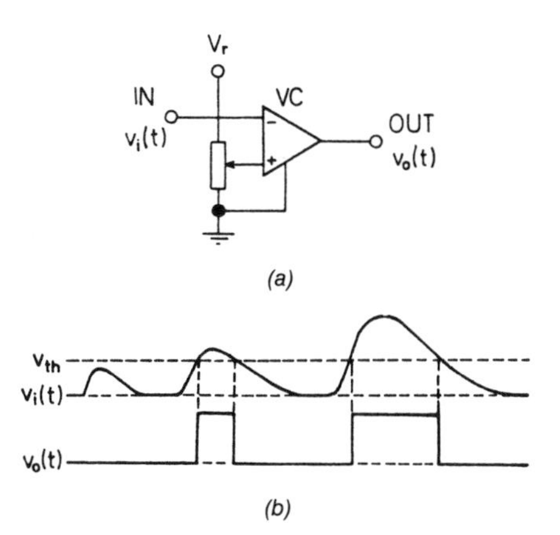

FIGURE 2.37. Basic configuration of an integral pulse-height discriminator (a) and illustration of its response (b). The symbols used: V_r—reference voltage, V_{th}—threshold (discrimination) level, VC—voltage comparator.

threshold can be simply derived by a voltage divider from a stable reference voltage.

Other more sophisticated microelectronic circuits can be used even more conveniently in designing integral as well as differential pulse-height discriminators. One such element can function as a charge-sensitive preamplifier, a charge-sensitive preamplifier together with a pulse-height discriminator, or a compact charge-sensitive preamplifier and linear amplifier (AMPTEK, 1990). These microelectronic elements were successfully tested in the processing and evaluation of signals from scintillation, proportional, and semiconductor detectors (Jiang et al., 1990).

Two integral discriminators can create a *differential discriminator* or *single-channel analyzer* (SCA) whose block diagram is shown in Figure 2.38. Its main function is in the production of a normalized output pulse for every input pulse, the pulse-height of which lies between the *upper* and *lower discrimination levels.* The difference between both these levels is called the *channel width* or the *discriminator window.* The logic unit accepts unified pulses (usually of different widths) from the ULD and LLD to generate a pulse when a pulse appears at the output of the LLD, but no pulse

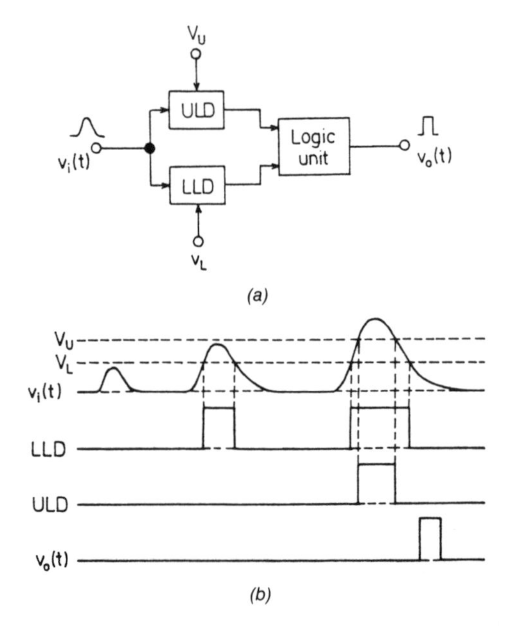

FIGURE 2.38. Single-channel analyzer: (a) block diagram, (b) illustration of its function (ULD—the upper-level discriminator, LLD—the lower-level discriminator, V_U—the upper discrimination level, V_L—the lower discrimination level).

is produced by the ULD. As long as both these discriminators produce pulses the logic unit will not respond.

Various methods may be chosen to design the logic unit, sometimes also referred to as the anti-coincidence gate. The techniques depend on the discriminators used and the requirements for the preservation of the time information (Sabol, 1974). Most of the earlier designed SCAs interrogated the logic circuitry when the input signal decreased below the lower discrimination level and produced a corresponding output pulse at that time, provided the maximum of the input signal occurred within the window (Milam, 1973). This type of signal can be used in simple counting systems but is not satisfactory when the SCA output signal is supposed to be correlated in time with the beginning of the input pulse.

In applications where conservation of the *time information* is important, the output pulse of a SCA must be derived carefully from a well-defined part of the detector or amplifier signal. Several time pick-off methods have been developed to produce time-correlated pulses.

(1) *Leading edge discrimination*—the time (logic, reference) pulse is generated in the time when the input pulse crosses a fixed discrimination level. Although the discrimination level is usually set very low, the shape and pulse-height of the input pulse always causes some timing uncertainties that may be quite small when detectors with very fast response are used.

(2) *Crossover timing* or *zero crossing trigger*—this virtually amplitude-independent timing is based on the detection of the time at which a bipolar pulse is changing its polarity, e.g., when it crosses from the positive to the negative side of the zero level.

(3) *Constant fraction timing*—the time pulse is picked up at a fixed time after the leading edge of the input pulse has reached a *constant fraction* of its maximum amplitude. This method of timing appears to produce the best results.

The parameters of a typical SCA can be summarized as follows: discrimination level range 0.02–10 V, integral non-linearity (defined similarly as the same parameter of linear amplifier) better than $\pm 0.25\%$, stability better than $0.005\%/°C$, pulse pair resolution < 0.5 μs.

In addition to pulse-height selection for particle counting, single-channel analyzers can be used in numerous applications including those measurements requiring time correlations and spectrometry based on scanning or sweeping the input pulse dynamic range by properly chosen discrimination levels. The latter function, however, is now performed almost entirely by *multichannel analyzers,* which are, thanks to advances in microelectronics, available in a variety of modifications, including computer-based versions.

After the pulse-height selection of detector or amplifier pulses containing information about energy (and also about the types of particles or the times of their interactions) in an analog form, uniform pulses are produced, and these pulses then carry the relevant information in a discrete or digital form. The presence of such a pulse at the output of a particular selector means that the parameters of the corresponding input meet the set criteria of this selector. In order to evaluate the number of such pulses, the standardized pulses obtained at the selector output are counted in devices called *counters* or *scalers.*

A *counter* is actually a registrator or recorder of pulses entering its input. For every pulse coming into the counter, its previous total is increased by one. The counter is active only during the selected measuring interval controlled externally or by a timer, which is usually an inherent part of any counter nowadays. In addition to the possibility of selecting the time interval for counting, one can also preselect the number of counts, which, when reached, will stop the counter and show the relevant time interval during which these counts were accumulated. Information about both the number of counts and time is visualized on a display. These data can also be printed by an attached printer, stored, or transmitted to other electronic blocks or to a computer.

A simplified diagram of an *n-decade* counter consisting of individual decade counters, decoders, and digital displays is shown in Figure 2.39 (Mann et al., 1980).

Individual blocks of a pulse counter are now available as compact integrated circuits. The counters built from digital monolithic elements are fast enough (their time resolution is usually much shorter than the dead time of

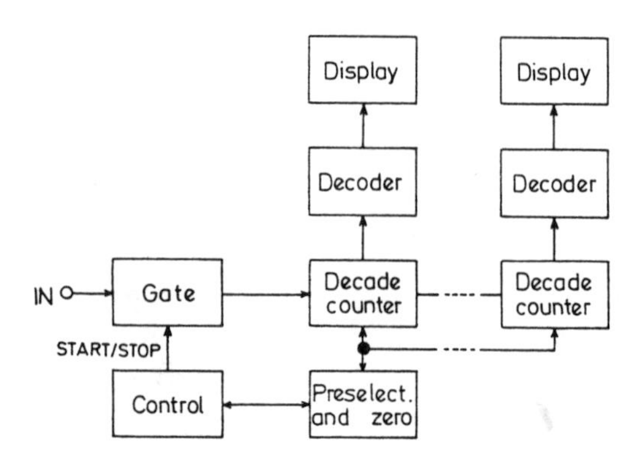

FIGURE 2.39. Basic configuration of a pulse counter.

detectors) so that they do not present any limitation regarding counting rates or timing. Anyway, in the case of low-level counting these aspects are usually irrelevant.

In principle, pulses can also be evaluated by *ratemeters*, which give information about the mean number of pulses per unit of time rather than number of pulses observed in a selected measuring time interval. Because of the low count-rate encountered in low radioactivity measurements, the use of ratemeters in such applications is very limited.

2.4.3.3 ENERGY SPECTROMETRY

Measurements of the energy distribution of detected particles are based on the evaluation of pulse-height spectra. To date pulse-height analysis is almost exclusively carried out by means of *multichannel analyzers* (MCA), which continuously select and store information about the height of input pulses.

The first MCAs were assembled using a series of integral discriminators to divide a given range of input pulse height into a number of fixed adjacent windows. Each of these windows represented one channel from which pulses were counted and stored. This technique did not allow the building of an analyzer with more than about twenty to fifty channel analyzers. The main problems of using this approach were the insufficient stability of discrimination levels and the complexity, which result in unreliable operation since the number of channels was practically proportional to the number of required electronic components.

The operation of the current multichannel analyzers is based on the principle of converting the pulse-height of an input analog signal into an equivalent digital number. This conversion process takes place in the most important part of the MCA–the *analog-to-digital converter* (ADC). There are two basic methods used for converting analog information that corresponds to the pulse-height into an adequate digital representation. The first method was devised by Wilkinson in 1949 and in a modified version is still widely used in most ADCs. The principle of the *Wilkinson type ADC* consists in the conversion of the pulse-height into a proportional time interval which is then digitalized. The other technique–the *successive approximation type ADC*–was developed later, mainly with the intention of reducing the conversion time and making ADCs faster.

The operating principle of a *Wilkinson type converter* is based on charging a storage capacitor by an input pulse to its maximum pulse-height and then, after some *stretching*, this capacitor is discharged by a constant current. The end of a discharge is monitored by a comparator. During the period of discharge, which is proportional to the peak value of the input pulse, a group of standard-shaped pulses is generated. The pulses are counted and their number is directly related to the amplitude of the input signal. One possible schematic illustration of such an operation is shown in Figure 2.40. A pulse-stretching circuit (PSC) ensures through a switch (Sw) that the voltage at the storage capacitor (C_s) is a flat-top for a short time. The end of this stretching will start the discharging process using a constant current source (CCS), and at the same time a bistable circuit is initiated from its *set* input. The end of the discharge is detected by a voltage comparator (VC), which also *resets* the BC so that the time interval T proportional to the maximum pulse-height V_m is produced.

The second type of ADC uses the technique of *successive approximation*, which is based on a series of comparisons of the input pulse-height with the reference voltage levels splitting the full input ADC range. Its function is illustrated in Figure 2.41. In the first logic operation the pulse-height of the input pulse is compared with a reference voltage, which is one-half of the ADC range (we may assume that its full range is 0–10 V). If it is above the reference (5 V), the reference is stored and the first bit is set to "1." Otherwise, it is discarded and the bit is set to "0." A further reference (2.5 V) is then added and again compared, after that another reference (1.25 V), and so on, in successively decreasing binary steps. At the end of this operation, the binary presentation will be a measure of the input pulse amplitude.

In comparison to the Wilkinson type, the conversion time of the successive approximation ADC is much shorter, but its linearity and stability is somewhat poorer. On the other hand, in the Wilkinson ADC, the conversion time is linearly proportional to the number of channels, while in the successive approximation converters this time increases as the logarithm of the number of channels.

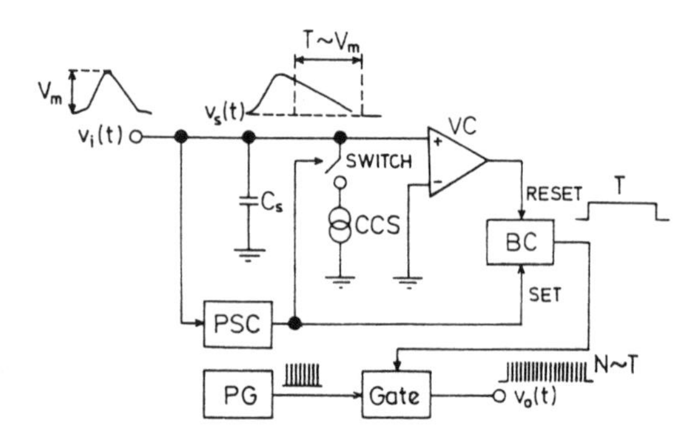

FIGURE 2.40. Possible principle configuration of a Wilkinson ADC (PSC—the pulse-stretching circuit, CCS—the constant current source, BC—the bistable circuit, VC—the voltage comparator).

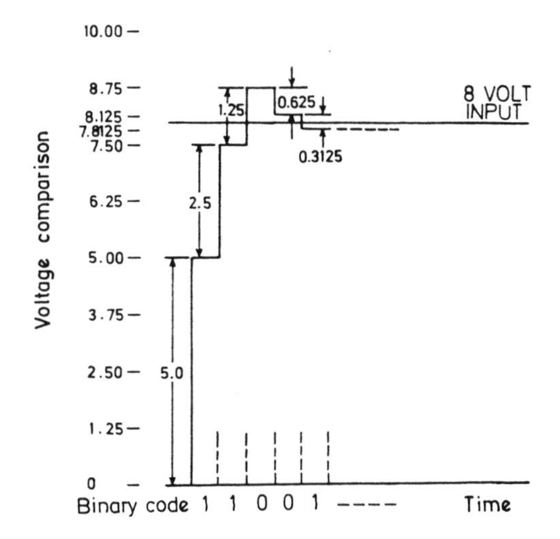

FIGURE 2.41. Principle of an ADC based on successive approximation. A series of comparisons between the input and reference voltage generates a binary code as the comparison voltage approaches the maximum pulse-height of the input signal. From Ross (1973).

The basic architecture of a conventional MCA in which all functions, from acquisition to display, are performed by dedicated hardware and software is shown in Figure 2.42. The input signal in such a MCA is first applied to a SCA, which decides whether the signal is in the region of interest. If its pulse-height falls in the selected region, the SCA produces a signal opening the linear gate through which the delayed input pulse passes to reach the ADC. The Wilkinson type ADC receives the input signal and generates a train of high-frequency pulses whose number is proportional to the pulse-height of the input signal. These pulses are counted by an address counter and *one* count is then added to the relevant channel in the memory. The data are processed, handled and organized in such a way that the results

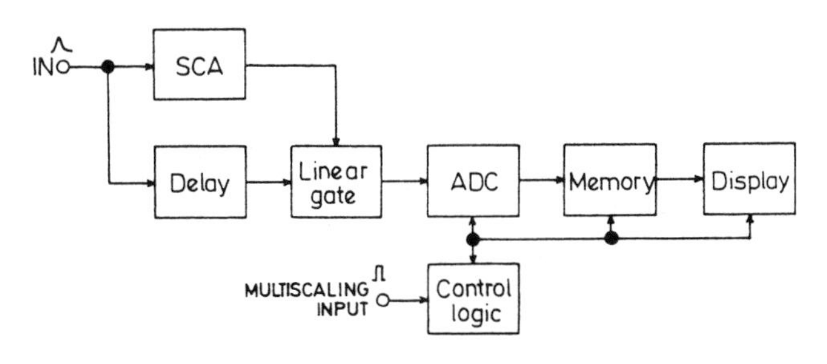

FIGURE 2.42. Block diagram of a conventional multichannel analyzer.

can be displayed as a pulse-height distribution. The desired information can be transmitted to a computer for further evaluation or to some accessories such as printers or plotters.

Most MCAs are usually provided with a *multichannel scaling mode,* which can be used, for example, for a direct measurement of radioactive decay curves. In this operation regime, each channel sequentially counts input pulses bypassing the ADC during a preset time interval. The memory locations are used here as scalers (counters), and their position can be automatically advanced by an internal clock or by external pulses.

The advanced MCAs offer various useful analysis functions, as, for example (Canberra, 1992):

(1) *Energy and time calibration*—based on a least squares fit, calibration in energy units (eV, keV, MeV) or in units of time (μs, ms, s) when multichannel scaling is used
(2) *Peak search*—the automatic analysis of unknown spectra using the calculated locations of peak centroid positions
(3) *Peak net area*—the calculation of the counts in the peak by subtracting a straight line background drawn between the end points of the region of interest set about the peak
(4) *Spectrum normalization*—increasing or decreasing the number of counts in every channel of a spectrum
(5) *Spectrum stripping*—based on the subtraction of a background or reference spectrum from another spectrum
(6) *Spectrum smoothing*—using a three-, five-, or seven-point method
(7) *Isotope identification*—by scanning an isotope library in order to find any entries that are within an energy window of the peak in the spectrum

Some of the main parameters of the advanced MCAs lie typically in the following range: Wilkinson type ADC with 100 MHz generator, number of channels 4096, conversion time better than 10–12 s at 1024 channels, integral non-linearity $< \pm 0.025\%$ and differential non-linearity $< \pm 1\%$ (both over the top 99% of the input range), temperature drift $< \pm 0.01\%/°C$, and long-term drift $< 0.05\%/24$ hours. The drifts in electronics or detector gain, if necessary, can be stabilized (Sabol, 1990).

The conventional (stand-alone, dedicated) MCAs have some disadvantages and limitations. First of all, they are usually one-purpose instruments that lack functionality outside their primary field of application. Most such analyzers cannot be programmed in a high-level language. The architecture used in the conventional MCAs is rather rigid, and it will be difficult and also expensive to expand these measuring systems in future.

Modern-day MCAs are computer-based systems, which are very similar to standard PCs supplied with appropriate ADCs and software. The first such concept and system was developed and introduced in 1983 by EG&G Ortec. Essentially, this approach consists of the separation of the MCA functionality into two parts (EG&G Ortec, 1983).

(1) Data acquisition and storage is performed by a hardware component, known as a *multichannel buffer* (MCB). There are now many types of these systems for both desk and portable PCs designed to meet the specific requirements of any spectrometry application. These systems usually include a digital stabilizer and they enable the computer control of all major functions as well as the operation and parameters of individual blocks of the whole measuring assembly (e.g., gain and shaping constants, bias supply—including protection of a detector, parameters of an ADC, a digital stabilizer or an attached sample changer, etc.). In addition to the external MCBs, *plug-in MCA cards* (ready to be used within a standard PC) consisting of ADC, microprocessor, program memory, and dual-ported data memory are now commercially available.

(2) The control and display function is carried out by software (usually referred to as *MCA Emulation Software*) running on a standard personal computer, which is connected to the MCB. The potential of software is virtually unlimited. This includes such operations as advanced spectral analysis, namely fast peak search, peak centroid and shape calculation, net areas and gross areas of peaks, spectrum sum, spectrum smooth, energy calibration, region-of-interest analysis, radionuclide identification, on-line activity calculation (even for data being acquired), spectral plotting, and also control of advanced hardware functions.

A principal configuration of a multichannel buffer and a PC is shown in Figure 2.43. During acquisition, no overhead is placed on the computer, and this allows for flexibility and expansion.

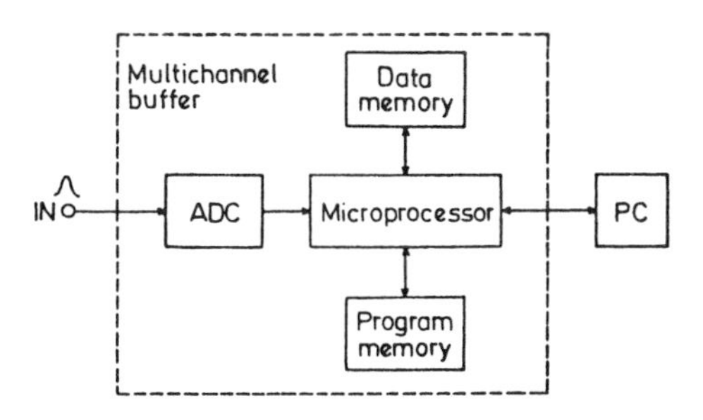

FIGURE 2.43. Functional block diagram of an advanced computer-based system for multichannel analysis. From EG&G Ortec (1993).

2.4.3.4 PULSE-SHAPE IDENTIFICATION

In addition to the information about the energy of particles contained usually in the pulse-height, the *shape* of the detector pulses can reveal, in some cases, the type of particles detected. The detector signal shape reflects to a certain extent an energy deposition pattern, which may be quite different for various particles. The energy deposition rate or the number of ionization events per unit of the particle track depends primarily on the particle mass and charge. Such information is usually encoded in the time profile of the detector output current. After its integration this information is preserved in the *rise time* of the resulting voltage pulse.

Illustrations of different shapes of the pulses from a scintillation detector based on an organic scintillator and a proportional counter are shown in Figure 2.44.

In the first case, Figure 2.44(a), the current pulse at the anode (or one of the last dynodes) of a photomultiplier follows the light emission rate, which in some organic scintillators can be approximated by a sum of two exponentials—one with a fast decay time and the other with a slow decay time. While the fast component is for all types of particles more or less the same, the slow part of the current signal depends primarily on the rate of energy loss dE/dl. Consequently, there is a difference in the current signal shape representing charged particles, such as alpha particles, protons, or electrons, with various energy deposition rates. The difference in the pulse shape is often used in the spectrometry and dosimetry of mixed *n*-gamma fields where it is important to distinguish between the response to neutrons (producing recoil protons) and gamma radiation (producing electrons).

Another example is the characteristic shape of the voltage pulse from a proportional counter where a difference in the rise time (leading edge) of pulses corresponding to different particles can be used for their identification [Figure 2.44(b)].

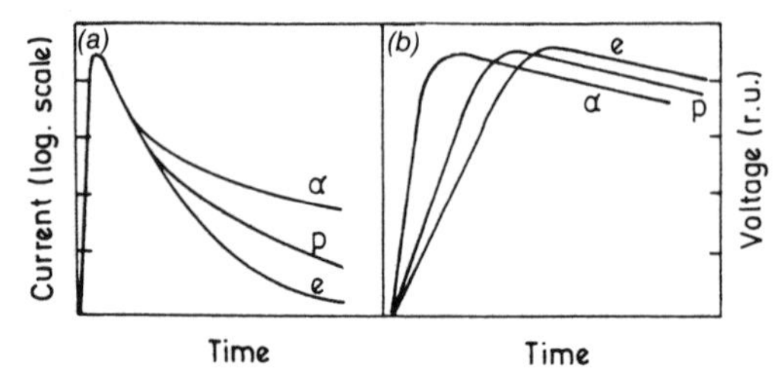

FIGURE 2.44. Characteristic response of detector signals suitable for pulse-shape analysis: (a) the current signal of a photomultiplier coupled to an organic scintillator, (b) the voltage response of a proportional counter to different types of particles.

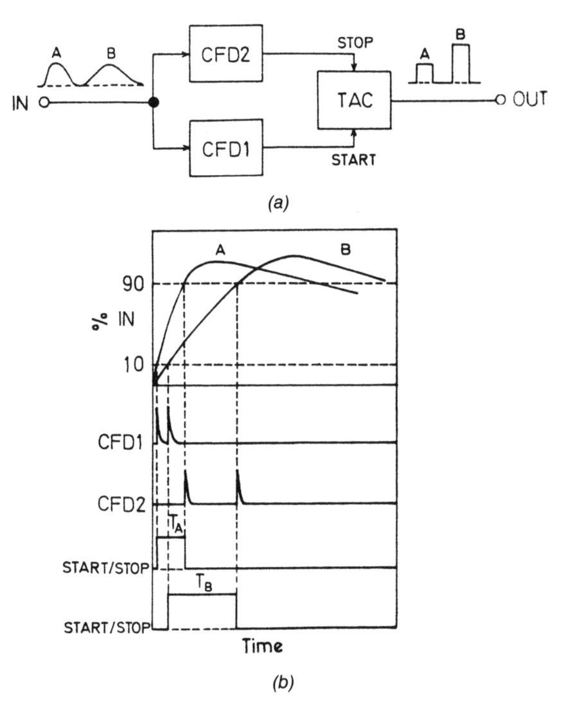

FIGURE 2.45. Block diagram of a typical pulse-shape analyzer (a) and illustration of its operation (b).

There are various methods of designing pulse-shaping analyzers or discriminators. Some of these systems are specially developed to meet specific requirements; others can be assembled from available electronic modules or units. One such measuring system, including a diagram illustrating its function, is presented in Figure 2.45. This pulse-shape analyżer is based on two constant-fraction discriminators set on two different fraction levels (say 10% and 90%, respectively). The output of the first discriminator is used as a *start pulse* for a *time-to-amplitude converter* (TAC). The pulse from a second constant-fraction discriminator serves similarly as a *stop pulse.* The TAC produces an output pulse whose amplitude is proportional to the time interval between the *start* and the *stop* pulses generated by the CFDs. This means that pulse-shape is converted into the pulse-height, which can then be further analyzed and selected using a SCA.

Compact *pulse-shape analyzers* are commercially available as a NIM (nuclear instrument module) block producing an output pulse with an amplitude proportional to the shape of the input signal. The output signal is independent of the amplitude of the input pulse and depends only on the shape of the input signal. Moreover, many types of universal and specialized constant-fraction discriminators, time-to-amplitude converters, as well as highly sensitive and stable timing single-channel analyzers are readily available in a modular form for designing and building various pulse-shape discriminators needed for enhanced selection of the desired type of radiation response or elimination of an unwanted, interfering response.

The pulse-shape discrimination method is a powerful tool in many radiation measurements. It can be used in the assay of low-level alpha-beta emitters by means of liquid-scintillation counting (Prichard et al., 1992). Although alpha-particle energies cover the 4–8 MeV range and beta-particles occur in the 0–2 MeV range, they cannot be distinguished by differences in their pulse-heights because alpha and beta particles interact with scintillation cocktail with such different efficiencies that the resulting light intensities are similar. This difficulty can be overcome by the use of a pulse-shape analyzer, which helps not only to separate alpha and beta responses, but also may substantially reduce the background pulse rate. Pulses produced by alpha particles are characterized by longer decay time than those generated by beta particles and this information is used in pulse-shape discrimination.

Other detectors, including semiconductor silicon (Pausch et al., 1992) and HPGe (Bemford et al., 1991) detectors, or scintillation detectors with inorganic scintillators (Doll et al., 1989; Harihar, 1992) also yield pulses with shapes dependent on the particle type.

2.4.3.5 TIME CORRELATION

Using reference time pulses correlated with the interactions of respective particles with a detector, various studies and operations can be carried out. We can measure the *distribution of time intervals* between successive pulses of a single detector or compare the time occurrence of pulses from two or more detection systems.

Since radioactive decay processes occur randomly in time, the same statistics can be applied to the emission of accompanying particles. Moreover, due to the fact that the interactions of individual particles with a detector are independent, in most cases the time distribution of pulses at the output of the detection system will also, in principle, follow the same statistical law—*Poisson distribution* (more details about the counting statistics can be found in Section 2.6). Comparing an experimentally determined pulse time interval distribution with that predicted on the basis of statistics, one can check whether a counting system is working properly or whether the output signal contains some additional pulses which do not fit the expected statistical behavior. Such pulses are usually of another origin (noise, interference of the ambient electric and magnetic fields, reflections in transmission lines connecting unmatched electronic blocks, etc.) than that of the interactions of measured radiation and are considered as spurious or unwanted pulses. They have to be eliminated from further counting or necessary steps should be applied to correct the final results. From the time interval distribution it is also possible to deduce information about the actual or effective dead time (Sabol, 1988a, 1988b, 1988c).

The measurements of time intervals can be based on their digitalization or conversion, using time-to-amplitude converters, into pulses with corresponding pulse-heights, which can then be evaluated by single-channel or multichannel analyzers.

The time correlation between the signals from two detectors can also be used for the energy measurement of particles. In such an arrangement, called a *time-of-flight method,* one detector gives a *start* signal, while the output pulse of the other detector is used as a *stop* signal. The purpose of the measurement is to determine the time interval (defined by the start and stop signals) it takes a particle to travel the known distance between these two detectors. Under certain circumstances, the time interval obtained is a measure of the speed of a particle and thus its energy.

Time correlation has very important applications, especially in measurements of the activity of radionuclides where signals from two or more detectors are evaluated by a *coincidence* or an *anti-coincidence* unit.

The principle of the operation of a coincidence unit is simple (Figure 2.46): it gives an output signal only when the time shift between input pulses is equal or less than its *resolution time* τ_r; otherwise, no signal at the output of the unit appears. The time resolution is the main parameter of the

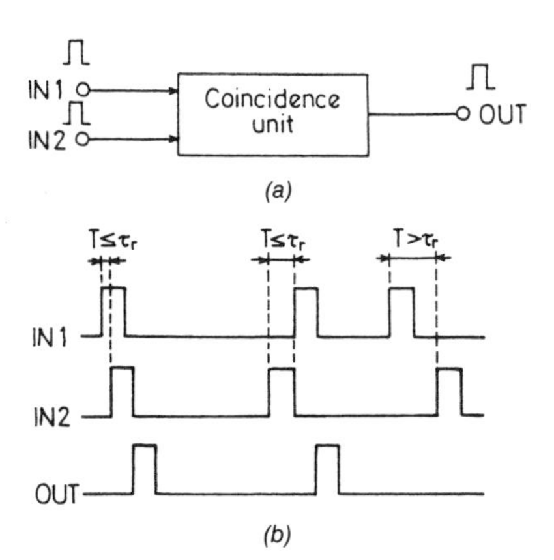

FIGURE 2.46. Principle of operation of a coincidence unit: (a) schematic diagram, (b) illustration of its response.

coincidence unit, and it can be set to an appropriate value consistent with the detectors used.

Although ideally the aim of the coincidence measurement would consist of counting *true* coincidence between two input pulses, in practice this seldom occurs because there are always some fluctuations and uncertainties in time correlations of *real* detectors. Because of this, the detector pulses corresponding to the detection events that took place at the same time in both detectors may not appear at the outputs of the respective systems at exactly the same time. In order to register such a time appearance of two pulses as a *true* coincidence, a certain fixed resolution time should be introduced so that pulses coming within this time interval will produce a signal at the output of the coincidence unit.

As long as input pulses entering the coincidence unit characterized by a resolution time τ_r are statistically independent, they will produce at its output a *random* (*falls* or *accidental*) *coincidence rate* n_r given by

$$n_r = 2n_1n_2\tau_r \tag{2.35}$$

where n_1 and n_2 are the input pulse rates.

A typical coincidence setup for the measurement of the activity of radionuclides emitting one beta particle and one photon gamma for each disintegration is shown in Figure 2.47. Since the detectors used for the detection of these different radiation components are usually of different types (for example, a proportional counter detects beta particles while a large-volume scintillation detector is used for the detection of gamma radiation), associated electronics blocks may differ as well. After amplification, relevant time pulses are produced in the timing SCAs whose discrimination levels are set in such a way as to accept for further time evaluation only pulses in a certain pulse-height range. The pulse rates in both beta and gamma channels (n_β and n_γ), as well as the pulse rate (n_c) at the output of the coincidence unit, are determined by the relevant counters.

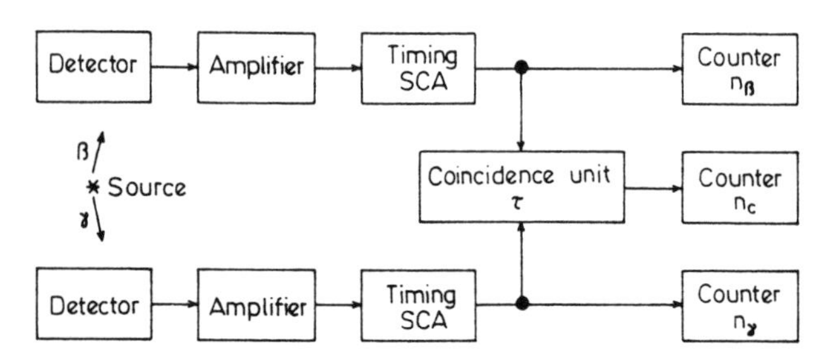

FIGURE 2.47. Block diagram of a basic coincidence counting system.

Consider a radioactive source with a simple and known decay scheme containing a single beta-gamma branch. Let A be the activity of the source placed between two detectors, one sensitive to beta particles, the other sensitive to gamma photons. For the counting rates n_β, n_γ and $n_{\beta\gamma}$ the following well-known equations can be written:

$$n_\beta = A\epsilon_\beta \tag{2.36}$$

$$n_\gamma = A\epsilon_\gamma \tag{2.37}$$

$$n_{\beta\gamma} = A\epsilon_\beta\epsilon_\gamma \tag{2.38}$$

and

$$A = \frac{n_\beta n_\gamma}{n_{\beta\gamma}} \tag{2.39}$$

where ϵ_β and ϵ_γ are detector efficiencies for beta and gamma radiation, respectively, and $n_{\beta\gamma}$ is the true coincidence rate that can be obtained from the total coincidence rate n_c by a simple subtraction of the random coincidence rate n_r. The activity of the source can then be expressed in the form

$$A = \frac{n_\beta n_\gamma}{n_c - 2n_\beta n_\gamma \tau_r} \tag{2.40}$$

In the derivation of these equations, it is assumed that the relevant detection efficiencies are constant and they include all effects that lead to the registration of the respective particles. In practice, however, this may not always be the case and appropriate corrections have to be introduced. Moreover, there are also some other effects which have to be considered, e.g., background count rates, dead time losses, and, in the case of short half-life radionuclides, the decay of the source during the measuring time interval must be taken into account as well.

In addition to a single beta-gamma decay scheme, radionuclides with a complex decay can also be measured by the coincidence method, but more corrections should be applied (NCRP, 1985). Besides beta-gamma coincidences, the coincidence between other particles or photons may also be used (gamma-gamma, alpha-beta, X-gamma, etc.). In these applications various types of detectors, often in modified versions, are employed. Beta particles and other charged particles or low energy X-rays are usually detected by gas-filled counters and also by thin plastic or liquid scintillation

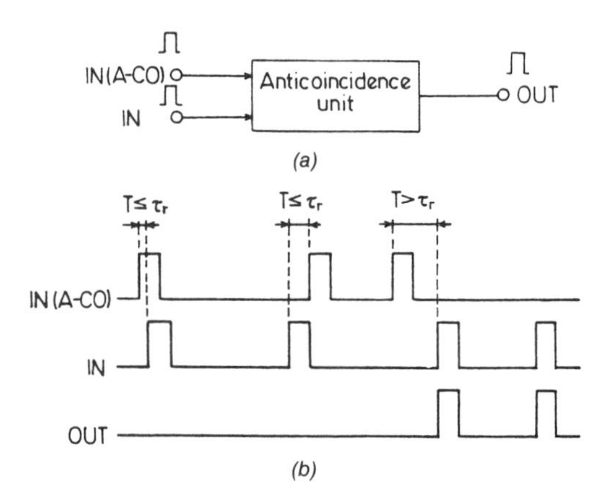

FIGURE 2.48. Principle diagram (a) and function (b) of a simple anti-coincidence unit.

detectors. Scintillation detectors, especially NaI(Tl), and recently HPGe detectors are also commonly used as photon detectors.

Anti-coincidence units are another type of useful block for time-related operations with pulses of two or more detectors. The function of a typical anti-coincidence unit with two inputs is illustrated in Figure 2.48. In this case, an output pulse appears as a response to an input pulse only when no pulse comes to another input within the *time resolution* of the unit.

The anti-coincidence and coincidence functions are now usually integrated in a single unit which can be used for both types of time correlation operations. Such universal coincidence units have four or more inputs whose mode (COINC—coincidence, or ANTI—anti-coincidence) can be set independently by toggle switches. The resolution (resolving) time is variable from a few tens of nanoseconds to a few microseconds.

Anti-coincidence units are frequently used in reducing the response of a main (sample) detector to background radiation by means of a surrounding guard detector. Its signal can eliminate those pulses which appear at the "same" time (depending on the set resolution time) at the output of the main detector. By this method the background rate of proportional counters may be reduced by a factor of five (Winkler and Rösner, 1989). Another very useful application of anti-coincidence gating is in environmental monitoring using a HPGe-NaI(Tl) Compton suppression spectrometer (Chung and Lee, 1988).

2.4.4 Computers in Low-Level Activity Measurements

Computers, thanks to their reasonable price, ever-improving performance, and versatility, are now commonly used in offices, laboratories, and field conditions to perform a variety of different and sometimes very complex tasks. Modern measuring procedures, methods, and instruments now utilize more and more the tremendous potential computers offer.

Applications of computers in the various areas of low-level radioactivity measurements and evaluations can be divided into three main groups.

(1) A computer can be used for the acquisition, storage, processing, handling, and evaluation of data representing various kinds of information or results of measurements. Many *calculations,* including the most sophisticated, can now be carried out by computers, which may be connected to computer networks through which they can have access to different data bases. These calculations, even with personal portable computers, may include a variety of mathematical simulations such as the penetration of radiation through matter and its interaction with materials of a given composition and geometry or the transport and behavior of radionuclides in the environment in miscellaneous situations.

(2) Computers can control and optimize measuring procedures, including the on-line evaluation of obtained results and their interpretations in line with the given requirements or criteria. A computer, for example, can take care of all the individual tasks essential for running a *low-level counting laboratory,* including communication with individual instruments, collection of measuring data, continuous registration of all relevant parameters, statistical treatment and analysis of data, applications of statistical models, error evaluation and reduction, introduction of necessary corrections, etc.

(3) Computers in a standard version equipped with some accessories or in a modified version that converts a computer into a *powerful measuring instrument* can be used for virtually any type of measurement. In such a way the computer can substantially increase the capability and performance of radiation detection systems. The computer can make most necessary calculations and even decisions with respect to the optimization of the measurement on the spot very efficiently, and, thanks to stored instructions and information, it can take into account many options and possibilities in order to choose the best solution under the changing circumstances and conditions. A special role in these applications of computers is played by computer-based multichannel analyzers, which have already been discussed. Such functions as the acquisition, storage and analysis of data, experiment monitoring, equipment calibration, and its quality control are now quite common not only in advanced MCAs, but also in many other instruments used in low-level activity measurements.

In the near future we may expect further advances in the applications of microelectronics in developing hardware with increased operational power and even more elaborate software, which will even more enhance the capabilities of nuclear instrumentation.

2.5 ELIMINATION OF THE EXTRANEOUS COUNTS

In view of the fact that the nuclear decay has a statistical character (Section 1.1.1), the measurement of low-level radioactivity generally requires long measurement intervals—sometimes even several days for one measurement. The length of the measurement interval is given by the counting rate and the required precision of the measurement (for more details, see Section 2.6).

A source of extraneous counts may be both the background counts and spurious pulses. The origin of background counts and their reduction has already been discussed (see Section 2.2.2). The reduction consists in the suppression of responses elicited in the detector by cosmic radiation and the sources of radionuclide radiation in the detector's surroundings and in the choice of high-purity radionuclide materials for all constructional materials.

False impulses are elicited by electric disturbances. Even a single such impulse can affect the results of the measurement of low-level radioactivity significantly; therefore, the basic requirement for measuring low-level radioactivities is perfect suppression of such events.

A simple testing of this suppressing in the measuring device may be carried out by connecting to the corresponding input of the starting member of the recording electronics (all inputs of which are perfectly shielded), for example, to a charge-sensitive preamplifier, a condenser with a capacity equal to the capacity of the detector at the working voltage, instead of the detector. When the lower discrimination level of the counting channel is set into the working position, no impulse should be recorded.

The reliability tests based on statistical evaluation of the measured values for the estimation of the effect of disturbances are given in Section 2.6.

The sources of disturbances can be divided, in principle, into the following basic groups:

- external electromagnetic field
- power supply net
- measuring electronic circuits, including connection and earth
- detector itself

Let us discuss these individual sources in greater detail, so that we can judge the possibilities for eliminating their effects.

Effect of the *external electromagnetic field* can be restricted mainly by limitation of sparking switches, controlled either manually (e.g., by a power supply switch) or automatically (e.g., in refrigerators). It is desirable that the number of connectors in laboratories for measurement of low activities should be minimal and that certain operational procedures would be excluded completely (e.g., testing the degree of evacuation of a glass apparatus by discharge). If the laboratory for the measurement of low activity is located in a building within a complex of further laboratories, similar measures should be carried out in the neighbouring laboratories as well. It is advisable to exchange fluorescent lamp illumination for bulb illumination in a laboratory intended for measurements of low activities.

Storms also have to be considered as sources, because in some cases they can manifest themselves by disturbances several hours before their arrival to the locality in question. The proximity of city ground transport is also very undesirable, for example, in streets adjoining the site of the measurement.

It should be stressed that the effect of the sources of an external electromagnetic field depends primarily on the level of the shielding of the input of the recording electronics (Section 2.4), especially at a high input sensitivity (e.g., in the case of semiconductor detectors, which are usually working without internal amplification). Some devices and frequently the entire laboratory designed for the measurement of low activity are built into a Faraday cage.

Disturbances in the *power supply net* can be limited only to a certain extent by stabilization and filtration. Laboratories designed for the measurement of low activity should have their own power source, or at least a distributor of the net voltage via its own transformer, in order to exclude the effect of the connection of electrical appliances with a large power input, which are connected to the same branch of the mains. Moreover, one's own feeding source can prevent better than the normal power net discontinuities connected with serious disturbances taking place in some parts of the measuring arrangements (e.g., photomultipliers of scintillation spectrometers should be fed continuously).

The *measuring arrangement* requires an installation that prevents crossing between the signal conductors, reflections in the transfer cables, loops permitting induction due to changing magnetic fields, and the eliciting of noise by mechanical, acoustic, or thermal effects during mounting or operation. The use of a voltage supply with a high stability and low ripple effect is important. The earth plays an important role. Its resistance should be lower than 0.5 Ω at a minimum number of earth and degenerative loops. No further devices should be grounded to the earth of the arrangement for measuring low activities. The requirements on partial electronic circuits of the measuring arrangement were mentioned in Section 2.4. if the source of disturbances is located directly in these circuits (noise, instability, etc.), it is useful, when looking for it, to proceed from the least sensitive terminal part toward the detector.

Disturbances in *low activity detectors* themselves may be due to improper construction, incorrect production technology, bad assembly, or inadequate operation conditions. The chances that these disturbances may occur in individual types of detectors (insulation, opaqueness in scintillation detectors, electronegative impurities in gas-filled detectors, etc.) are evident from their description (Section 2.3).

In addition to the measures following from the analysis of these four groups of possible disturbances, various other methods are used to decrease their effect. These methods are used mainly in cases when all the primary required measures cannot be realized. Thus, for example, in times when it is impossible to obtain one's own power supply in the radiochronological laboratory (Šilar and Tykva, 1991), taking measurements during the night hours proved to be a practical alternative, taking care that the reading of the measured counts was carried out at shorter time intervals. The effect of nighttime disturbances, when compared with those occurring during the day measurements in a single time interval (agreeing with the sum of partial night intervals) could thus be decreased distinctly.

Electric disturbances should always be taken into consideration when measuring low activities, and their elimination should be carried out carefully, according to available techniques.

2.6 COUNTING STATISTICS AND ERRORS

2.6.1 Statistics of Radioactive Decay

Radioactive decay is a random, stochastic process. Radioactive atoms (or rather nuclei) undergo disintegrations occurring with a certain probability depending on the decay constant of the radionuclide under consideration. The number of radioactive atoms N_0, initially present at time $t = 0$, due to decay will be reduced to $N(t)$ at time t, following the known equation

$$N(t) = N_0 e^{-\lambda t} \tag{2.41}$$

where $N(t)$ and N_0 are considered to be the *expectation values* (or true values) of the relevant parameters. This means that on average we may expect that the number of radioactive atoms will decrease with time exponentially. In any actual individual observation, however, the observed number of atoms may differ from the corresponding expectation value. The observed value will in fact fluctuate around the expectation value.

Since $N(t)$ represents the number of radioactive atoms at time t, obviously the number of atoms which have decayed in the time interval $(0,t)$ will be equal to $N_0 - N(t)$. Taking into account Equation (2.41), the probability that a radioactive atom will not decay within time t can be expressed as the ratio

$$\frac{N(t)}{N_0} = e^{-\lambda t} \tag{2.42}$$

and similarly, the probability p that a single atom will decay in time t is given by

$$p = \frac{N_0 - N(t)}{N_0} = 1 - e^{-\lambda t} \tag{2.43}$$

As long as the half-life of a radionuclide is sufficiently long (much longer than the observation interval t), then the number of decayed atoms $M = N_0 - N(t)$ will be very small in comparison to a large number N_0. Under these conditions, the probability $P(M)$ of observing exactly M atoms decaying in time t can be expressed by the *binomial distribution*

$$P(M) = \frac{N_0!}{M!(N_0 - M)!} p^n (1 - p)^{N_0 - M} \tag{2.44}$$

Its derivation can be found in many books dealing with radioactivity and counting statistics, for example, Evans (1955) or more recently, NCRP (1985).

Since the number N_0 is usually very large (about 10^{15} or more), the binomial distribution is computationally cumbersome and in nuclear counting applications it is more practical to use either the *Poisson distribution* or the *normal distribution* which can approximate the binomial distribution.

The *Poisson distribution* can be written in the following form

$$P(M) = \frac{(pN_0)^M}{M!} e^{-pN_0} \tag{2.45}$$

where $P(M)$ is the probability of observing (measuring) M decays in the time interval t during which the expected number of decays is equal to the expectation value pN_0.

The Poisson distribution may be presented in a more appropriate form

$$P(M) = \frac{(nt)^M}{M!} e^{-nt} \tag{2.46}$$

where n is the expectation value (in practice, it is approximated by a mean value) of decays in a unit of time.

The *normal* or, as it is sometimes called, *Gaussian distribution* is often used to approximate the Poisson distribution under the condition that $p \ll 1$. This distribution is given as

$$P(x) = \frac{1}{\sqrt{2\pi}\,\sigma} \exp\left(\frac{-(x - \bar{x})^2}{2\sigma^2}\right) \tag{2.47}$$

where x is a continuous variable (as distinct from the binomial or Poisson distributions defined only for integer values), $\bar{x}$ is its true (mean) value, and σ is the standard deviation which is related to the Poisson distribution parameters by the expression

$$\sigma^2 = pN_0 = nt \tag{2.48}$$

The normal distribution is commonly used to interpret nuclear counting experiments and it is useful to summarize here some of the important properties of this distribution. The normal distribution is plotted in Figure 2.49, which illustrates its various features.

The normal distribution is symmetric around $\bar{x}$, is defined uniquely by

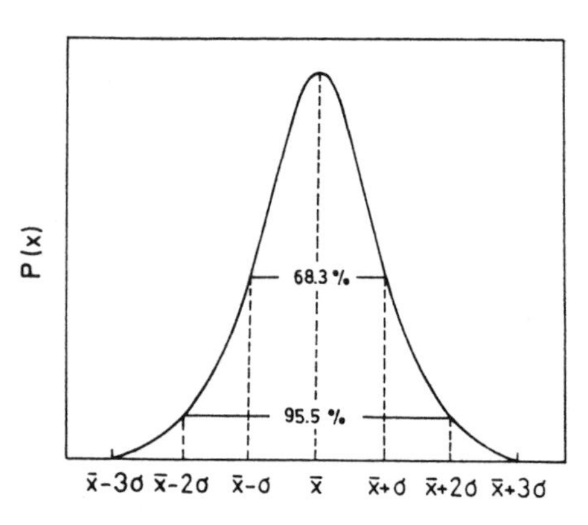

FIGURE 2.49. Normal distribution and some of its characteristic parameters.

the two parameters σ and $\bar{x}$, and extends x from $-\infty$ to $+\infty$. The probability of obtaining the value x from the interval (x_1, x_2) is equal to

$$P(x_1, x_2) = \int_{x_1}^{x_2} P(x)\mathrm{d}x \quad (2.49)$$

The infinite integral of the distribution satisfies the condition

$$P(x) = \int_{-\infty}^{+\infty} P(x)\mathrm{d}x = 1 \quad (2.50)$$

The mean value $\bar{x}$, the variance $V(x)$, the standard deviation σ, and the full width at half maximum (FWHM) are given by

$$\bar{x} = \int_{-\infty}^{+\infty} xP(x)\mathrm{d}x \quad (2.51)$$

$$V(x) = \int_{-\infty}^{+\infty} (x - \bar{x})^2 P(x)\mathrm{d}x = \sigma^2 \quad (2.52)$$

$$\sigma = \sqrt{V(x)} \quad (2.53)$$

$$\mathrm{FWHM} = (2\sqrt{2 \ln 2})\,\sigma = 2.35\sigma \quad (2.54)$$

2.6.2 Distribution of Detector Pulses

Pulses at the output of a detector, or more accurately, pulses produced by the discriminator follow essentially the same distribution as radioactive decay. Therefore, as long as the dead time of the detector and electronics can be neglected, pulses registered by a counter obey the Poisson law, which for this purpose can be written in the form

$$P(M) = \frac{N^M}{M!}e^{-N} = \frac{(n \cdot t)^M}{M!}e^{-n \cdot t} \quad (2.55)$$

where $P(N)$ is the probability of observing (measuring) M pulses during time t, N is the expected (true) number of pulses in the same interval, and

n is the expected pulse rate. Both N and n are in practice replaced by their corresponding mean values $\bar{N}$ and $\bar{n}$, i.e.,

$$\bar{N} = \frac{\sum_{i}^{K} N_i}{K} \tag{2.56}$$

$$\bar{n} = \frac{\sum_{i}^{K} n_i}{K} = \frac{\bar{N}}{t} \tag{2.57}$$

and

$$N = \lim_{K \to \infty} (\bar{N}) \tag{2.58}$$

$$n = \lim_{K \to \infty} (\bar{n}) \tag{2.59}$$

In some counting applications it is important to know the distribution of time intervals between successive pulses. The relevant distribution can be derived using the probability that the next pulse will occur in the infinitesimal time interval $\mathrm{d}t$ following the time t which elapsed since the appearance of the previous pulse. Obviously, this probability, $P(t)\mathrm{d}t$, will be equal to the product of the probability $P(0)$ that no pulse appears in the interval $(0, t)$, and the probability $P_1(\mathrm{d}t)$ that a pulse occurs during $\mathrm{d}t$. These probabilities are given by

$$P(0) = \frac{(n \cdot t)^0}{0!} e^{-n \cdot t} = e^{-n \cdot t} \tag{2.60}$$

$$P_1(\mathrm{d}t) = n \cdot \mathrm{d}t \tag{2.61}$$

which after the substitution results in the equation representing the distribution function for time intervals between adjacent detector pulses

$$P(t) = n e^{-n \cdot t} \tag{2.62}$$

The time interval distribution has a simple exponential shape with the most probable interval being zero.

The mean or average time interval can be calculated using a standard formula

$$\bar{t} = \frac{\int tP(t)\mathrm{d}t}{\int P(t)\mathrm{d}t} = \frac{\int tne^{-n \cdot t}\mathrm{d}t}{\int ne^{-n \cdot t}\mathrm{d}t} = \frac{1}{n} \tag{2.63}$$

which is what can intuitively be expected.

When the dead time τ of a counting system is taken into consideration the time interval probability density is be given by

$$P_\tau(t) = \begin{cases} 0 & \text{for } t < \tau \\ ne^{-n(t-\tau)} & \text{for } t \geq \tau \end{cases} \tag{2.64}$$

and the mean time interval in this case is given by

$$\bar{t} = \tau + \frac{1}{n} \tag{2.65}$$

The situation is illustrated in Figure 2.50, which shows that the time interval distribution in the presence of a certain dead time is a shifted (just by τ) modification of the distribution of a system having zero dead time.

In low-level counting, where measuring time is usually very long and each sample is measured only once, the simultaneous evaluation of the time interval distribution may give us further information which can be of use in eliminating false pulses.

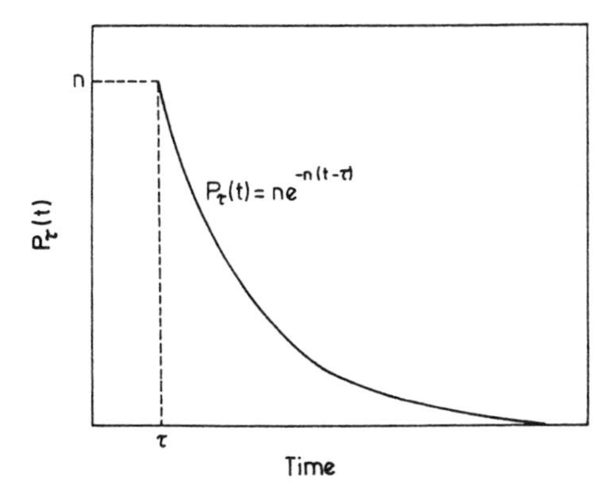

FIGURE 2.50. Time interval distribution of pulses from a counting system with a dead time.

2.6.3 Instrumental and Statistical Errors

Errors due to serious malfunctioning of the detectors and associated electronics will usually be noticed. They affect the results of the measurements in such a way that we can detect them relatively easily and take necessary steps to eliminate the source of the difficulties. However, it is generally more difficult to detect *small* deviations and fluctuations and to identify the possible reasons or problems behind them. In present-day instrumentation this task is usually performed by computers or microprocessor-based blocks, which can continuously check, control, and introduce a necessary compensation to correct any malfunction of the relevant units in the measuring assembly. Computerized quality control and assurance contribute enormously to increasing the reliability and accuracy of nuclear instrumentation.

There are, however, some uncertainties and fluctuations which are caused by the statistical nature of radioactive decay. *Statistical errors* differ principally from *instrumental errors;* the first can never be fully eliminated, but they can be predicted, evaluated, and taken into account in the presentation and interpretation of final results.

The most important parameter used for the assessment of statistical fluctuations is the *standard deviation,* σ, which is calculated from the following equation:

$$\sigma = \sqrt{\frac{\sum_{i}^{K} (N_i - N)^2}{K - 1}} \tag{2.66}$$

where i refers to a given observation, N_i is the value of a random variable referring to the ith observation, $\bar{N}$ is the mean value, and K is the number of observations (measurements). The meaning of the standard deviation can be interpreted in terms of the probability that a single value N_i will lie inside the range $(N - k\sigma, N - k\sigma)$ where N is the true value and k is a coefficient defining the so-called *confidence limits* or intervals. The probability or *confidence level* that a value N_i will fall within these limits is often expressed as percentage. Some limits are specified by special names of relevant errors. Values of a confidence level as a function of the confidence interval together with error names commonly used are presented in Table 2.18.

The interpretation of the confidence levels related to a chosen confidence interval is self-explanatory: for example, the *probable error*—interval

TABLE 2.18. Confidence Intervals (Limits) and Corresponding Confidence Levels Including Commonly Used Error Terminology. The Confidence Level Represents the Probability That Values Will Lie within Confidence Interval Limits (k Is the Number of Standard Deviations).

Confidence Interval $\pm k\sigma$	Name of Error	Confidence Level (%)
$\pm 0.675\ \sigma$	Probable error	50
$\pm 0.841\ \sigma$	–	60
$\pm 1.000\ \sigma$	Standard deviation	68.3
$\pm 1.038\ \sigma$	–	70
$\pm 1.150\ \sigma$	–	75
$\pm 1.281\ \sigma$	–	80
$\pm 1.645\ \sigma$	9/10 error	90
$\pm 1.960\ \sigma$	95% error	95
$\pm 2.000\ \sigma$	(95% error)	95.4
$\pm 2.240\ \sigma$	–	97.5
$\pm 2.575\ \sigma$	99% error	99
$\pm 3.000\ \sigma$	–	99.73
$\pm 3.291\ \sigma$	–	99.90
$\pm 4.000\ \sigma$	–	99.99

$(\bar{x} \pm 0.675\sigma)$ – is equally likely to be exceeded or not, since the confidence level is just 50%. This means that 50% of the observations will fall within the limits $(\bar{x} \pm 0.675\sigma)$. Similarly, the *standard deviation* represents the 68.3% probability of finding a result x (including x) within the interval from $(\bar{x} - \sigma)$ to $(\bar{x} + \sigma)$ where x is the true mean value.

The confidence level (CL) for any confidence interval $(\bar{x} \pm k\sigma)$ can be calculated as

$$CL = P(\bar{x} - k\sigma, \bar{x} + k\sigma) = \int_{\bar{x}-k\sigma}^{\bar{x}+k\sigma} P(x)\mathrm{d}x \qquad (2.67)$$

where $P(x)$ is the distribution given by Equation (2.45).

It is important to realize that the standard deviation σ represents an error of a single measurement. Such standard deviation – it may be better to use the symbol $\sigma(N_i)$ for it – can be calculated on the basis of experimental values N_i using Equation (2.66). In that measurement we obtain K values of N_i from which the mean value is determined. Now suppose that we were to repeat this measurement L times, getting L values of $\bar{N}_j$ $(j = 2,3,4,\ldots,L)$, and we are interested in finding the standard deviation

$\sigma(\bar{N}_j)$ of these mean values $\bar{N}_j$. Generally, for this standard deviation the following expression can be derived

$$\sigma(\bar{N}_j) = \sqrt{\frac{\sum_i^K (N_i - N)^2}{L(K - 1)}} \tag{2.68}$$

In practice, it is often necessary to use and combine results of various measurements which have different statistical errors. In this case the final mean value should be calculated by weighting the results of individual measurements with respect to their standard deviations. The weighted mean value can be found using the following equation (Bevington, 1969):

$$\bar{N} = \frac{\sum_i^K \frac{N_i}{\sigma_i^2}}{\sum_{i-1}^K \frac{1}{\sigma_i^2}} \tag{2.69}$$

Many quantities and parameters involving radiation or radiation sources rely on the counting of pulses. Often only one such measurement is performed, giving the result in the form of a single value, say M_1. In this case we have no other option but to consider the value obtained as the best estimate of the true mean value. Consequently, its standard deviation will be $\sqrt{M_1}$, and the result will be given as $M_1 \pm \sqrt{M_1}$.

Results of the measurement are sometimes reported in terms of the *relative standard error,* δ, which is often given in percentage. Assuming a single value M_1 in a counting experiment, the relative error can be expressed as

$$\delta = \frac{\sigma(M_1)}{M_1} = \frac{\sqrt{M_1}}{M_1} = \frac{1}{\sqrt{M_1}} \tag{2.70}$$

Using the above equation one can assess how many counts have to be accumulated in order to keep the relative standard error below the selected level. The number of counts required for a given value of the relative error are listed in Table 2.19.

TABLE 2.19. The Minimum Number of Counts Needed to Keep the Relative Standard Error at the Stated Level.

Number of Counts M_1	Corresponding Relative Error δ $\left(\delta = \frac{100\%}{\sqrt{M_1}}\right)$
100	10
1,000	3.16
10,000	1.0
100,000	0.316
1,000,000	0.1

2.6.4 Propagation of Errors

In most experiments the measured quantity or parameter is a function of more than one random variable. For example, a low-level counting may involve two or more mathematical operations such as addition, subtraction, multiplication, or division of measured mean values and their standard deviations. In such situations, it is desirable to express the resulting error of the complex quantity be means of the errors of the individual random variables.

Let us consider a general case where the quantity to be determined is a function of n independent variables $x_1, x_2, x_3, \ldots, x_n$ whose both mean values x_i and standard deviations $\sigma(x_i)$ are known from the measurements. Under these conditions the quantity represented by a function f can be reported as

$$f = f(x_1, x_2, x_3, \ldots, x_n) \pm \sigma_f \tag{2.71}$$

where the standard deviation σ_f is given by the following *error propagation formula:*

$$\sigma_f^2 = \left(\frac{\delta f}{\delta x_1}\right)^2 \sigma^2(x_1) + \left(\frac{\delta f}{\delta x_2}\right)^2 \sigma^2(x_2) + \left(\frac{\delta f}{\delta x_3}\right)^2 \sigma^2(x_3) + \ldots + \left(\frac{\delta f}{\delta x_n}\right)^2 \sigma^2(x_n) \tag{2.72}$$

We may apply the error propagation formula to some common functions with independent (uncorrelated) variables:

(1) The *sum or difference of variables* multiplied by constants a_1 and a_2:

$$f(x_1, x_2) = a_1\sigma(x_1) \pm a_2\sigma(x_2) \tag{2.73}$$

$$\sigma_f = \sqrt{a_1^2 \cdot \sigma^2(x_1) + a_2^2 \cdot \sigma^2(x_2)} \tag{2.74}$$

(2) The *multiplication or division by a constant:*

$$f(x_1) = a_1 x_1 \tag{2.75}$$

$$\sigma_f = a_1 \cdot \sigma(x_1) \tag{2.76}$$

or

$$f(x_1) = \frac{x_1}{a_1} \tag{2.77}$$

$$\sigma_f = \frac{\sigma(x_1)}{a_1} \tag{2.78}$$

(3) The *multiplication or division of variables:*

$$f(x_1, x_1) = x_1 \cdot x_2 \tag{2.79}$$

$$\sigma_f = \sqrt{x_1^2 \cdot \sigma^2(x_1) + x_2^2 \cdot \sigma^2(x_2)} \tag{2.80}$$

or

$$f(x_1, x_1) = \frac{x_1}{x_2} \tag{2.81}$$

$$\sigma_f = \sqrt{\left(\frac{\sigma(x_1)}{x_1}\right)^2 + \left(\frac{\sigma(x_2)}{x_2}\right)^2} \tag{2.82}$$

2.6.5 Optimization of Measurements

Although in most counting experiments either the number of counts in the preselected time interval is measured or the time interval needed to accumulate a preselected number of counts is determined by a timer; the results of these measurements are usually converted into *counting (pulse) rates.*

In activity measurements one of the simplest procedures consists of counting the total number of pulses M_{sb} (source + background) during the time interval T_{sb} with the radioactive source (sample) present, and then counting the background pulse M_b during the interval T_b when there is no source involved. The corresponding counting rates are as follows

$$m_{sb} = \frac{M_{sb}}{T_{sb}} \tag{2.83}$$

$$m_b = \frac{M_b}{T_b} \tag{2.84}$$

where m_{sb} and m_b are the *total (gross) counting rate* and the *background counting rate,* respectively.

For evaluation of the activity of the sample we need the *net counting rate* m_s (related only to the source) which can be obtained from the results of the measurements described as

$$m_s = m_{sb} - m_b = \frac{M_{sb}}{T_{sb}} - \frac{M_b}{T_b} \tag{2.85}$$

Applying the propagation error formula [Equation (2.72)] to the counting rate m_s leads to the following general expression for the standard deviation $\sigma(m_s)$ of the net counting rate

$$\sigma(m_s) = \left(\frac{\delta m_s}{\delta M_{sb}}\right)^2 \sigma^2(M_{sb}) + \left(\frac{\delta m_s}{\delta T_{sb}}\right)^2 \sigma^2(T_{sb}) + \left(\frac{\delta m_s}{\delta M_b}\right)^2 \sigma^2(M_b) + \left(\frac{\delta m_s}{\delta T_b}\right)^2 \sigma^2(T_b) \tag{2.86}$$

Since advanced electronic timers can measure the time intervals very accurately so that the standard deviations $\sigma(T_{sb})$ and $\sigma(T_b)$ are much smaller than the errors in the counting M_t and M_b the above expression may be reduced to

$$\sigma(m_s) = \sqrt{\left(\frac{\delta m_s}{\delta M_{sb}}\right)^2 \sigma^2(M_{sb}) + \left(\frac{\delta m_s}{\delta M_b}\right)^2 \sigma^2(M_b)} \tag{2.87}$$

Taking into account that

$$\sigma(M_{sb}) = \sqrt{M_{sb}} \text{ and } \sigma(M_b) = \sqrt{M_b} \tag{2.88}$$

we obtain finally this often used formula in a common form

$$\sigma(m_s) = \sqrt{\frac{M_{sb}}{T_{sb}^2} + \frac{M_b}{T_b^2}} \tag{2.89}$$

In all counting measurements it is very important to evaluate the resulting net counting rate with *minimum possible error.* Although we always try to keep the background counting rate to the lowest levels, it cannot be completely eliminated and its presence should be taken into consideration. As has already been mentioned, the activity of a sample is usually measured by using a detection system with known efficiency, first with the source in position and then with the source removed. In order to reproduce the scattering of the background radiation in the detector by the measuring sample, the sample is usually replaced during the background measurement by an inactive dummy sample positioned in place of the active sample (Mann et al., 1991). Bearing in mind Equation (2.89), the effect of the background radiation is obvious: it will tend to broaden the distribution of the net counting rate m_s.

In the measurements of sources having low levels of activity the total measuring time, i.e., $T_t = T_{sb} + T_b$, is usually limited, and it is necessary to decide how the time available should best be subdivided in order to minimize the standard deviation of the net counting rate $\sigma(m_s)$.

Considering $\sigma(m_s)$ as a function of T_{sb} and T_b, we square Equation (2.89) and find the minimum by differentiation

$$2\sigma(m_s) \cdot \mathrm{d}\sigma(m_s) = -2\frac{M_{sb}}{(T_{sb})^3}\mathrm{d}T_{sb} - 2\frac{M_b}{(T_b)^3}\mathrm{d}T_b \tag{2.90}$$

by setting the differential $\mathrm{d}\sigma(m_n)$ to zero and taking into account that the time interval $T_t = T_{sb} + T_b$ is constant, i.e.,

$$\mathrm{d}T_t = 0 = \mathrm{d}T_{sb} + \mathrm{d}T_b \tag{2.91}$$

the result is then obtained in the form

$$\frac{T_{sb}}{T_b} = \sqrt{\frac{m_s + m_b}{m_b}} = \sqrt{\frac{m_{sb}}{m_b}} \tag{2.92}$$

Since the decision is based on the ratio m_{sb}/m_b, we have to make an estimate of these counting rates in a short preliminary measurement.

In order to compare the effects of various arrangements of the experiment, including the type of detector, source-detector configuration, and settings of electronic blocks, it is useful to introduce the concept of the *figure of merit* which will be equal to the inverse of the total time $1/T$. The figure of merit, required to determine the net counting rate m_s within a given relative standard deviation $\delta(m_s)$, should be made as large as possible. After some rearrangements the figure of merit can be expressed as

$$\frac{1}{T} = \frac{m_s}{(\sqrt{m_{sb}} + \sqrt{m_b})^2}\delta^2 \qquad (2.93)$$

where

$$\delta = \delta(m_s) = \frac{\sigma(m_s)}{m_s} \qquad (2.94)$$

Regarding the figure of merit, two extreme cases may be of special interest in practical applications (Mann et al., 1991):

(1) The background counting rate is much lower than the net counting rate ($m_s \ll m_s$). In this case the figure of merit can be approximated as

$$\frac{1}{T} = m_s \cdot \delta^2 \qquad (2.95)$$

which implies that the counting rate m_s must be as large as possible. In other words, since the rate m_s depends primarily on the total detection efficiency, this parameter should be as high as is achievable under the given circumstances.

(2) The background rate is high compared with the net counting rate. Applying this condition to Equation (2.93) one can get the relationship

$$\frac{1}{T} = \frac{m_s^2}{4m_b}\delta^2 \qquad (2.96)$$

In this situation the highest value of the figure of merit is obtained when the ratio $m_s^2/4m_b$ is maximized.

The performance of instruments for the activity measurement is sometimes evaluated using the concept of the *minimum detectable activity* which may arbitrarily be defined in many different ways depending on the choice of the acceptable maximum error of the measured result. For many applications a simple approach is sufficient. For example, the minimum detectable activity can be defined as that activity which gives rise to a counting rate of three times the background standard deviation (ICRU, 1972; Mann et al., 1991). In the case of low-level counting and some other related fields more elaborate and refined definitions of the minimum detectable activity were introduced (Watt and Ramsden, 1964; Currie, 1968; Sumerling, 1983; Curie, 1984; Cline, 1990).

Although there is no general rule as to the selection of the maximum tolerated or acceptable standard deviation $\sigma(m_s)$, an upper limit is set from the requirement that $\sigma(m_s) \leq m_s$ (Tsoulfanidis, 1983). Usually, only measurements satisfying the following condition will be accepted

$$\sigma(m_s) \leq f \cdot m_s \text{ where } f < 1 \tag{2.97}$$

This condition can be rewritten in the form

$$\sqrt{\frac{M_{sb}}{T_{sb}^2} + \frac{M_b}{T_b^2}} \leq f \cdot m_s \tag{2.98}$$

and from the substitution and rearrangement we obtain the resulting formula for the net counting rate m_s as

$$m_s \geq \frac{1 + \sqrt{1 + 4f^2 m_b T + 4f^2 T^2 \sigma_b}}{2f^2 T} \tag{2.99}$$

where T is the counting time with the sample present and σ_b is the standard deviation of the background counting rate, i.e., $\sigma(m_b)$. The above equation indicates that the minimum detectable activity is determined by the following three factors:

(1) The factor f, which is selected by the investigator
(2) The background counting rate m_b and its standard deviation $\sigma(m_b)$—parameters essentially depending on the measuring system including its shielding
(3) The counting time T, which is limited by the long-term stability of the measuring system and also, in the case of short-lived radioactive sources, by the half-life of the radionuclide involved

2.6.6 Quality of Fit and Statistical Tests

So far we have only considered fluctuations and uncertainties in the experimental counting data originating in the statistical nature of radioactive disintegrations. In real measurements there are, however, some other factors and causes that may also affect the distribution of counts obtained in the evaluation of radioactive sources. Since the number of repeated measurements in low-level counting is usually very limited, it is not easy to distinguish between statistical fluctuations and extraneous disturbances responsible for false, often abnormal counts which may introduce a considerable error into the final results.

Although we may be able in some cases to avoid and reject suspected erroneous data by simply using common sense, it is often necessary to check the reliability and soundness of experimental results by statistical methods. The main aim of such tests is to find whether the results of measurements are compatible with the valid statistical models. Many different test procedures where a characteristic *test parameter* based on experimental data is checked and compared with the theoretically predicted value have been developed.

From numerous statistical tests used to verify experimental data the following three methods are often applied for the treatment of results related to radioactivity measurements: *Student's t-test, F-test,* and *chi-squared test,* each based on a theoretical frequency distribution (Student's t distribution, the F distribution, and the chi-square distribution). Detailed descriptions of these tests can be found in standard references and many other publications (Evans, 1955; Parratt, 1961; Dixon and Massey, 1969; NBS, 1969; Weise, 1971; CRC, 1985).

Perhaps the most commonly used statistical test to check the quality of data from counting experiments is the *chi-squared* (χ^2) *test,* also called the Pearson test, based on mean-square deviations between the measured and predicted values in different subdivided groups of the experimental results.

In this test the value of

$$\chi^2 = \frac{\sum_{i}^{K} (\bar{N} - N_i)^2}{\bar{N}} \tag{2.100}$$

is compared to the theoretical distribution of this quantity. The value χ^2 is calculated from K measurements of N_i and then it is compared with the probability values given in the relevant tables used with the χ^2 test. Table 2.20 may serve as an example of such criteria. The range of acceptable χ^2 values is selected around the value 0.5 (the best fit), usually in the interval from 0.1 to 0.9. If the value χ^2 corresponds to this probability range the assumed distribution has a high probability of fitting the experimental data. The cases with lower or higher probabilities of χ^2 lying outside the region

TABLE 2.20. Probability Values for the χ^2 Test.

Degree of Freedom (K–1)	Probability						
	0.99	0.95	0.90	0.50	0.10	0.05	0.01
2	0.020	0.103	0.211	1.386	4.605	5.991	9.210
3	0.115	0.352	0.584	2.366	6.251	7.815	11.345
4	0.297	0.711	1.064	3.357	7.779	9.488	13.277
5	0.554	1.145	1.610	4.351	9.236	11.070	15.086
6	0.872	1.635	2.204	5.348	10.645	12.592	16.812
7	1.239	2.167	2.833	6.346	12.017	14.067	18.475
8	1.646	2.733	3.490	7.344	13.362	15.507	20.090
9	2.088	3.325	4.168	8.343	14.684	16.919	21.666
10	2.558	3.940	4.865	9.342	15.987	18.307	23.209
11	3.053	4.575	5.578	10.341	17.275	19.675	24.725
12	3.571	5.226	6.304	11.340	18.549	21.026	26.217
13	4.107	5.892	7.042	12.340	19.812	22.363	27.688
14	4.660	6.571	7.790	13.339	21.064	23.685	29.141
15	5.229	7.261	8.547	14.339	22.307	24.996	30.578
16	5.812	7.962	9.312	15.338	23.542	26.296	32.000
17	6.408	8.672	10.085	16.338	24.769	27.587	33.409
18	7.015	9.390	10.865	17.338	25.989	28.869	34.805
19	7.633	10.117	11.651	18.338	27.204	30.144	36.191
20	8.260	10.851	12.443	19.337	28.412	31.410	37.566
21	8.897	11.591	13.240	20.337	29.615	32.671	38.932
22	9.542	12.338	14.041	21.337	30.813	33.924	40.289
23	10.196	13.091	14.848	22.337	32.007	35.172	41.638
24	10.856	13.848	15.659	23.337	33.196	36.415	42.980
25	11.534	14.611	16.473	24.337	34.382	37.382	44.314
26	12.198	15.379	17.292	25.336	35.653	38.885	45.642

mentioned should be considered with suspicion. There may be a number of reasons why the data failed the test within the preselected range. Some possible reasons can be summarized as follows (Tsoulfanidis, 1983):

(1) Unstable measuring instruments are liable to give inconsistent results due to spurious counts or other changes of their parameters during the measurements.

(2) Interference from the ambient environment may result in additional pulses.

(3) Lack of uniformity in sample preparation and treatment may contribute to widely scattered results.

(4) Some results (the so-called *outliers*) may fall far away from the average. Since the data follow the normal statistical distribution, which extends from $-\infty$ to $+\infty$, in principle, all results are possible. In

practice, however, this is always suspicious and warrants attention. A similar situation may arise when the experimental data show results that are statistically too good (e.g., counts are more or less the same).

The outliers are sometimes rejected from the processed experimental data using different criteria. Some experimenters use an arbitrary criterion and reject any result differing from the mean value by more than three or four standard deviations (Mann et al., 1980). There are many other criteria for data rejection such as, for example, *Chauvenet's criterion* suggested by W. Chauvenet more than ninety years ago. His criterion takes into account also the number of observations K and postulates that a result may be rejected if it has a standard deviation from the mean greater than that corresponding to the $1-1/(2K)$ error.

2.7 STANDARDIZATION AND CALIBRATION IN LOW-LEVEL RADIOACTIVITY MEASUREMENT

2.7.1 International Character of Activity Measurements

As in other fields of science, technology, medicine, and commerce, in areas relating to the measurement and monitoring of radioactivity there should be internationally agreed upon and maintained standards as well as measuring procedures, assuring worldwide conformity of radioactivity measurements. This has long been recognized by the International Bureau of Weights and Measures, where already in 1911 the first international standard of radioactivity was prepared (Rytz, 1978). In order to improve the accuracy and uniformity of measurements throughout the world, international comparisons of radionuclides and other radiation sources are periodically organized. In these intercomparisons many national laboratories and institutions take part including, more recently, participants from developing countries. In addition to such national laboratories as NIST (USA), PTB (Germany), or NPL (UK), a very active role in radioactivity standardization is played by international organizations, agencies, and commissions, especially the International Atomic Energy Agency, the World Health Organization, the International Commission on Radiation Units and Measurements, and the International Committee on Radionuclide Metrology.

Although in the beginning efforts in radionuclide metrology were focused only on a few individual radionuclides without attempting to cover all ranges of activity, during the last decade more intensive work has concentrated on standardization and calibration addressing the special needs and

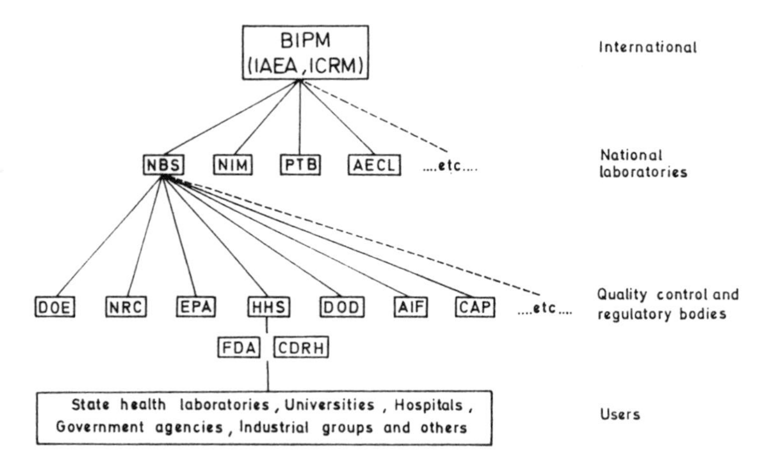

FIGURE 2.51. Hierarchy of radioactivity-standardization laboratories and organizations. From Mann et al. (1991).

requirements of low-level radioactivities. The main tasks in this field are directed toward *radioactivity standards* (reference and calibration sources), *traceability*, and *quality assurance*.

There are many different types of radioactivity standards; some of them represent individual radionuclides, usually of the highest purity and known impurities, or more complex sources such as natural-matrix standard reference materials used for the calibration of equipment measuring environmental radioactivity. Ideally, these standards should be *traceable* to the relevant international standards within the hierarchical *international reference system,* whose diagram is shown in Figure 2.51 (Mann et al., 1991). Traceability in radioactivity must be based primarily on periodic demonstrations of the reliability of radiochemical procedures, the correct calibration of the measuring instruments, and the performance of the laboratory with test samples. To achieve these goals, NIST, for example, prepares a number of classes of (a) test samples to check the calibrations of instruments, (b) test sources with chemical and radioactivity interferences to check the adequacy of radiochemical procedures and the competence of the technicians, and (c) natural-matrix materials to test sample-handling and dissolution techniques (Inn et al., 1984).

2.7.2 Sources for Energy Calibrations

Probably the first calibration of any detection system for the measurement of activity consists in the conversion of pulse-heights into the energy scale. In the majority of cases a response is a linear function of the energy deposited in the detector and, in principle, two sufficiently different energy points are usually enough to draw a calibration line. For measuring systems with non-linear detector responses and also for the more precise calibration of systems with a linear or rather a quasi-linear energy-response relationship, more *energy calibration sources* are required. The preparation of these sources should be done in such a way as to preserve the energy of emitted particles as much as possible. Consequently, *gamma sources* tend to have a "point" form to avoid attenuation and scattering within the source, while *alpha* and *beta* sources are usually prepared as disks on which a very thin layer of active material is deposited.

Recommended sources, including energy values, for the calibration of photon detectors can be found in various references. A good information source for such data is NCRP Report 58 (NCRP, 1985), which contains a comprehensive table of gamma rays as reference standards for energy calibration. Details of some *gamma reference sources* are given in Table 2.21 (Liden and Holm, 1985).

TABLE 2.21. Radionuclides Suitable for the Preparation of Reference Gamma Point Sources Used for the Energy Calibration of Photon Detectors.

Radionuclide	Half-Life	*E* (keV)	Photons per Decay
^{241}Am	433 ± 2 y	26.35	0.0258 ± 0.0022
		59.54	0.363 ± 0.004
^{109}Cd	453 ± 2 d	88.04	0.0373 ± 0.0006
^{57}Co	270.9 ± 0.6 d	122.06	0.8559 ± 0.0019
^{139}Ce	137.65 ± 0.05 d	165.85	0.8006 ± 0.0013
^{203}Hg	46.59 ± 0.05 d	279.19	0.815 ± 0.008
^{51}Cr	27.704 ± 0.002 d	320.08	0.0980 ± 0.0010
^{113}Sn	114.9 ± 0.1 d	391.69	0.6490 ± 0.0020
^{85}Sr	64.85 ± 0.03 d	513.99	0.980 ± 0.010
^{207}Bi	38 ± 0.3 y	569.67	0.978 ± 0.005
		1063.61	0.74 ± 0.03
		1770.22	0.073 ± 0.004
^{137}Cs	30.0 ± 0.2 y	661.65	0.899 ± 0.004
^{94}Nb	(2.3 ± 0.16)10^4 y	702.63	1
		871.10	1
^{54}Mn	312.5 ± 0.5 d	834.83	0.999760 ± 0.000002
^{88}Y	107 ± 1 d	898.83	0.934 ± 0.007
		1836.04	0.9935 ± 0.0003
^{65}Zn	244.1 ± 0.2 d	1115.52	0.5075 ± 0.0010
^{60}Co	5.271 ± 0.001 y	1173.21	0.99900 ± 0.00020
		1332.46	1
^{22}Na	2.602 ± 0.002 y	1274.54	0.99940 ± 0.00020

TABLE 2.22. Calibration Energies for Electrons (C.E.—Conversion Electron).

Source	Energy (keV)	Electrons per Decay (%)	Radiation
^{125}I	3.7	79.3	C.E.-K
^{55}Fe	4.9–5.2	48.5	Auger-K
	5.7–5.9	11.0	
	6.3–6.5	0.9	
^{133}Ba	17.2	10.6	C.E.-K
^{125}I	21.8–23.0	13.1	Auger-K
	25.8–27.4	6.0	
	29.8–31.7	0.8	
	30.5–31.2	10.7	C.E.-L
^{133}Ba	45.0	45.2	C.E.-K
^{131}I	45.6	3.5	C.E.-K
^{199}Au	125.1	5.5	C.E.-K
^{203}Hg	193.6	13.5	C.E.-K
^{131}I	329.9	1.5	C.E.-K
^{207}Bi	481.7	1.6	C.E.-K
^{137}Cs	624.1	8.1	C.E.-K

Based on Mann et al. (1991).

A very useful radionuclide for the energy calibration of gamma spectrometry systems is ^{152}Eu with a half-life of 13.2 years.

The energy calibration of beta detection systems can best be performed using radionuclides emitting *conversion* as well as *Auger electrons* which are essentially monoenergetic. Some data relevant to such calibration standards are summarized in Table 2.22 (Mann et al., 1991).

The energy calibration of detection and spectrometry systems for measuring *alpha-emitting radionuclides* requires specially prepared alpha sources with minimum distortion of their monoenergetic properties. The most often used alpha calibration sources are ^{241}Am, and also ^{210}Po, ^{239}Pu, and ^{244}Cm. In addition to energies, radionuclide impurity, chemical stability, and accompanying radiation should also be considered. Parameters of some alpha sources suitable for calibration are listed in Table 2.23.

2.7.3 Special Radioactivity Standards

2.7.3.1 ENVIRONMENTAL NATURAL-MATRIX STANDARDS

Environmental radioactivity measurement for use in analytical intercomparison and as standard reference materials, requires very large homogeneous samples of a variety of matrices, each naturally contaminated by a number of long-lived radionuclides, at several different ranges of concentrations (Bowen, 1978).

A number of standard environmentally quasi-equilibrated natural-matrix materials to test sample preparation and analytical techniques for radionuclides in soil, sediments, animal and human organs have been prepared and produced. These include such matrices as fresh water, sea water, river sediment, lake sediment, ocean sediment, milk powder, soils of various origin, animal bones, human lung, human liver, and other materials.

It is important that natural-matrix materials contain "useful" activities of all the artificial radionuclides having half-lives of about one year and longer, and which are now known or expected to be of environmental concern. They may include especially the following types of radionuclides (Bowen, 1978):

(1) *Activation products,* in particular ^{55}Fe, ^{54}Mn, ^{60}Co, ^{65}Zn, ^{110m}Ag, ^{134}Cs and ^{3}H
(2) *Fission products,* notably ^{90}Sr and ^{137}Cs, but also ^{106}Ru, ^{126}Sb, ^{144}Ce, ^{147}Pm, and ^{155}Eu
(3) *Transuranium nuclides,* especially ^{238}Pu, ^{239,240}Pu and ^{241}Am, but also ^{244}Cm and ^{237}Np

TABLE 2.23. Parameters of Some Alpha Sources Used in Energy Calibration (ΔE_α—Uncertainty).

Source	Half-Life	E_α (MeV)	ΔE_α (keV)	Alpha per Decay (%)
^{232}Th	1.2×10^{10} y	4.012	±5	77
		3.953	±8	23
^{238}U	4.5×10^{9} y	4.196	±5	77
		4.149	±5	23
^{230}Th	7.7×10^{4} y	4.6875	±1.5	76.3
		4.6210	±1.5	23.5
^{239}Pu	2.4×10^{4} y	5.1554	±0.7	73.3
		5.1429	±0.8	15.1
		5.1046	±0.8	11.5
^{210}Po	138 d	5.30451	±0.07	>99
^{241}Am	433 y	5.48574	±0.12	85.2
		5.44298	±0.13	12.8
^{238}Pu	88 y	5.49921	±0.20	71.1
		5.4565	±0.4	28.7
^{244}Cm	18 y	5.80496	±0.05	76.4
		5.762835	±0.03	23.6

It seems that special arrangements will have to continue to be made for such radionuclides as ^{14}C, ^{85}Kr, ^{99}Tc, and ^{129}I, either because their analysis requires exceedingly large sample aliquots, or because their preservation in samples demands extremely special handling and manipulation.

According to Bowen (Bowen, 1978) the following operational classification of materials in terms of the activity concentrations (per g of solid) of their major constituents may be used as natural-matrix standards:

(1) *High-level*—37 to 7400 Bq/kg (materials contaminated by releases from fuel-reprocessing plants or from solid-waste dumps)
(2) *Medium-level*—1.9 to 185 mBq/g (materials contaminated by leakage from nuclear-power reactors)
(3) *Low-level*—3.7 to 370 mBq/kg (materials contaminated by world-wide fallout)
(4) *Blank (very low-level)*—involves materials that have been protected either by their location or history from the introduction of man-made radionuclides (their contamination by artificial radionuclides is below 1.8 mBq/kg)

Radioactive fallout due to the Chernobyl accident will presumably affect all four categories. Since both this accident and nuclear weapons tests have delivered fallout mainly to the northern hemisphere, it appears likely that some southern areas will prove to yield sediment, soil, or biota samples which may be used as blanks for most environmental measuring applications.

A limited variety of specific standard reference materials is available from IAEA, NIST and some other big laboratories. Some of these materials, including the main radionuclides, are listed in Table 2.24 (Debertin and Helmer, 1988; NIST, 1991). These standards present to the users a broad sampling of radiochemistry problems commonly encountered in environmental radioactivity measurements. The standard reference materials were prepared for use in tests of measurements of environmental radioactivity contained in matrices similar to the sample, for evaluating analytical methods, or as a generally available calibrated "real" sample matrix in interlaboratory comparisons. A mixture of these materials may be prepared gravimetrically by the users and presented as an unknown to the analyst to test measurement capability.

The following are brief descriptions of some reference material available from NIST (1991):

- *Peruvian soil*—contains radioactivity concentrations for many fallout radionuclides and can be used as a blank or for sensitive

TABLE 2.24. Some Standard Reference Materials Available from IAEA and NIST.

Material	Radionuclides	Producer
Marine sediment	^{60}Co, ^{90}Sr, ^{137}Cs, ^{238}Pu, $^{239,240}Pu$, ^{241}Am	IAEA
Marine algae	^{40}K, ^{54}Mn, ^{60}Co, ^{90}Sr, ^{99}Tc, ^{137}Cs, ^{226}Ra, $^{239,240}Pu$	IAEA
Homogenized fish	^{90}Sr, ^{134}Cs, ^{137}Cs transuranics	IAEA
Calcined animal bone	^{90}Sr, ^{266}Ra	IAEA
Milk powder	^{90}Sr, ^{137}Cs	IAEA
Peruvian soil	^{137}Cs, ^{230}Th, ^{232}Th, $^{239,240}Pu$, ^{241}Am	NIST
River sediment	^{60}Co, ^{137}Cs, ^{252}Eu, ^{254}Eu, ^{226}Ra, ^{238}Pu, $^{239,240}Pu$, ^{241}Am	NIST
Rocky Flats soil	^{40}K, ^{90}Sr, ^{137}Cs, ^{226}Ra, ^{228}Ac, ^{230}Th, ^{232}Th, ^{234}U, ^{238}U, ^{238}Pu, $^{239,240}Pu$, ^{241}Am	
Human lung	^{232}Th, ^{234}U, ^{238}U, $^{239,240}Pu$, $^{238}Pu/(^{239,240}Pu)$	
Human liver	^{238}Pu, $^{239,240}Pu$, ^{241}Am	

tests of radioanalytical procedures at *low-level radioactivity* concentrations for other radionuclides.

- *River sediment*—was collected downstream from a nuclear reactor facility. Concentrations of fission and activation products are elevated over typical world-wide levels. Radionuclides $^{239,240}Pu$ and ^{241}Am are very homogeneously distributed through the sample and are in acid-leachable forms. Inhomogeneity is 3% or better for other radionuclides.
- *Rocky Flats soil*—was collected within 13 cm of the soil surface at Rocky Flats, Colorado. The concentrations of $^{239,240}Pu$ and ^{241}Am are about an order of magnitude higher than typical world-wide levels. Approximately 10% of the plutonium is in an acid-resistant form. The material also contains "hot" particles and a statistical method is provided for dealing with these. Inhomogeneities, excluding hot particles, are on the order of 3% or better for other radionuclides.
- *Human lung*—contains activity concentrations on the order of 10^{-4} Bq/g. It has been freeze-dried, cryogenically ground, homogenized, and packed in a glass bottle under vacuum. There is significant inhomogeneity in $^{239,240}Pu$ concentration, which is unavoidable because plutonium was taken into the lungs in particulate form. Assessments of accuracy of measurement technique can be improved by averaging over several samples.
- *Human liver*—contains activity concentrations on the order of 10^{-4} Bq/g. Similar to the human lung material, it has also been freeze-dried, cryogenically ground, homogenized, and packed in a glass bottle under vacuum.

2.7.3.2 *ARTIFICIAL MATRIX MATERIALS*

Some manufacturers and producers of calibration sources offer solid reference sources in various shapes and sizes (Debertin, 1991). These calibration sources are filled with a plastic material containing the selected radionuclides. The density of the plastic varies from 0.7 to 2.5 g/cm^3. The containers used are usually Marinelli beakers and polyethylene or glass bottles. In addition, containers supplied by the customers can be filled according to special requirements. It is assumed that the attenuation properties of the plastic material are similar to those of a sample material of the same density.

It is also possible to use readily available standard solutions of accurately known radionuclide concentrations to prepare one's own calibration sources. In this case the material under study is spiked with known amounts of the calibrated solution. Using such a procedure one can produce a calibration source in the same matrix as the actual sample. River sediment reference sources were prepared by this method and described by Mundschenk (1990).

2.7.3.3 OTHER TYPES OF REFERENCE SOURCES AND MATERIALS

Some manufacturers supply custom-made radionuclide calibration standards where there is an increasing demand, especially for mixed gamma sources. These sources are usually traceable to recognized national laboratories such as NIST and they are tailored to meet the users' individual needs (custom geometries, activities and matrices, etc.).

Mixed gamma standards for efficiency calibrations of germanium detectors containing the mixture of eight to ten radionuclides covering a range from 59.5 keV to 1836 keV are readily available in different forms (Analytics, 1992). The matrices include liquids, point sources, single air filters, composite air filters, charcoal cartridges, silver zeolite cartridges, water equivalent solids, continuous filter paper, simulated gases, as well as sand, dirt, or vegetation in closed containers.

Other special sources may include:

- *Simulated gas standards*—are used for the calibration of gamma-ray detectors for radioactive gas counting. These sources are usually prepared directly in standard counting containers which require no transfers and are leakproof.
- *Mixed gas standards*—allow each user to calibrate virtually any gas counting container using a true gas mixture of such radionuclides as ^{85}Kr, ^{133}Xe, and ^{127}Xe. In addition, unpressurized as well as pressurized single radionuclide gas standards are also available (Analytics, 1992).
- *Radionuclide liquid standards*—are supplied in various activities and volumes as single radionuclide standards and special mixtures.
- *Plate sources*—are available for calibrating surface contamination monitors. These standards are prepared from durable, rigid, and low-scattering plastic backing made of opaque Plexiglass with a 1.7 mg/cm^2 mylar face for beta and a 0.5 mg/cm^2 mylar face for alpha sources.
- *Large area gamma standards*—are designed for the calibration of germanium and sodium iodide detectors. They are especially

suitable for detectors monitoring contaminated surfaces. These standards can be used flat, for ground contamination monitoring, or rolled, to simulate contaminated pipes.

- *Continuous filter paper standards*—are available with one or more radionuclides which may include ^{241}Am, ^{36}Cl, ^{60}Co, ^{137}Cs, ^{147}Pm, ^{89}Sr, ^{99}Tc, and ^{204}Tl.
- *Planchet standards*—are provided on the specified type of planchet, with required active area and activity.
- *Whole body counting standards*—are available in bottles or they may have a special shape to meet the requirements of different whole body monitors.
- *Radon and thoron calibrated sources and standards*—are used for calibration of radon and thoron monitors as well as radon and thoron progeny working level or concentration meters. The standards contain either a radium (^{226}Ra) or thorium (^{228}Th) source which emanates radon or thoron gas, respectively (Pylon, 1992).

2.8 SAMPLE TREATMENT

2.8.1 Sample Collection, Transportation, Storage, and Measurement

In addition to the laboratory processes proper, sample treatment includes, in the case of low-level activities, indispensable keeping of specific activity

- during the sample collection
- during the transportation from locality to laboratory processing
- during the storage to the time of processing

If the sample treatment was carried out incorrectly, between the sample withdrawal and the measurement, considerable distortion of the true value of the specific activity may take place, either by an increase or decrease.

The reason for the increase is because of the contamination of the sample from the surrounding medium, for example, radioactive fallout from air, radionuclide contaminated soil, natural concentrations of radionuclides in the surrounding material or chemicals used, etc. Therefore it is indispensable to protect the sample perfectly and, if possible, permanently from contact with the outside medium during all operations (for example) by using separate working tools, by selecting suitable or previously adjusted protecting packings, etc.).

In contrast to this a decrease of specific activity can be caused mainly by sorption of the radionuclide labeled molecule, or a considerable part of it (e.g., after dissociation in solution) onto the walls of the vessel in which the sample is stored, transported, and measured. The importance of this effect is, of course, affected both by the nature of the labeled compound and the medium in which it is kept, and by the material on which sorption takes place. We have shown (Rexa and Tykva, 1990) that, depending on the character of the sample, sorption from solution may be combined with sedimentation to the bottom, which also contributes to a decrease in the volume of activity. The simplest methods leading to a decrease of sorption are usually an increase in the acidity of the solution and selection of a suitable material for the container (teflon may be used with preference, but polyethylene may also do, while untreated glass surface is usually unsuitable for a direct contact).

A prolonged measurement of low activity requires a guarantee of stable measurement conditions. Therefore, it must also be guaranteed that no change takes place in the sample until the end of the measurement. As an example let us mention liquid scintillators, where under certain conditions we observed sorption of the dissolved sample on the wall of the measuring vial, or its partial precipitation followed by sedimentation. This undesirable phenomenon can be prevented by adequate preparation of the sample (Tykva, 1980).

The reasons why sample treatment is needed for low-level radioactivities may be divided as follows:

- The first fundamental reason consists in the very different character of the sample: fractions from separation micromethods (e.g., after thin-layer chromatography or gel electrophoresis), cell suspensions, plant explantates, animal organs, parts of insects, wood, soil, polluted water, bones, fragments of meteorites, honey, milk, and many other different samples. For a certain method of radioactivity measurement various forms of the starting sample have to be converted to a form required by the selected method of measurement, e.g., enabling a reproducible introduction into a liquid scintillator. This first reason is substantially valid for all radioactivity measurements.
- Sample withdrawal, transport, storage, and preparation for the measurement must be always carried out in such a manner as to prevent an increase of the measured value by introduction of radionuclides from the medium into the sample (e.g., by the chemicals used). This reason in radioactivity assay is more or less general, but for low levels it is indispensable to follow it very strictly.

- The third reason for sample treatment is quite specific for low levels. It is treatment leading to an increase of specific activity of the sample. A smaller sample requires a smaller detector, with a corresponding decrease of the background and a diminishing of the shielding arrangement. In many cases the measurement of low-level radioactivity without this radionuclide enrichment cannot be done by a certain type of detection; or it cannot be carried out at all. An increase of specific activity may be also achieved using isolation of the measured radionuclide, in which, however, the original chemical composition of the sample is not preserved, as in the case of enrichment.

From the facts mentioned so far in this section it is evident that the conditions for sample treatment are specific for various forms of sample and that they also differ with the analyzed radiation. The description of generally utilized radiochemical procedures, such as precipitation, extraction, sorption, crystallization, evaporation in the case of non-volatile samples, distillation in the case of volatile samples, ion exchange during the passage through a bed of ion exchanger or chromatographic separation, may be found in more detail in the corresponding literature (Cheronis, 1954; Bubner and Schmidt, 1966; Faires and Parks, 1973; Shono, 1990).

An example of a parallel sample treatment of very different nature is the low-level measurement of natural radionuclides around a coal-fired power plant (Rösner et al., 1984). The authors compared the specific activities of radionuclides ^{137}Pb and ^{137}Po in soil, fly ash, and vegetation (grass, cereals, and corn). Each sample of the soil was composed of four subsamples, weighing about 10 kg. All subsamples were transported in plastic bags to the laboratory, then dried in plastic trays and stored in polyethylene bottles. Fly ash samples were collected from the electrostatic precipitator several times within a year at full load. Four vegetation subsamples, each about 200 g fresh weight, were dried in an oven and the loss of weight was determined. The samples were milled and then stored in polyethylene bottles. The laboratory adjustment of the sample for measurement, which was different for each of the the three mentioned groups, is described in detail in the mentioned paper.

As an example of the gradual use of several working procedures for the transfer of the analyzed radionuclide from an environmental sample into the detector, the transfer of krypton dissolved in water into internal gas counter may be used (Held et al., 1992). A water sample of about 200 l was flushed with helium to extract the dissolved gases. Krypton was then extracted from the mixture by repeated adsorption on charcoal, fractionated by desorption and subsequent gas chromatography.

The use of several subsequently repeated working procedures is common for low-level activities, while an independently used single operation is only rarely used for low-level sample treatment.

2.8.2 Laboratory Procedures

Among individual laboratory procedures, the *separation* of the measured radionuclide from the rest of the sample is most commonly used. Its independent use can be demonstrated by the separation of strontium from a large amount of calcium to determine the low level of ^{90}Sr in environmental materials, especially in sea water (Bojanowski and Knapinska-Skiba, 1990). As an example of separation by precipitation from the gas phase, the determination of the specific activity of $^{14}CO_2$ in air may be mentioned, used for the assessment of the exposure to ^{14}C–produced in the nuclear fuel cycle in radiation protection (Tschurlovits et al., 1982). While *chemical precipitation* to barium carbonate is used for this separation of carbon dioxide, *electrostatic precipitation* may also be used, made possible by the fact that small dust particles are electrically charged. The preparation of the sample for silicon detector counting of post radon decay products may serve as an example of such a procedure (Pereira and da Silva, 1989).

To the most frequently applied preparations of low-level samples belongs further *concentration* used both independently–for example in measurements of ^{137}Cs in lacustrine sediment and its pore water (Das, 1992)–and also as preconcentration applied for measuring low specific activities before separation–for example during the determination of plutonium levels in water (Yu-Fu et al., 1991). In some cases the concentration is achieved by passing the sample through a suitable absorption reagent, as evident from the arrangement for the passage of 2 m^3 of sea water in the determination of ^{51}Mn, ^{60}Co, ^{65}Zn, and ^{144}Ce (Hashimoto et al., 1989).

In order to transfer low activities of a substance into solution, either from a solid sample or from another solution, *extraction* is used efficiently. This procedure may be illustrated by the extractional system used for the determination of ^{90}Sr in milk (Kimura et al., 1980). If the extraction process is performed in a continuous and multi-stage manner on a column, it represents *extraction chromatography* (Braun and Ghersini, 1975).

The mentioned determination of ^{90}Sr in milk may also be achieved relatively rapidly by cation exchange, by concentrating the relevant cations in the milk sample (Suomela and Wallberg, 1989). The frequently used *ion exchange* is often employed in the form of *ion-exchange chromatography,* as for example in the determination of uranium series radionuclides ^{226}Ra and ^{231}Pa (Saarinen and Suksi, 1992).

Another important set of methods includes *sorption* procedures. For measurement of ^{60}Co concentration in sea water, a special device was used consisting of a pumping system combined with cartridges packed with the adsorbent (Hashimoto et al., 1989). In each experiment 2 m^3 of sea water was treated for 360 minutes. Detectable concentration was nearly 3.7 mBq m^{-3}. Chemisorption based on differing distribution ratios is used to a smaller extent, which may be increased by adjustments of a suitable solid phase, as was shown, e.g., for radioidine by using $AgNo_3$ for Fe_2O_3 and $BaSO_4$ (Kepák, 1966).

In order to complete enumeration of the basic radiochemical treatments let us also mention a few modifications of standard laboratory procedures for processing samples: *evaporation* to increase volume activity or the preparation of a solid dry residue, *filtration* to capture particles dispersed either in the gaseous or the liquid medium, and eventually the little-used separation by *distillation* based on various boiling points of the labeled component and other components of the analyzed system.

In view of the character of the interaction of photons with matter (Section 1.2.3), the terminal part of the preparation of the sample with gamma radionuclides before their movement into the detector remains without bigger complications and it usually requires only homogenization and subsequent transfer, into a plastic tube (Hedvall and Pettersson, 1987). In contrast to this, low-level samples for alpha spectrometry and the measurement of low specific activity of low-energy beta nuclides require special treatment before their introduction into the detector.

To prepare samples for alpha spectrometry, which requires very thin uniform samples owing to the very high specific ionization of the analyzed particles, *electrodeposition* from suitably prepared solution is used, which was described, e.g., for the determination of isotopic composition of enriched or depleted uranium samples (Sánchez et al., 1992). Such electrodeposition is frequently preceded by several preparatory steps, e.g., in the determination of plutonium and americium in environmental samples it involves preconcentration of the nuclides by coprecipitation and ion exchange (Pillai and Matkar, 1987) or by a pretreatment of bones during the quantitative analysis of the content of environmental levels of uranium, thorium, and plutonium (Singh et al., 1984).

For low levels of weak beta emitters, especially the most frequently used, ^3H and ^{14}C, a number of special methods and operation procedures was developed for the adjustment of the sample to a state required by the detector, most frequently by using a liquid scintillation spectrometer (see Section 2.3.4.4) or an internal gas-filled counter (Section 2.3.3.4). Since the aim of these detectors is to eliminate self-absorption of the emitted low-energy electrons and to achieve a counting geometry approaching 4π sr, the goal of sample treatment is a homogeneous dispersion of the sample in the liquid scintillation medium (most frequently dissolution, less frequently dispersion in the scintillation emulsion or gel), or introduction of a gas filling, prepared from the sample, into the internal gas counter.

The initial samples for the measurement occur most frequently in solid, but also often in liquid phase. As an introductory step for the preparation of the filling of the internal gas counter various types of *oxidation* are usually used (Šilar and Tykva, 1977). For liquid scintillation counting, oxidation followed by *absorption* of the carbon dioxide formed in a quaternary ammoniacal base was used mainly in previous years (Tykva, 1964). For scintillation counting improving so-called scintillation cocktails are presently produced by a number of firms and commercially available. These make possible a homogeneous dispersion of many samples of biological material (Tykva, 1976) in the scintillation medium. The use of oxidation is limited mainly to the preparation of carbon dioxide as the starting substance for the *synthesis of benzene* as a component of a liquid scintillator at very low levels of specific activity (Polach et al., 1983), or to samples which decrease detection efficiency distinctly (for example some plant tissues) after introduction to the scintillation cocktail or to those which can be introduced into the scintillation medium only with difficulty (for example samples containing chitin, like insect cuticules, or bones). In contrast to this, some low-level samples may be introduced directly into the liquid scintillator without any special preparation for counting, as for example in ^{14}C-estimations in ethyl alcohol or vinegar (Schoenhofer, 1991).

Regarding the filling of an internal gas counter, the water vapor formed by oxidation cannot be used for the determination of tritium, because the reactive hydroxyl groups $-O^3H$ cause a distinct contamination of surfaces, especially in some materials, e.g., glass. In the determination of ^{14}C the carbon dioxide formed may be used for filling, but in view of its high affinity to electrons, it is important to decrease the presence of electronegative admixtures before its introduction into the detector, using special traps (Šilar and Tykva, 1991). For these reasons hydrocarbons (e.g., methane) are frequently used as the detector filling. Moreover, they enable the determination of lower specific activity while using the same volume of the measured gas, with respect to the increased number of atoms of the corresponding element in the molecule (e.g., four atoms of hydrogen in molecules of methane). As an example of such a *synthesis of hydrocarbons,* we can examine the preparation of propane via hydrogenation of propadiene, for measurement of the low activity of tritium (Wolf and Singer, 1991). If it is necessary to choose the filling of an internal gas counter, it is desirable to evaluate the necessary equipment, the complexity and the duration of the preparation in a careful manner, as well as the possible volume of the measured gas (detector volume and the filling pressure) and the requirements on the measurable specific activity of the radionuclide. The safety measures also play a certain role in the preparation of some hydrocarbons.

Before the preparation of the filling itself (either of the scintillation vial or of the gas counter), *enrichment* is sometimes used to increase the specific activity of the sample. In the case of ^3H, enrichment is common especially for measurement of the specific activity of samples in which the value of the specific activity achieves values about 1 TU (ratio $^3H/^1H = 10^{-18}$), e.g., in non-contaminated surface waters. If such samples have to be introduced into the scintillation medium directly in the form of water, different electrochemical properties of H_2O and HOT molecules for *electrolysis* are made use of. In a relatively simple cooled electrolytic arrangement (Moser and Rauert, 1980) it is possible to increase the specific activity of the tritium in the electrolyte at the preferential dissociation of H_2O molecules, currently twenty times for ten to twelve samples at the same time, with several standard samples. In laboratories using liquid scintillation counting of water samples for ^3H-dating or control measurements of surface and underground waters of a level of 1 TU, this procedure of enrichment of distilled samples is widespread.

For higher specific activities water is usually introduced into the scintillation vial without enrichment and both gels and solutions formed with the scintillation medium are applied (Takiue et al., 1989). To measure the background, "dead water" is used, which does not contain tritium (Morishima et al., 1987).

During the measurement of low specific activity of ^{14}C, enrichment is used quite exceptionally. *Thermodiffusion* in the form of carbon monoxide permits radionuclide dating up to approximately 70,000 years (Haring et al., 1958), and practically no other developed methods (Freyer and Wagener, 1967; Thiemann, 1969) are in use.

The difference between the qualitatively wide spectrum of sample treatments and relatively low number of methods of measurements is due to the large variety of the characters of samples, and hence to differing quality and quantity of individual types of low level samples. Due simply to their treatments, the broad assortment of samples is reduced to only several basic types, suitable for individual types of detectors.

In this sense the variability of various sample treatments corresponds in the case of low levels to the preparation of compounds labeled with radionuclides for tracer experiments. Both groups contain a large number of very different methodical procedures and their combinations, while as far as measurement is concerned, only relatively few methods are used.

2.9 REFERENCES

Aglietta, M. et al. 1992. "Study of the Low Energy Background Radiation and the Effect of the ^{222}Rn in the LSD Underground Experiment," *Nucl. Phys. B., Proc. Suppl.*, 28A:430–434.

Aitken, M. 1985. *Thermoluminiscence Dating*. New York: Academic Press.

AMPTEK, Inc. 1990. *Catalog and Application Notes*. Angelo Drive 5, Bedford, MA 01730, USA.

Analytics. 1992. *Catalog*. Analytics, Inc., 1380 Seaboard Industrial Boulevard, Atlanta, GA 30318.

Arnold, J. R. 1987. "Decay Counting in the Age of AMS," *Nucl. Instrum. Methods Phys. Res.*, B29:424–426.

Arnold, L. D. 1983. "An Approach to Low-Level H3 and C14 Analysis at the Alberta Environmental Centre," in *Advances in Scintillation Counting*, S. A. McQuarrie, C. Ediss and L. I. Wiebe, eds., Edmonton: University of Alberta, pp. 442–455.

Arthur, R. J., J. H. Reeves and H. S. Miley. 1987. "Use of Low-Background Germanium Detectors to Preselect High-Radiopurity Materials Intended for Constructing Advanced Ultra-Low Detectors," *Nuclear Science Symposium*, October 21–23, San Francisco, INIS, DE 8800 5136, p. 14.

Bamford, G. J., A. C. Rester, R. L. Coldwell, et al. 1991. "Neutron, Proton and Gamma-Ray Event Identification with a HPGe Detector through Pulse Shape Analysis," *IEEE Trans. Nucl. Sci.*, NS38(2):200–208.

Baxter, M. S. and R. D. Scott. 1987. *Resonance Ionisation Mass Spectrometry. A Review of Its Status and Potential as a Radioanalytical Technique*. London: Department of the Environment.

Beer, J. et al. 1981. "Carbon-14 Dating of Ice with Accelerator-Based Mass Spectrometry," in *Methods of Low-Level Counting and Spectrometry*. Vienna: International Atomic Energy Agency, pp. 419–430.

Bertolini, G. and A. Coche, eds. 1968. *Semiconductor Detectors*. Amsterdam: North Holland Publ.

Bertsche, K. J., C. A. Karadi and R. A. Müller. 1991. "Radiocarbon Detection with a Small Low Energy Cyclotron," *Nucl. Instrum. Methods. Phys. Res.*, A301:358–371.

Bevington, P. R. 1969. *Data Reduction and Error Analysis for the Physical Sciences*. New York, NY: McGraw-Hill.

Bigu, J. 1992. "Contamination of Alpha-Particle Detectors by Desorption of ^{222}Rn Progeny from Metal Surfaces," *Appl. Radiat. Isot.*, 43:747–751.

Birks, J. B. 1964. *Theory and Practice of Scintillation Counting*. Oxford: Pergamon Press.

Bojanovski, R. and D. Knapinska-Skiba. 1990. "Determination of Low-Level ^{90}Sr in Environmental Materials: A Novel Approach to the Classical Method," *J. Radioanal. Nucl. Chem. Articles*, 138:207–218.

Bowen, V. T. 1978. "Natural Matrix Standards," *Environment Internat.*, 1:35–39.

Bransome, E. D., ed. 1970. *The Current State of Liquid Scintillation Counting*. New York: Grune and Stratton.

Braun, T. and G. Ghersini, eds. 1975. *Extraction Chromatography*. Budapest: Akadémiai Kiadó.

Bray, G. A. 1960. "A Simple Efficient Liquid Scintillator for Counting Aqueous Solution in a Liquid Scintillation Counter," *Anal. Biochem.*, 1:279–285.

Brodzinski, R. L. et al. 1988. "Achieving Ultralow Background in a Germanium Spectrometer," *J. Radioanal. Nucl. Chem. Articles*, 124:513–521.

Brodzinski, R. L., H. S. Miley and J. H. Reeves. 1990. "Further Reduction of Radioactive

Backgrounds in Ultrascusitive Germanium Spectrometers," *Nucl. Instrum. Methods Phys. Res.*, A292:337–342.

Bubner, M. and H. Schmidt. 1966. *Die Synthese Kohlenstoff-14-markierter organischer Verbindungen.* Leipzig: Georg Thieme Verlag.

Butterfield, D. and H. Polach. 1983. "Effects of Vial Holder Materials and Design on Low Level ^{14}C Counting," in *Advances in Scintillation Counting*, S. A. McQuarrie, C. Ediss and L. I. Wiebe, eds., Edmonton: University of Alberta, pp. 468–477.

Cameron, J. R., N. Suntharalingam and G. N. Kenney. 1968. *Thermoluminiscent Dosimetry.* Madison: University of Wisconsin Press.

Canberra. 1992. "Canberra Instruments Catalog," Canberra Industries, Inc., 800 Research Parkway, Meriden, CT 064450, USA.

Carles, A. G., M. T. Martin-Casallo and A. G. Malonda. 1991. "Spectrum Unfolding and Double Window Methods Applied to Standardization of ^{14}C and ^{3}H Mixtures," *Nucl. Instrum. Methods Phys. Res.*, A307:484–490.

Chereji, I. 1992. "^{222}Rn (^{226}Ra) Determination in Water by Scintillation Methods," *J. Radioanal. Nucl. Chem.*, 165:263–267.

Cheronis, N. D. 1954. "Micro and Semimicro Methods," *Technique of Organic Chemistry, Vol. VI*, A. Weissberger, ed., New York: Interscience Publ., pp. 1–409.

Chung, C. and C. J. Lee. 1988. "Environmental Monitoring Using a HPGe-NaI(Tl) Compton Suppression Spectrometer," *Nucl. Instrum. Meths. Phys. Res.*, A273:436–440.

Chwaszczewska, J. and W. Przyborski. 1967. "Low-Background Counting with Silicon Detectors," in *Radiocarbon Dating and Methods of Low-Level Counting.* Vienna: International Atomic Energy Agency, pp. 711–720.

Cline, J. E. 1990. "Comparison of Detection-Limit Computations for Four Commercial Gamma-Ray Analysis Programs," *Nucl. Instrum. Meths. Phys. Res.*, A286:421–428.

CRC. 1985. *Handbook of Chemistry and Physics.* Boca Raton, FL: The Chemical Rubber Co.

Crook, M. A. and P. Johnson, eds. 1976. *Liquid Scintillation Counting, Vol. V.* London: Heyden.

Currie, L. A. 1968. "Limits for Qualitative Detection and Quantitative Determination," *Analytical Chem.*, 40(3):586–593.

Currie, L. A. 1984. "Lower Limit of Detection: Definition and Elaboration of a Proposed Position for Radiological Effluent and Environmental Measurements," *NUREG/ CR-4007.* Gaithersburg, MD: National Institute for Standards and Technology.

D'Amico, A. and F. Mazzetti, ed. 1986. *Noise in Physical Systems and 1/f Noise.* Amsterdam: North-Holland Publ.

Das, H. A. 1987. "The Advantage of Anti-Compton Counting in the Measurement of Low-Level Radioactivity by Gamma-Ray Spectrometry," *J. Radioanal. Nucl. Chem. Articles,* 115:159–173.

Das, H. A. 1992. "Release of ^{137}Cs from Anoxic Lacustrine Sediment Measurement and Formulation," *J. Radioanal. Nucl. Chem. Articles,* 156:129–149.

Das, H. A. and R. N. J. Comans. 1990. "On the Limits of Low Level Measurements of ^{137}Cs as a Natural Radiotracer," *J. Radioanal. Nucl. Chem. Articles,* 138:407–416.

Dearnaley, G. and D. C. Northrop. 1966. *Semiconductor Counters for Nuclear Radiation, Second Edition.* London: Spot.

Debertin, K. 1991. "Efficiency Calibration in Gamma-Ray Spectrometry with Germanium

Detectors," in *Low-Level Measurements of Man-Made Radionuclides in the Environment,* M. García-León and G. Madurga, eds., Singapore: World Scientific, pp. 3–14.

Debertin, K. and R. G. Helmer. 1988. *Gamma- and X-Ray Spectrometry with Semiconductor Detectors.* Amsterdam: North-Holland.

Delibrias, G. and J. L. Rapaire. 1967. "Radioactive Dating and Methods of Low-Level Counting," *Proceeding Series of IAEA.* Vienna: IAEA, pp. 603–612.

Dixon, W. J. and F. J. Massey. 1969. *Introduction to Statistical Analysis.* New York, NY: McGraw-Hill.

Doll, P., G. Fink, M. Haupenthal, et al. 1989. "Pulse Shape Discrimination with a Large NaI Crystal," *Nucl. Instrum. Meths. Phys. Res.,* A285(3):464–468.

EG&G Ortec. 1993. "EG&G Ortec Nuclear Instruments," *Catalog 1993.* EG&G Ortec, 100 Midland Road, Oak Ridge, TN 37831-0895, USA.

Evans, R. D. 1955. *The Atomic Nucleus.* New York, NY: McGraw-Hill Book Company, Inc.

Fairbank, W. M., Jr. 1987. "Photon Burst Mass Spectrometry," *Nucl. Instrum. Methods Phys. Res.,* B29:407–414.

Faires, R. A. and B. H. Parks. 1973. *Radioisotope Laboratory Techniques.* London: Butterworth.

Fischer, H. and G. Werner. 1971. *Autoradiographie.* Berlin: Walter de Gruyter.

Fleischer, R. L., P. B. Price and R. M. Walker. 1975. *Nuclear Tracks in Solids.* Berkeley: University of California Press.

Freyer, H. and K. Wagener. 1967. "Electrochemical Processes for Isotope-Enrichment" (in German), *Angew. Chem. Ed. Ger.,* 79:734–744.

Galbraith, W. 1958. *Extensive Air Showers.* London: Butterworth.

Gfeler, C. and H. Oeschger. 1962. "Estimation of Measurement Limits of Lower Activities" (in German), *Compte rendu de la rénnion de la Societé suiss de Physique,* 35:307–313.

Grinberg, B. and Y. Le Gallic. 1961. "Basic Characteristics of a Laboratory Designed for Measuring Very Low Activities" (in French), *Int. J. Appl. Radiat. Isotopes,* 12:104–117.

Harihar, P., W. Stapor, A. B. Campbell, et al. 1992. "Nuclear Spectroscopy Using Risetimes in Cerussite Scintillator," *Nucl. Instrum. Meths. Phys. Res.,* A322:40–42.

Haring, A., A. E. de Vries and H. de Vries. 1958. "Radiocarbon Dating up to 70,000 Years by Isotopic Enrichment," *Science,* 128:472–473.

Hashimoto, T. et al. 1989. "Measurement of Radionuclides in Sea Water," in *Proc. 15. Regional Congress of the International Radiation Protection Association (IRPA) on the Radioecology of Natural and Artificial Radionuclides.* Köln, Germany: Verlag TUEV Reinland, pp. 480–483.

Haviland, R. T. and L. L. Bieber. 1970. "Scintillation Counting of ^{32}P without Added Scintillator in Aqueous Solutions and Organic Solvents and on Dry Chromatographic Media," *Anal. Biochem.,* 33:323–334.

Hedvall, R. and H. Pettersson. 1987. "Gamma-Spectrometric Determination of Uranium Isotopes in Biofuel Ash," *J. Radioanal. Nucl. Chem. Articles,* 115:221–216.

Hedvall, R., H. Pettersson and S. Erlandsson. 1987. "Gamma Spectrometric Determination of Uranium Isotopes in Biofuel Ash," *J. Radioanal. Nucl. Chem. Articles,* 115:211–216.

Held, J., S. Schuhbeck and W. Rauert. 1992. "A Simplified Method of ^{85}Kr Measurement for Dating Young Groundwaters," *Appl. Radiat. Isot.,* 43:939–942.

Heusser, G. et al. 1989. "Construction of Low-Level Ge-Detector," *Appl. Radiat. Isot.*, 40(5):393–395.

Horrock, D. L. 1980. "Effect of Impurity and Color Quenching upon the Liquid Scintillation Pulse Height Distributions," in *Liquid Scintillation Counting: Recent Applications and Development, Vol. I,* C. T. Peng, D. L. Horrocks, E. L. Alpen, eds., New York: Academic Press, pp. 173–186.

Hötzl, H. and R. Winkler. 1984. "Experiences with Large Area Frisch Gird Chambers in Low-Level Alpha Spectrometry," *Nucl. Instrum. Methods Phys. Res.*, 223:290–294.

Houtermans, F. G. and H. Oeschger. 1958. "Proportional Counter for Measurement of Lower Activities of Weak Beta Emitters" (in German), *Helv. Phys. Acta,* 31:117–126.

Hubert, P. et al. 1986. "Alpha-Rays Induced Background in Ultra Low-Level Counting with Ge Spectrometers," *Nucl. Instrum. Methods Phys. Res.*, A252:87–90.

Huimin, J. et al. 1989. "On-Line Measurement of Low-Level Gamma Radioactivity in Large-Volume Water Using NaI(Tl) Detector," *At. Ener. Sci. Technol.*, 23(1):20–25.

Hutchison, N. J. 1984. "Hybridisation Histochemistry: *In situ* Hybridisation at the Electromicroscope Level," in *Immunolabelling for Electromicroscopy,* A. Polak and E. Varnedell, eds., Amsterdam: Elsevier, p. 341.

ICRU. 1972. *Measurement of Low-Level Radioactivity,* ICRU Report 22. Bethesda, MD: International Commission on Radiation Units and Measurements.

Inn, K. G., P. A. Mullen and J. M. R. Hutchinson. 1984. "Radioactivity Standards for Environmental Monitoring. II," *Environment Internat.*, 10:91–97.

Janossy, L., ed. 1950. *Cosmic Rays, Second Edition.* Oxford: Clarendon Press.

Jelley, J. V. 1958. *Čerenkov Radiation and Its Applications.* London: Pergamon Press.

Jiang, S. H., M. C. Horng and C. I. Hsu. 1990. "Radiation Instrumentation with the Microelectronical Technique," *Nucl. Instrum. Meths. Phys. Res.*, A299:298–301.

Jisl, R. and R. Tykva. 1986. "Correction of the Continual Scanning Record of Radioactivity Distribution. II. Deconvolution," *Nucl. Instrum. Methods Phys. Res.*, A251:166–171.

Johnson, S. J. 1967. "4π Low-Level Beta Counter Using Two Si-Au Detectors," in *Radioactive Dating and Methods of Low-Level Counting.* Vienna: International Atomic Energy Agency, pp. 721–728.

Kamikubota, N. et al. 1989. "Low-Level Radioactive Isotopes Contained in Materials Used for Beta-Ray and Gamma-Ray Detectors," in *Radiation Detectors and Their Uses,* M. Miyajima, S. Sasaki and T. Doke, eds., Ibaraki (Japan): Tsukuba, pp. 19–23.

Kasharov, L. L. and A. V. Fisenko. 1987. *Low-Level Counting and Spectrometry,* P. Povinec, ed., Bratislava: Veda, pp. 71–74.

Kemmer, J. 1968. "Ge(Li) Gamma Spectrometer of Low Level Gamma Activities" (in German), *Nucl. Instrum. Methods,* 64:268.

Kepák, F. 1966. "Sorption of Small Amounts of Radioiodine as Iodide Anions on Hydrated Ferric Oxide Containing Silver," *Coll. Czech. Chem. Commun.*, 31:1493–1500.

Kilgus, U., R. Kotthaus and E. Lange. 1990. "Prospects of CsI(Tl) Photoiode Detectors for Low-Level Spectroscopy," *Nucl. Instrum. Methods Phys. Res.*, A297:425–440.

Kimura, K., M. Sawada and T. Shono. 1980. "Poly(crownether) Catalysed Derivatisation of Lower Fatty Acids for Gas Chromatography," *Fresenius Z. Anal. Chem.*, 301:250.

Klucke, H. and J. Beetz. 1987. "The Optimization of an Anticompton Spectrometer for the Analysis of Biological Samples," in *Low-Level Counting and Spectrometry,* P. Povinec, ed., Bratislava: Veda, pp. 149–154.

Knoll, G. F. 1979. *Radiation Detection and Measurement.* New York: John Wiley and Sons.

Kopeštanský, J., R. Tykva and S. Staněk. 1991. "SIMS Study of Semiconductor-Oxid-Metal Structure: $AuSiO_xSi(111)$ and $AlSiO_xSi(111)$," *Progr. Surf. Sci.*, 35:215–218.

Kovalchuk, E. L. et al. 1982. "U 235, Th 222, K 40 along the Main 3 km Adit of the Baksan Neutrino Observatory," in *Natural Radiation Environment,* W. G. Vohra et al., eds., New York: John Wiley and Sons, pp. 201–205.

Kowalski, E. 1970. *Nuclear Electronics.* New York, Heidelberg: Springer Verlag.

Kuhn, L. 1968. "The Measurement of the Lowest Activities by Semiconductor Detectors" (in German), *Atompraxis,* 14(9/10):1–3.

Liden, K. and E. Holm. 1985. "Measurement and Dosimetry of Radioactivity in the Environment," in *The Dosimetry of Ionizing Radiation, Vol. I,* K. R. Kase, B. E. Bjärngard and F. H. Attix, eds., Orlando: Academic Press.

Lindstrom, L. M. and J. K. Langland. 1990. "A Low-Background Gamma-Ray Assay Laboratory for Activation Analysis," *Nucl. Instrum. Methods Phys. Res.*, A299:425–429.

Mann, W. B., R. L. Ayres and S. B. Garfinkel. 1980. *Radioactivity and Its Measurement, Second Edition.* Oxford: Pergamon Press.

Mann, W. B., A. Rytz and A. Spernol. 1991. *Radioactivity Measurements, Principles and Practice.* Oxford: Pergamon Press.

Matyjek, M. et al. 1988. "Evaluation of Alpha Spectrometers with Surface-Barrier Detectors for Low Level Measurements," in *Report on the Consultans Meeting on Rapid Instrumental and Separation Methods for Monitoring Radionuclides in Food and Environmental Samples.* Vienna: International Atomic Energy Agency, pp. 83–92.

May, H. and L. B. Marinelli. 1964. *The Natural Radiation Environment,* J. A. S. Adams and W. M. Lowder, eds., Chicago: University of Chicago Press, p. 463.

McQuarrie, S. A., L. I. Wiebe and C. Ediss. 1980. "Observations of the Performance of ESP and H# in Liquid Scintillation Counting," in *Liquid Scintillation Counting: Recent Applications and Development, Vol. I,* C. T. Peng, D. L. Horrocks, E. L. Alpen, eds., New York: Academic Press, pp. 291–300.

Milam, J. K. 1973. "Single-Channel Analyzers," in *Instrumentation in Applied Nuclear Chemistry,* J. Krugers, ed., New York, NY: Plenum Press.

Miley, H. S., R. L. Brodzinski and J. H. Reeves. 1991. *Abstracts of 2nd International Conference on Methods and Applications of Radioanalytical Chemistry,* Kona, Hawaii, American Nuclear Society, p. 28.

Minobe, M. et al. 1987. "Development of Low-Level Alpha Particle Counting System," (in Japanese) *Sumitomo Kagatu,* No. 1: 92–99.

Moljk, A., R. W. P. Drever and S. C. Curran. 1957. "The Background of Counters and Radiocarbon Dating," *Proc. Royal Soc. London,* 239:433–455.

Mondspiegel, K., J. Sabol and R. Tykva. 1990. "Contribution to the Evaluation of Radiation Hazard in the Surrounding of the Uranium Ore Mill MAPE Mydlovary" (in Czech), *Report of the Institute of Landscape Ecology.* České Budějovice, Czech Republic: Academy of Sciences.

Morishima, H. et al. 1987. "Preparation of 'Dead Water' for Low Background Liquid Scintillation Counting" (in Japanese), *Radioisotopes-Tokio-Japan,* 36:126–128.

Moser, H. and W. Rauert. 1980. *Isotopenmethoden in der Hydrologie.* Berlin: Gebrüder Borntraeger, p. 40.

Müller, G. et al. 1990. "Low-Background Counting Using Ge(Li) Detectors with Anti-Muon-Shields," *Nucl. Instrum. Methods Phys. Res.*, A295:133–139.

Mundschenk, H. 1990. "Vergleichsanalyse Radionuklide im Sediment," *Report BfG-0525* (Bundesanstalt für Gewässerkunde, Koblenz, Germany).

Murase, Y. et al. 1989. "Effect of Air Luminiscence Counts on Determination of ^{3}H by Liquid Scintillation Counting," in *Proc. 15. Regional Congress of the International Radiation Protection Association (IRPA) on the Radioecology of Natural and Artificial Radionuclides.* Köln, Germany: Verlag TUEV, Reinland, pp. 509–513.

Murray, A. S. and M. J. Aitken. 1988. "Analysis of Low-Level Natural Radioactivity in Small Mineral Samples for Use in Thermoluminiscence Dating, Using High-Resolution Gamma Spectrometry," *Appl. Radiat. Isot.*, 39:145–158.

Murray, A. S. et al. 1987. "Analysis for Naturally Occurring Radionuclides at Environmental Concentrations by Gamma Spectrometry," *J. Radioanal. Nucl. Chem. Articles,* 115:263–288.

NBS. 1969. "Statistical Concepts and Procedures," in *Precision Measurements and Calibration, Vol. 1,* NBS Spec. Publ. 300, H. H. Ku, ed., Gaithersburg, MD: National Bureau of Standards.

NCRP. 1985. *A Handbook of Radioactivity Measurements Procedures,* NCRP Report No. 58. Bethesda, MD: National Council for Radiation Protection and Measurements.

Nicholson, P. W. 1974. *Nuclear Electronics.* London: John Wiley and Sons.

Niese, S., W. Helbig and H. Kleeberg. 1989. "Multi-Sample Beta-Gamma Coincidence Spectrometry in an Underground Laboratory," *J. Radioanal. Nucl. Chem.*, 129:387–391.

NIST. 1991. *Radioactive Standard Reference Materials Catalog.* National Institute of Standards and Technology, Gaithersburg, MD 20899.

Noujaim, A. A. et al., eds. 1976. *Liquid Scintillation Science and Technology.* New York: Academic Press.

Oikari, T. et al. 1987. "Simultaneous Counting of Low Alpha- and Beta-Particle Activities by Liquid Scintillation Spectrometry and Pulse-Shape Analysis," *Appl. Radiat. Isot.*, 38(10):875–878.

Parratt, L. G. 1961. *Probability and Experimental Errors in Science: An Elementary Survey.* New York, NY: John Wiley.

Pausch, G., W. Bohne, H. Fuchs, et al. 1992. "Particle Identification in Solid-State Detectors by Exploiting Pulse Shape," *Nucl. Instrum. Meth. Phys. Res.*, A322:43–52.

Pereira, E. B. and H. E. da Silva. 1989. "Atmospheric Radon Measurements by Electrostatic Precipitation," *Nucl. Instrum. Methods Phys. Res.*, A280:503–505.

Pillai, K. C. and V. M. Matkar. 1987. "Determination of Plutonium and Americium in Environmental Samples and Assessment of Thorium in Bone Samples from Normal and High Background Areas," *J. Radioanal. Nucl. Chem., Articles,* 115:217–229.

Plaga, R. 1991. "Silicon for Ultra Low Level Detectors and ^{32}Si," *Nucl. Instrum. Methods Phys. Res.*, A309:598–599.

Polach, H. A. 1987. "Perspectives in Radiocarbon Dating by Radiometry," *Nucl. Instrum. Methods Phys. Res.*, B29:415–423.

Polach, H. A. et al. 1983. "An Ideal Vial and Cocktail for Low-Level Scintillation Counting," in *Advances in Scintillation Counting,* S. A. McQuarrie, C. Edis and L. I. Wiebe, eds., Edmonton, Canada: University of Alberta, pp. 508–525.

Povinec, P. 1980. "Proportional Chambers for Low Level Counting," *Nucl. Instrum. Methods*, 176:111–117.

Prichard, H., M. Venco and C. L. Dodson. 1992. "Liquid-Scintillation Analysis of ^{222}Rn in Water by Alpha-Beta Discrimination," *Radioactivity & Radiochemistry*, 3(1):28–36.

Pylon. 1992. *Catalog*. Pylon Electronics, Inc., 147 Colonnade Road, Ottawa, Ontario, Canada.

Rancoita, P. G. and A. Seidman. 1982. "Silicon Detectors for High-Energy Physics," *Rivista Nuovo Cimento*, 5(7).

Revzan, K. L. et al. 1991. "Modeling Radon Entry into Houses with Basements: Model Description and Verification," *Lawrence Berkeley Laboratory Rep.* LBL–27 742.

Rexa, R. and R. Tykva. 1990. "The Problem of Instability in Liquid Scintillation Counting," *Appl. Radiat. Isot.*, A41:867–871.

Rogers, A. W. 1979. *Techniques of Autoradiography, Third Edition.* Amsterdam: Elsevier.

Rösner, G. 1981. "Measurements of Actinide Nuclides in Water Samples from the Primary Circuit of a Nuclear Power Plant," *J. Radioanal. Chem.*, 64(1):55–64.

Rösner, G. et al. 1984. "Low Level Measurements of Natural Radionuclides in Soil Samples around a Coal-Fired Power Plant," *Nucl. Instrum. Methods Phys. Res.*, 223:585–589.

Ross, W. A. 1973. "Multichannel Analyzers," in *Instrumentation in Applied Nuclear Chemistry*, J. Krugers, ed., New York, NY: Plenum Press.

Rossi, B. and H. Staub. 1949. *Ionization Chambers and Counters.* New York: McGraw-Hill.

Rytz, A. 1978. "International Coherence of Activity Measurements," *Environment Internat.*, 1:15–18.

Rytz, A. 1979. "New Catalog of Recommended Alpha Energy and Intensity Values," *At. Data Nucl. Data Tables*, 23:507.

Saarinen, I. and J. Suksi. 1992. "Determination of Uranium Series Radionuclides ^{231}Pa and ^{226}Ra Using Liquid Scintillation Counting (LSC)," *Nuclear Waste Commission of Finnish Power Companies, Rep. VJT-92-20,* Helsinki.

Sabol, J. 1974. "Simple Anticoincidence Gates for S-C Pulse Height Selectors," *Rev. Sci. Instrum.*, 45:464.

Sabol, J. 1988a. "Another Method of Dead Time Correction," *J. Radioanal. Nucl. Chem., Letters*, 127:389–394.

Sabol, J. 1988b. "Approximation of the Nonparalyzable Dead Time by the Shortest Time Interval," *Rev. Sci. Instrum.*, 59(9):2086–2087.

Sabol, J. 1988c. "Dead-Time Corrections and Effects of Dead-Time on the Counting Statistics of G-M Counters," *Radiat. Prot. Dosim.*, 23:445.

Sabol, J. 1990. "Reduction of Additional Fluctuations of a Stabilized Scintillation Spectrometer Based on Radioactive Reference Source," *Rev. Scient. Instrum.*, 61:1255.

Sánchez, A. M. et al. 1992. "A Rapid Method for Determination of the Isotopic Composition of Uranium Samples by Alpha Spectrometry," *Nucl. Instrum. Methods Phys. Res.*, A313:219–226.

Sayre, E. V. et al. 1981. "Small Gas Proportional Counters for the ^{14}C Measurement of Very Small Samples," in *Methods of Low Level Counting and Spectrometry.* Vienna: International Atomic Energy Agency, pp. 393–407.

Schoenhofer, F. 1991. "Low-Level Measurements by Liquid Scintillation Counting," in *Second International Conference on Methods and Applications of Radioanalytical Chemistry, Abstracts.* Kona, Hawaii: American Nuclear Society, p. 61.

Schoenhofer, F. and F. Henrich. 1987. "Recent Progress and Application of Low Level Liquid Scintillation Counting," *J. Radioanal. Nucl. Chem. Articles,* 115: 317–333.

Shizuma, K., K. Iwatani and H. Hasai. 1987. "Gamma-Ray Scattering in the Low-Background Shielding for Ge Detector," *Radioisotopes Tokyo Japan,* 36:465–468.

Shono, T. 1990. *Electroorganic Synthesis.* London: Academic Press.

Šilar, J. and R. Tykva. 1977. "Radiocarbon Dating Laboratory on the Charles University, Prague: Methods and Results," in *Proc. Int. Conf. Low-Radioactivity Measurements and Applications,* P. Povinec and S. Usačev, eds., Bratislava: Slovenské pedagogické nakladatelstvo, pp. 331–334.

Šilar, J. and R. Tykva. 1991. "Charles University, Prague: Radiocarbon Measurements I," *Radiocarbon,* 33(1):69–78.

Sing, N. P. et al. 1984. "Quantificative Determination of Environmental Levels of Uranium, Thorium and Plutonium in Bone by Solvent Extraction and Alpha Spectrometry," *Nucl. Instrum. Methods Phys. Res.*, 223:558–562.

Sumerling, T. J. 1983. "Calculating a Decision Level for Use in Radioactivity Counting Experiments," *Nucl. Instrum. Meths.*, 206:501.

Suomela, J. and L. Wallberg. 1989. "Rapid Determination of ^{89}Sr and ^{90}Sr in Milk," in *Proc. 15. Regional Congress of the International Radiation Protection Association (IRPA) on the Radioecology of Natural and Artificial Radionuclides.* Köln, Germany: Verlag TUEV Reinland, pp. 465–467.

Swailem, F. M. and M. B. Moussa. 1985. "Studies of Factors Affecting Emission of Čerenkov Radiation in Transparent Media," *Isotopenpraxis,* 21:5–8.

Takiue, M., H. Fujii and Y. Homma. 1989. "Reliability of the Low Level ^{3}H Activity Determined by Liquid Scintillation Measurement," in *Proc. 15. Regional Congress of the International Radiation Protection Association (IRPA) on the Radioecology of Natural and Artificial Radionuclides.* Köln, Germany: Verlag TUEV Reinland, pp. 503–508.

Takiue, M. et al. 1990. "A New Approach to Analytical Radioassay of Multiple Beta-Labelled Samples Using a Liquid Scintillation Spectrometer," *Nucl. Instrum. Methods Phys. Res.*, A293:596–600.

Thiemann, W. 1969. "Enrichment of Heavy Carbon Isotopes by Electrolysis" (in German), *Z. Naturforsh,* 24a:830–835.

Tschurlovits, M., K. J. Pfeiffer and D. Rank. 1982. "A Comparison of Different Methods for Determination of $^{14}CO_2$ in Air," *Atomkernenergie-Kerntechnik,* 40:267–269.

Tsoulfanidis, N. 1983. *Measurement and Application of Radiation.* Washington: Hemisphere Publ. Comp.

Tykva, R. 1964. "Measurement of ^{14}C-Radioactivity of Not Scintillator-Soluble Substances by Liquid Scintillation or in Internal Gas Counter" (in German), *Coll. Czech. Chem. Commun.*, 29:680–689.

Tykva, R. 1967. "Analysis of Errors and Accuracy of Simultaneous ^{3}H and ^{14}C Assay in Organic Substances by Means of an Internal Gas Counter," *Int. J. Appl. Radiat. Isotopes,* 18:45–56.

Tykva, R. 1976. "Exclusion of the Sample Sorption Effect on the Instability of Liquid Scintillation by Emulsion Counting," *Anal. Biochem.*, 70:621–623.

Tykva, R. 1977. "Semiconductor Detectors for Low-Levels of Alpha, Beta or Gamma Nuclides: Counting, Spectrometry and Applications," in *Low-Radioactivity*

Measurements and Applications, P. Povinec and S. Usacev, eds., Bratislava: Slovenské pedagogické nakladatelstvo, pp. 213–217.

Tykva, R. 1980. "Limits of Beta Counting due to Sample Sorption and Procedures for Exclusion of the Counting Rate Instability," *Liquid Scintillation Counting: Recent Applications and Development, Vol. I,* C.-T. Peng, D. L. Horrocks and E. L. Alpen, eds., New York: Academic Press, pp. 225–233.

Tykva, R. 1987. "Radiomicroscopy," *Nucl. Instrum. Methods Phys. Res.*, A251:488–490.

Tykva, R. 1989. "The Present State of Surface-Barrier Silicon Detectors (SBSD)," in *Proc. 3rd Int. Symp. on High Energy Experiments and Methods,* P. Reimer, M. Suk and V. Šimák, eds., Prague: Institute of Physics, Czech Academy of Sciences, pp. 71–75.

Tykva, R. and R. Jisl. 1986. "Correction of the Continual Scanning Record of Radioactivity Distribution. I. Smoothing," *Nucl. Instrum. Methods Phys. Res.*, A251:160–165.

Tykva, R. and L. Kokta. 1967. "Gas Proportional Counter for Determination of Low-Level ^{3}H and ^{14}C in Small Samples," *Nucl. Instrum. Methods,* 55:381–382.

Tykva, R. and B. Pavlů. 1973. "Photodensitometric Investigation of the Conditions of Tritium Determination on Paper Chromatograms by Fluorography," *Coll. Czech Chem. Commun.*, 38:25–28.

Tykva, R. and J. Veselý. 1965. "Determination of the Efficiency of Autoradiographic Detection of Tritium on Stripping Film Using an Internal Gas Counting Tube," *Coll. Czech Chem. Commun.*, 30:898–899.

Tykva, R. et al. 1987. "High-Sensitivity Semiconductor Spectrometry of Charged Particles in the Intershock Project," in *Proc. of 20. International Cosmic Ray Conference, Vol. 4. International Union of Pure and Applied Physics.* Moscow: Nauka, pp. 418–421.

Tykva, R. et al. 1990. "Special Silicon Detectors for the Actinide Assay in Snow Fields in High Tatras," in *20 Journée de Actinides, Abstract.* Prague: Charles University, pp. 76–77.

Tykva, R. et al. 1992. "A Topographic Method for Studying Uptake, Translocation and Distribution of Inorganic Ions Using Two Radiotracers Simultaneously," *J. Exp. Bot.*, 43:1083–1087.

Unterricker, S. et al. 1988. "Investigations on the Location and Construction of a Low Background Gamma-Spectrometer," *J. Radioanal. Nucl. Chem. Articles,* 122:271–278.

Unterricker, S. et al. 1989. "Possibilities of γ-Spectrometric Determination of the Amount of Natural Activities in Shielding and Construction Material," *Isotopenpraxis,* 25:247–252.

Vapirev, E. I. and A. V. Hristova. 1991. "Quantitative Analysis of a Mixture of Beta-Emitters," *Nucl. Instrum., Methods Phys. Res.*, A307:126–131.

Verplancke, J. 1992. "Low Level Gamma Spectroscopy: Low, Lower, Lowest," *Nucl. Instrum. Methods Phys. Res.*, A312:174–182.

Veselý, J. and R. Tykva. 1968. "Correlation of Simultaneous ^{3}H and ^{14}C Autoradiography with Gas Counting," *Int. J. Appl. Radiat. Isotopes,* 19:705–707.

Walford, G. V., J. A. Cooper and R. N. Keyser. 1976. "Evaluation of Standardized Ge(Li) Gamma Ray Detectors for Low Level Environmental Measurements," *IEEE Trans. Nucl. Sci.*, NS23(1):734–737.

Walton, J. E. and E. E. Haller. 1982. "Silicon Radiation Detectors—Materials and Applications," *Lawrence Berkeley Lab, Rep. No. 14909,* Berkeley.

Watt, D. E. and D. Ramsden. 1964. *High Sensitivity Counting Techniques.* Oxford: Pergamon Press.

Weise, L. 1971. *Statistische Auswertung von Kernstahlungs-messungen.* Munich: Oldenbourg.

Weller, R. L. 1964. *The Natural Radiation Environment,* J. A. S. Adams and W. M. Lowder, eds., Chicago: University of Chicago Press, p. 567.

Wilkinson, D. H. 1950. *Ionization Chambers and Counters.* London: Cambridge University Press.

Winkler, R. and G. Rösner. 1989. "Proportional Counter for Low-Level ^{241}Pu Measurement," *Nucl. Instrum. Methods. Phys. Res.*, A274:359–361.

Winn, W. G. 1987. "Ultra-Sensitive Examination of Environmental Samples with SRL Underground Counting Facility," *Annual Meeting of the American Nuclear Society,* June 7–12, Dallas, INIS, DE 8700 6384, p. 6.

Winn, W. G., W. G. Bowman and A. L. Boni. 1988. "Ultra-Clean Underground Counting Facility for Low-Level Environmental Samples," *Sci. Tot. Envir.*, 69:107–144.

Wirdzek, Š. and D. Kažimir. 1985. "Radioactivity of Frequently Used Building Materials in Slovakia," in *Low-Level Counting and Spectrometry, Proc. Third Int. Conference,* P. Povinec, ed., Bratislava: Veda, pp. 275–380.

Wogman, N. A. 1981. "Design and Environmental Applications of an Ultra-Low Background, High-Efficiency Intrinsic Ge Gamma-Ray Spectrometer," in *Methods of Low-Level Counting and Spectrometry.* Vienna: International Atomic Energy Agency, pp. 15–29.

Wogman, N. A. and J. C. Laul. 1982. "Natural Contamination in Radionuclide Detection Systems," in *Natural Radiation Environment,* W. G. Vohra et al., eds., New York: John Wiley and Sons, pp. 384–390.

Wolf, M. and C. Singer. 1991. "Synthesis of Propane via Hydrogenation of Propadiene, Propyne or Propene, for Low-Level Measurement of Tritium" (in German), in *Annual Report 1990 of the GSF Institute of Hydrology No. GSF-HY-1/91.* Neuherberg, Germany: Forschungszentrum fuer Umwelt und Gesundheit, pp. 150–153.

Yamaya, T. et al. 1981. "Fano Factor for Proton in Silicon at 87 K," in *Proc. Int. Symp. Nuclear Radiation Detectors,* K. Husimi and Y. Shida, eds., Tokyo: Institute for Nuclear Study, University of Tokyo, pp. 221–223.

Yu-Fu, Yu., H. E. Bjoernstad and B. Salbu. 1992. "Determination of Plutonium 239, Plutonium 240 and Plutonium 241 in Environmental Samples Using Low-Level Liquid Scintillation Spectrometry," *Analyst,* 117(3):439–442.

Yu-Fu, Yu., B. Salbu and H. E. Bjoernstad. 1991. "Recent Advances in the Determination of Low Level Plutonium in Environmental and Biological Materials," *J. Radioanal. Nucl. Chem. Articles,* 148:163–174.

Zastawny, A. and B. Rabsztyn. 1986. "A Needle Gas Counter for Measurements of Low Beta Radioactivity Solid Emitters, First Measurements," *Isotopenpraxis,* 22(6):193–197.

Zdesenko, Yu. G. et al. 1985. "Preliminary Results Neutrinoless Double Beta-Decay of ^{76}Ge" (in Russian), *Izv. Akad. Nauk SSSR, Ser. fiz.*, 49:862–867.

CHAPTER 3

Selected Fields of Low-Level Radiation

3.1 STARTING DATA

As was shown in Chapter 1, low levels of radionuclides are present practically everywhere, as natural components of the universe or from man-made sources. In this chapter we present, only briefly, some further examples of low-level fields.

One very extensive field for the evaluation of low levels of radionuclides is their dissipation from sources (Chapter 1) in the environment. In addition to the study of a number of physical, physico-chemical, biochemical, and biological dependencies, the aim of these measurements is the *protection of man,* mainly from *internal contamination* caused by radionuclides present in inhaled air and from contamination of the food chain (Desmet and Myttenaere, 1988).

Therefore, the study of the conditions of transport of radionuclides in the environment, combined with low-level analyses, is carried out in many laboratories and is the subject of many publications. The basic data on individual types of transport are extensive (atmospheric, aquatic, and terrestrial) and they are surveyed in the appropriate monographs (Eisenbud, 1963; Kathren, 1991). Fundamental geophysical data on individual systems can also be found in these monographs. In addition to various modes of propagation of radionuclides in the environment, some knowledge has also been collected concerning the endangering of the food chain by animal products contaminated with radionuclides (Wirth and Kaul, 1989).

From the point of view of the evaluation of low-level radioactivities we can divide the themes into the following three parts:

- transport in the atmosphere and its influence on the earth's surface

TABLE 3.1. Environmental Level of Some Transuranic Nuclides.

Sample		^{238}Pu	$^{239,240}Pu$	^{241}Am
Soil (Bq/kg)		0.7	0.1–7	0.02
Herbage (Bq/kg)	Plants	$4.5 \cdot 10^{-4}$	0.3–2	
	Lichen		4–10	0.7–2
	Grain, vegetables	$(0.2–14) \cdot 10^{-5}$	$(4–89) \cdot 10^{-4}$	
Water (μBq/L)	Sea		0.7–52	0.2–8
	Lake		0.1–29	0.46

Based on Yu-Fu, Salbu and Bjornstad (1991).

- transport in fresh water and marine ecosystems, accumulation in sediments, and passage into aquatic animals
- transport from the soil into the food chain, especially into plants

The requirements on low-level assay are demonstrated by the example of transuranium elements in Tables 3.1 and 3.2, which illustrate both their proportion in selected samples and their detection limits. From the values presented it is evident that such determinations require not only detectors with high counting efficiency and low background, but also highly sophisticated sample preparation (Yu-Fu et al., 1991).

Other selected fields include radiochronology, activation analysis, whole body counting, monitoring, and radon assessment. The radiotracer methodology as a source of low levels was already discussed in Section 1.1.4.

3.2 TRANSPORT OF RADIONUCLIDES IN THE ENVIRONMENT

3.2.1 Transport in the Atmosphere

In lower layers of air, the concentration of radionuclides in many places on the earth has been investigated for more than twenty years as part of radioecological studies dealing with *seasonal and long-term variations,* to assess the radiation exposure of the population. An example of a low-level source of radionuclides spreading through the atmosphere is a coal-fired power plant (Table 3.3). It is estimated that about 1400 tons of uranium were released into the atmosphere in 1974 from coal-fired electric generating stations in the United States alone (Tadmor, 1986). From a review surveying the results of several authors (Tadmor, 1986), it is evident that the content of radionuclides in various coals is very different (see also Table 1.47), but in all cases the highest activity is released into fly-ash and thus into the atmosphere. The dependence of their relative content on escaping fly-ash particles is given in Table 3.4. Although no influence of these emissions on the natural activity in soil from the plant surroundings was found for the tested type of coal and the conditions used (Rösner et al., 1984), the radionuclides released in this way contributed to low-level measurements of time and space variation of their concentrations (Rangarajan et al., 1986; Hötzl and Winkler, 1987).

The movement of radionuclides in the atmosphere is represented in addition to small particles of solid phase (the mentioned fly-ash, aerosols, etc.) by various gaseous components as well, for example, ^{222}Rn, ^{3}H, $^{14}CO_2$, ^{85}Kr, and others, which come into the atmosphere from various sources (Chapter 1). In addition to concentrations of these radionuclides in the atmosphere, their deposition in the soil and also in plants, which enter the food chain, is also a subject of investigation. According to a high radiation hazard, a special interest is devoted to radon and its decay products (Section 3.7). So, e.g., recently we found that cigarette smoke increases the measured equivalent volume activity of ^{222}Rn considerably (Tučková and Tykva, 1994).

We found useful (Tykva et al., 1990) *integral information* for a given time interval to determine the content of actinides in the atmosphere using low level alpha spectrometry of snow samples collected at the same site (approximately 2000 m above sea level) and in the same month (Table 3.5).

Many data concerning the transport of radionuclides in the atmosphere in dependence on various factors were obtained from the Chernobyl reactor accident (Section 1.1.3.5), the role of which in the contamination of agricultural and natural ecosystems is still being continuously evaluated and which will undoubtedly be a reason for control in the future from some viewpoints. The data (see Tables 1.33–1.37 and Figures 1.20–1.26) on measured radionuclides, as well as the isotopic ratio at various places on earth, give

TABLE 3.2. Detection Limits of Transuranium Elements in the Environment.

Sample	Detection Limit
Soil, bottom sediments	$37 \cdot 10^{-3}$ Bq/kg
Bottom sediments	$3.7 \cdot 10^{-5}$ Bq per sample
Fallout	$3.7 \cdot 10^{-3}$ Bq/m^3
Atmospheric dust	$2 \cdot 10^{-7}$ Bq/m^3
Natural water	$1.8 \cdot 10^{-5}$ Bq/L
Biological materials (raw mass)	$3.7 \cdot 10^{-3}$ Bq/kg
Biological ash	$1 \cdot 10^{-4}$ Bq/g

Based on Yu-Fu, Salbu and Bjornstad (1991).

TABLE 3.3. Concentrations of Elements and Radionuclides in Coal.

Element, Radionuclide	U (ppm)	^{238}U ($Bq \cdot g^{-1}$)	Th (ppm)	^{232}Th ($Bq \cdot g^{-1}$)	^{226}Ra ($Bq \cdot g^{-1}$)	^{210}Po ($Bq \cdot g^{-1}$)	^{210}Pb ($Bq \cdot g^{-1}$)
Interval of conc.	0.4–1800	0.01–0.16	0.25–79	0.002–0.1	0.01–2.6	0.01–0.03	0.01–0.03

Adapted from Tadmor (1986).

TABLE 3.4. Variation of the Enrichment Factor of Radionuclides as a Function of the Size of the Escaping Fly-Ash Particles.

Particle Size[a] (μm)	Enrichment Factor Related to Mineral Matter of Coal			
	^{238}U	^{226}Ra	^{228}Th	^{210}Pb
2	2.8	2	1.2	4.8
10	1.6	1.3	1.1	2.1
17	1.3	1.1	1.1	1.4

[a]Particles sizes of < 10 μm are considered to be within the respirable range.
Based on Tadmor (1986).

some insight not only into the operating conditions prior to the accident and into the processes during its individual phases, but, especially, into the atmospheric transport of the debris. As an example let us mention the measurement of the content of ^{131}I, ^{137}Cs, ^{103}Ru, and ^{132}Te-^{132}I in the rain falling over the period of ten days in May 1986 and the measurements of corresponding activity values, deposited on one m^2 of ground (Martin et al., 1988). The authors then followed further transport of radionuclides into vegetables, milk, and meat (this transport is the subject of Section 3.2.3).

3.2.2 The Movement of Radionuclides in Water

The determination of the movement of radionuclides in surface and underground waters including sea water, comprises – from the point of view of measurements – not only water alone, but marine fauna and flora and the analysis of sediments as well. In a review of the transfer of long-lived radionuclides through the marine food chain (Belot, 1986), an objective example is mentioned: the radionuclides concentrated in the sediment are transferred to the detrital worms, which are eaten by crustaceans, themselves eaten by squids, in turn eaten by fish, and finally ingested by man.

The measurement of low activities thus contributes to the solution of three thematic areas. The first one is the *distribution of radionuclides in water* and the differences in their behavior under certain conditions. Thus,

TABLE 3.5. Measured Activity ($mBq \cdot m^{-2}$) in Snow Samples.

	1985	1986	1987	1988
^{234}U	2414 ± 139	3474 ± 113	8188 ± 256	1025 ± 58
^{238}U	1785 ± 137	3789 ± 120	8022 ± 254	835 ± 54
$^{239,240}Pu$	1936 ± 101	414 ± 31	259 ± 49	685 ± 43

for example, it was observed whether the changes in pH and/or concentrations of naturally occurring colloidal organic carbon cause differences in adsorption properties between Am and Cm (Nakayama and Nelson, 1988). To clarify these effects batch adsorption experiments were conducted on natural water-sediment systems to which americium and curium were added concurrently.

The second thematic area is the *passage of radionuclides from water into biological objects* living in water (Poletilo, 1990). In most fresh water systems in the southeastern United States, the mosquitofish is an abundant prey species for many kinds of larger fish and wildlife. Moreover, it is an appropriate subject for the study of contaminant distribution on a microgeographic scale and the distribution of contaminants between different sex and/or size classes (Newman and Brisbin, 1990). The measured whole body concentrations of ^{137}Cs in three locations of a former cooling reservoir of a nuclear reactor (Table 3.6) suggest that these values may vary between sex or size, as well as within microgeographic scales.

By measurement of ^{241}Pu in the food chain lichen reindeer man during 1961–1975 it was found (Holm and Persson, 1977) that a maximum in lichen occurred in 1963–1966. This maximum was in coincidence with the $^{239+240}$Pu maximum and was caused by the extensive nuclear weapon tests during 1961–1962.

Important information may be obtained by measuring low levels of *radionuclides in sediments*. In the South Spanish marine environment it has been found (Manjón et al., 1992) that local man-made sources are the main contributors to the activity of the sediment taken from two rivers (Table 3.7). The comparison of suspended matter and corresponding contamination of the surface sediments (Petit et al., 1987) are shown in Table 3.8 for ^{210}Pb, ^{226}Ra, and some heavy metals. The distinct gradient of specific activity in the sediment was shown in the measurements of the total alpha and beta activity in dried samples of 30-mm diameter (Mondspeigel et al., 1990) after withdrawal from the Bezdrev pond in South Czech Republic, in the proximity of the mouth of the creek, which comes from the nearby uranium ore mill at Mydlovary (Table 3.9).

3.2.3 Transport of Radionuclides from the Soil

Radionuclides that come to the ground from the atmosphere are deposited on the earth's surface, and from there a substantial part of them comes into the upper layers of soil. At the same time there are three pathways for radionuclide transport: *resuspension into the atmosphere, migration within the soil,* and *uptake by the roots of plants.* The greatest attention is paid to the study of the conditions for the intake of radionuclides by plants and their

TABLE 3.6. Sex, Wet Weight and ^{137}Cs Concentration in Mosquitofish Collected from Tree Locations in a Former Reactor Cooling Reservoir.

			Wet Wt (g)		^{137}Cs ($Bq \cdot g^{-1}$ wet wt)		Concentration Factor
Loc.	Sex	N	Median	Range	Median	Range	[($Bq \cdot g^{-1}$ dry wt)/($Bq \cdot liter^{-1}$)]
1	F	28	0.26	0.12–1.20	1.68	0.80–3.37	7,895
	M	28	0.16	0.10–0.23	1.73	0.74–2.94	8,755
2	F	55	0.31	0.15–0.96	1.61	0.90–2.98	8,474
	M	29	0.17	0.10–0.28	1.72	0.71–2.76	10,287
3	F	31	0.28	0.13–0.54	1.51	0.84–3.66	7,359
	M	19	0.16	0.10–0.26	2.23	0.69–5.02	11,285

Adapted from Newman and Brisbin (1990).

TABLE 3.7. U-Isotopes and ^{210}Po Activity Concentrations for Sediment Samples Taken at Two Rivers (A, B).

Sample	Activity Concentration (mBq·g^{-1} of sediments)			
	^{238}U	^{235}U	^{234}U	^{210}Po
A1	250.0 ±1.7	11.0 ±1.2	260.0 ±1.7	252.0 ±10.0
A2	277.0 ±13.0	10.3 ±0.8	290.0 ±14.0	341.0 ±12.0
B1	1060.0 ±75.0	48.0 ±4.9	1060.0 ±75.0	1194.8 ±112.2
B2	42.0 ±3.9	2.80 ±0.66	45.0 ±4.2	41.6 ±2.7
B3	1100.0 ±130.0	55.0 ±10.0	1200.0 ±130.0	820.0 ±50.2
B4	160.0 ±11.0	7.7 ±1.0	170.0 ±12.0	615.2 ±28.0

Adapted from Manjón, Martínez-Aguirre and García-León (1992).

TABLE 3.8. Radionuclides (^{210}Pb, ^{226}Ra) and Heavy Metals (Pb, Zn, Cd, Cu) in Suspended Matter of the River Meuse and in Surface Sediments (0–2 cm Depth).

Radionuclides	Sediments	Suspended Matter
^{210}Pb$_{tot}$ (Bq·g^{-1})	276 ± 12	258 ± 12
^{226}Ra (Bq·g^{-1})	129 ± 10	131 ± 10
Pb (ppm)	115 ± 5	124 ± 5
Zn (ppm)	772 ± 31	837 ± 33
Cd (ppm)	4.8 ± 0.2	4.8 ± 0.2
Cu (ppm)	50 ± 2	80 ± 3

Based on Petit, Thomas and Lamberts (1987).

TABLE 3.9. The Depth Profile of Measured Counting Rate in Sediments at a Certain Site of the Pond in the Surrounding of the Uranium Ore Mill.

Depth from Sediment Surface	Dry Sample (g)	Rate (cpm·g^{-1})
0–10 cm	4.48	44.1
20–30 cm	4.85	9.4
30–40 cm	4.30	7.1
40–50 cm	4.56	4.9

possible further transport into the meat or milk of animals designed for human nutrition.

Low-level counting techniques provided proof that for radium, polonium, lead, and calcium in different plants, soils, and environmental conditions, the plant uptake response to substrate concentration is a non-linear function (Simon and Ibrahim, 1987). There are many soil factors influencing their transfer, such as organic matter content, pH, clay fraction, type of clay minerals, soil moisture content, and the amount of soluble and exchangeable potassium, calcium, and ammonium, and also plant parameters, such as growth and development, species, and variety. The effect of some soil parameters on transfer of ^{134}Cs, ^{137}Cs, ^{85}Sr, ^{89}Sr, and ^{90}Sr from soil to edible plant parts were investigated in detail (van Bergeijk et al., 1992). The soil organic matter content has a great impact on the transfer of these two elements: transfer of Cs increases while that of Sr decreases with increasing organic matter content. In the pH 3.9–8.4 range the transfer of Cs was not affected by soil pH but the transfer of Sr decreased by a factor of 1.7 when soil pH increased from pH 4.5 to 7.4. The study of the impact of aging and climatic conditions on transfer of Cs and Sr showed several differences (Noordijk et al., 1992). The transfer of Cs to edible parts of several plant species gradually decreased as a result of aging, while that of Sr did not change when its residence time in the soil increased. After correction for the effect of aging, annual fluctuations of up to a factor of ten were found, depending on the plant species. These fluctuations were attributed to the effects of climatic conditions.

Special attention was devoted to the relationship between the concentrations of the radionuclides in the soil and corresponding representation of vegetal products, forming human food. The non-linear transfer, mentioned earlier, is evident from the results in Table 3.10 for two-year-old container-grown blueberry plants. The tracer methodology (^{125}I, ^{75}Se, ^{134}Cs), neutron activation, atomic absorption, or inductively coupled plasma spectroscopy was used.

The movement of the deposit of radionuclides in plants was investigated in the study of the transfer of ^{134}Cs, ^{137}Cs, ^{131}I, and ^{103}Rn from flowers to honey and pollen (Bunzel et al., 1988). The results suggest that the radionuclides were taken up by the plant leaves and transported to nectar and pollen.

Good models for such purposes were given by the Chernobyl accident. Thus, e.g., samples of pasture grass and fresh farm milk were measured. Using the measured data (Table 3.11), influenced by the conditions applied (Dreicer and Klusek, 1988), a milk transfer coefficient of 0.001 $d \cdot liter^{-1}$ was calculated. No significant difference was found using the total grass as compared to only its upper portion. A larger number of similar studies were

TABLE 3.10. Summary of the Leaf/Soil Concentration Ratio (CR) and Berry/Leaf Transfer Ratio (TR).

Element	Ratio	Sample	Number of Samples	Range
I	CR	June	9	0.009–4
		August	4	0.4–6
		Fall 1	6	0.6–1
	TR	All	3	0.41–2.3
Se	CR	June	18	0.22–26
		August	18	0.31–16
		Fall 1	14	0.31–10
		Fall 2	14	0.22–13
	TR	All	12	0.14–0.5
Cs	CR	June	6	0.64–9
		August	18	3.7–20
		Fall 1	17	3.2–16
		Fall 2	18	1.4–4.4
	TR	All	20	0.5–5.5
Pb	CR	All	14	0.00008–0.6
	TR	All	6	0.066–3.1
U	CR	June	4	0.016–0.10
		August	5	0.0038–0.035
		Fall 1	19	0.0008–0.040
		Fall 2	19	0.0004–0.14
	TR	All	7	0.042–4.0

Adapted from Sheppard and Evenden (1988).
Fall 1 and Fall 2 mean fall of year 1 and fall of year 2.

carried out, for example, for the environment of a nuclear fuel reprocessing plant (Hauschild and Aumann, 1989).

The results of low level measurements are sometimes not sufficient for assessing the environmental and radiological consequences. Such difficulties are overcome by recourse to mathematical models which describe radionuclide dispersion in the environment and transport to man (Jackson et al., 1987). Such predicted concentrations of selected radionuclides after one year of continuous deposition are presented as ratios relative to milk in Table 3.12.

Finally, the effect of tissue cooking on decreasing radionuclide concentrations (an effect that depends upon their chemical bindings) should be noted. Thus, significant decreases in ^{137}Cs, ^{134}Cs, ^{60}Co, and ^{110m}Ag concentrations in the carcass and liver samples of mallard ducks *(Anas platyrhynchos)* released at reactor radioactive leaching ponds were detected (Halford, 1987).

TABLE 3.11. Concentration of ^{131}I in Grass and Fresh Milk Collected at Chester after Chernobyl Accident.

Sampling Date[a]	Grass (Bq·kg^{-1} dry wt)		Fresh Farm Milk (Bq·L^{-1})
	Top	Total	
May			
9	14.8 ± 6.7	9.6 ± 6.7	
10	40.7 ± 12.2	29.2 ± 14.7	
12	102.0 ± 11.2	75.5 ± 11.9	1.07 ± 0.07
13	101.0 ± 13.1	67.0 ± 13.5	1.45 ± 0.07
14	67.3 ± 8.7	51.1 ± 15.6	1.35 ± 0.07
15	89.0 ± 14.2	72.2 ± 17.1	1.12 ± 0.06
16	80.7 ± 12.1	69.9 ± 15.9	1.35 ± 0.08
17	93.5 ± 16.8	64.4 ± 17.8	1.47 ± 0.08
19	53.5 ± 9.1	65.1 ± 24.5	0.88 ± 0.07
20			0.77 ± 0.07
21	17.4 ± 4.4	17.4 ± 5.2	
22	40.5 ± 2.0	55.1 ± 14.6	
23	26.5 ± 6.4	32.9 ± 12.0	0.62 ± 0.15
26	43.8 ± 10.5	52.9 ± 16.6	0.47 ± 0.08
27			0.38 ± 0.06
28	31.3 ± 12.8	36.3 ± 19.2	0.29 ± 0.06
29	25.1 ± 7.3	30.7 ± 10.6	0.37 ± 0.08
30	30.8 ± 8.9	20.7 ± 10.9	0.23 ± 0.08
June			
3	39.9 ± 11.6	46.6 ± 16.8	0.28 ± 0.06
4			0.32 ± 0.08
5			0.26 ± 0.04
6			0.35 ± 0.08

[a]Date of collection for grass; for milk the date of milking.
Based on Dreicer and Klusek (1988).

TABLE 3.12. Ratio of Predicted Concentrations in Various Foodstuffs to That for Milk Following Continuous Deposition (for the Food Products at the Time of the First Harvest and for Milk at 365 days).

Food Products	Radionuclide			
	^{137}Cs	^{129}I	^{241}Pu	^{241}Am
Root vegetables	0.5	0.8	1400	38
Leafy-green vegetables	4.7	8.4	7400	590
Fruits	2.8	17	7	20
Cereals	2.0	38	27	6.5
Beef	6.0	2.2	15	1.3
Beef offal	1.7	–	4200	240

Based on Jackson, Coughtrey and Crabtree (1987).

3.3 RADIOCHRONOLOGY

The determination of age by means of *the measurement of specific activity* of corresponding natural radionuclides is based on the half-life (Figure 1.1) of suitable original radionuclides in the environment (Table 3.13). The basis of radiochronology is Equation (1.4). Since we know the specific activity of the corresponding radionuclide at the starting time $t = 0$ (for example for ^{14}C at the time of death of organic material and the cessation of metabolism), we can determine the specific activity at the time of the actual measurement and calculate, on the basis of these two values, the time necessary for a given decrease of activity and thus the age of the sample. It is evident that in radiochronology we seek to determine values *lower* than the natural concentration of the radionuclide; therefore, the determination of the age represents one of the sources of extremely low levels.

The most widespread method of radiochronology is ^{14}C-dating (Libby, 1955), which has been used since the 1950s in many countries and which has contributed to the development of low-level techniques in the fields of measurement and sample treatment. This spread has been caused both by the position of carbon in nature, and by the half-time of its decay (5730 years). Correlation with a number of other methods of age determination (e.g., dendrochronology, stratigraphy, etc.) has shown that the ^{14}C-dating is reliable. As limits on the ends of the dating scale, the international stan-

TABLE 3.13. Radiochronological Methods and Their Application.

Method	Radionuclide Used	Decay Product	Application
Radiocarbon method	^{14}C	^{14}N	Up to order of magn. of 10^4 y
Rb-Sr method	^{87}Rb	^{87}Sr	Young Tertiary-Precambrian
K-Ar method	^{40}K	^{40}Ar	Pleistocene-Precambrian
Tritium method	^{3}H	^{3}He	Up to order of magn. of 10^3 y
Krypton method	^{85}Kr	^{85}Rb	Up to order of magn. of 10^3y
Decay series	^{238}U	^{206}Pb	Tertiary-Precambrian
methods	^{234}U	^{230}Th	Holocene-Pleistocene
	^{235}U	^{207}Pb	Tertiary-Precambrian
	^{232}Th	^{208}Pb	Tertiary-Precambrian
He methods	^{238}U	^{4}He	Cenosoic-Precambrian
	^{235}U	^{4}He	Cenosoic-Precambrian
	^{232}Th	^{4}He	Cenosoic-Precambrian
Re-Os method	^{187}Re	^{187}Os	Tertiary-Precambrian
Th-Pb method	^{230}Th	^{226}Ra	Early-Middle Pleistocene
Pa method	^{231}Pa	^{227}Ac	Würm-Glacial
K-Ca method	^{40}K	^{40}Ca	Pleistocene-Precambrian

dards "modern" oxalic acid or sucrose and "nonradioactive" anthracite have proven valuable.

In the current low-level arrangements the determination of age is carried out most frequently up to approximately 20,000 years for various fields (Šilar and Tykva, 1991). Thus, for example, within the frame of the expedition of the Czech Institute of Egyptology, Charles University, we analyzed the royal cemetery of the 5th dynasty on the edge of the Western Desert above Abusir (29°54′ N, 31°13′ E), 30 km south of Cairo on the west bank of the river Nile, using different kinds of wood (*Ficus* sp., or *Acacia arabica,* respectively), and the linen of mummy wrappings, as well as samples from other analogous sites, textiles, charcoal, etc. (the interval from 1500 ± 100 to 7240 ± 150 years). In addition to archeological samples of organic origin we also dated geological samples, using in some instances the necessary parallel correction measurement $\delta^{13}C$ by mass spectrometry (e.g., the measurements of tuffs from the Holocene stratotype profile, which have been correlated using previous paleontologically estimated ages). We thus determined the oldest age of 28,500 ± 1500 years in corals of the Jamanitas Formation. Age determination by means of ^{14}C was also used for further studies, for example in organic chemistry for controlling the origin of certain natural compounds, especially for pharmaceutical industry: the specific activity of ^{14}C in a substance extracted from contemporary plants is distinctly higher than it is in the synthetic compound, because the chemicals used are prepared from coal or crude oil. This has also been used in the analysis of ethanol.

Other dating methods are meant predominantly for very old samples in geology and space research (e.g., meteorites) with the exception of the decay of the radionuclides ^{3}H and ^{85}Kr used for younger samples (Table 3.13). Tritium is used relatively frequently, because of its half-life of 12.26 years, predominantly for studies in hydrology (Moser and Rauert, 1980) and also for the needs of the food industry (e.g., the control of wine). Another radionuclide used for dating young groundwaters is ^{85}Kr (Held et al., 1992), in the use of which the distortion of tritium concentration in the environment due to nuclear explosions (see Section 1.1.3) is eliminated.

Both the decay series (lead) and the helium methods are based on the decay of primordial radionuclides (Section 1.1.2.2). The helium method is especially suitable for magnetites containing uranium, because they prevent the escape of the helium formed. For the determination of a very great age, two methods are widely used, which directly afford a stable radionuclide on decomposition: the Rb-Sr and K-Ar methods. The use of other methods is infrequent. A more detailed survey of the dating methods is to be found in the literature (Kathren, 1991).

3.4 ACTIVATION ANALYSES

As mentioned in Section 1.1.4, stable atoms may be converted to emitters of alpha, beta, gamma, or X radiation by transmutation of their nuclei. The concentration of non-radioactive atoms is thus determined by means of low-level spectrometry of radionuclides (Niese, 1985; Petra et al., 1990). For transmutation, particles are used incident on the sample with the necessary energy, i.e., either neutrons in the *irradiation channels of the nuclear reactor* (NAA) or *accelerated charged particles* at the outlet of the accelerator (e.g., PIXE).

This methodical approach finds an application both for biological material in trace element research, having a fundamental importance for studying metabolism in living organisms (Mills, 1985), and in the research of admixtures in inorganic materials (Niese, 1985). In the latest studies impurities in semiconductor silicon were determined using NAA (Cr, Co, As, Mo, Ag, Sb, W), the concentration of which was in the interval of 10^{11} to 10^{12} atoms per cm^3. The analysis was carried out by low-level gamma coincidence spectroscopy after irradiation of 1 g of sample for about 90 hours at 8×10^{13} neutrons cm^{-2} s^{-1} and appropriate radiochemical sample treatment (separation and purification).

A further broad field of use of activation analysis is environmental measurement. Thus, concentrations of ^{129}I and natural ^{127}I in soils, food crops, and animal products collected in the environment of the nuclear fuel reprocessing plant were determined by NAA (Robens and Aumann, 1988). The ^{125}I tracer was added to the dried samples to provide an estimate of the chemical yield of the pre-irradiation separation of iodine. Iodine was separated by combustion in a stream of O_2 and collected on activated charcoal by heating and finally trapped in a quartz tube. The quartz tube was sealed and used as an irradiation ampoule. One or more of the ampoules were irradiated with reactor neutrons [thermal neutron flux: 1.4×10^{14} cm^{-2} s^{-1}, epithermal neutron flux (fluence rate): 1.3×10^{13} cm^{-2} s^{-1}, equivalent fission neutron flux: 5.2×10^{13} cm^{-2} s^{-1}] for eight to ten hours together with a comparative standard containing known amounts of ^{127}I and ^{129}I. The neutron-induced nuclear reactions used for the activation analysis of the iodine isotopes are $^{129}I(n,\gamma)$ ^{130}I ($T_{1/2}$ = 12 h) and $^{127}I(n,2n)$ ^{126}I ($t_{1/2}$ = 13 d). After irradiation, the contents of the ampoules were subjected to further purification by solvent extraction and distillation steps. The ^{130}I and ^{126}I activities produced in the samples and comparative standards during irradiation were measured by gamma-ray spectrometry providing a measure of the ^{129}I and ^{127}I contents, respectively. A sum-coincidence spectrometry technique was used for measuring very small amounts of ^{130}I in the presence of high ^{126}I activities. The measured results are summarized in Table 3.14.

TABLE 3.14. ^{129}I and ^{127}I Concentrations and $^{129}I/^{127}I$ Isotope Ratios in Soil (Dry Wt) and Fruit (Wet Wt) from the Environment of the Reprocessing Plant.

Sample	Date of Collection (1983)	Water Content (%)	^{127}I (10^{-7} g·g^{-1})	^{129}I (10^{-13} g·g^{-1})	^{129}I (μBq·g^{-1})	$^{129}I/^{127}I$ (10^{-6})
Soil (0–30 cm)	13 July		7.5 ± 1	8.0 ± 1	5.2 ± 0.8	1.07 ± 0.2
Plums	8 Sept.	82.1	0.55 ± 0.2	0.34 ± 0.09	0.23 ± 0.06	0.6 ± 0.3
Pears	8 Sept.	78.3	0.080 ± 0.02	0.041 ± 0.01	0.027 ± 0.007	0.51 ± 0.2
Raspberries	13 July	85.5	0.38 ± 0.08	0.19 ± 0.05	0.12 ± 0.03	0.50 ± 0.2
Soil (0–30 cm)	13 July		12 ± 3	5.9 ± 1	3.8 ± 0.8	0.49 ± 0.2
Rhubarb	13 July	90.0	0.32 ± 0.08	0.19 ± 0.05	0.12 ± 0.03	0.59 ± 0.2
Cherries	13 July	84.7	0.20 ± 0.06	0.09 ± 0.04	0.061 ± 0.02	0.45 ± 0.2
Soil	21 July		16 ± 3	8.5 ± 0.9	5.6 ± 0.6	0.53 ± 0.1
Apples	8 Sept.	82.8	2.8 ± 0.4	0.038 ± 0.01	0.025 ± 0.007	0.014 ± 0.004

Based on Robens and Aumann (1988).

An analogous procedure was also used for estimation of the $^{128}I/^{127}I$ ratios in various Japanese environmental samples (Katagiri et al., 1990).

Determination of thallium in complex biological matrices by NAA has been described employing substoichiometric extraction of thallium and low-level beta counting (Itawi and Turel, 1987). Two samples and a standard can be processed within two hours.

In connection with the use of platinum-containing drugs in cancer therapy and the introduction of platinum catalysts in car exhaust systems, the limit of platinum determination in biological material by NAA has been investigated (Xilei et al., 1991). After irradiating for one hour at 5×10^{13} $cm^{-2}\ s^{-1}$ the ultimate limit of detection is approximately 30 $pg \cdot g^{-1}$.

The construction and performance of a HPGe-NaI(Tl) Compton suppression spectrometer for NAA determination of nanogram quantities of arsenic in different samples of biological material has also been described without chemical separations (Petra et al., 1990). The detection limits were between 1–4 $ng \cdot g^{-1}$, respectively, while for citrus leaves the limit was 50 $ng \cdot g^{-1}$.

In collaboration with Vrije Universiteit in Amsterdam (Dr. R. D. Vis, head of the laboratory) we have developed a new methodical approach to study the putative binding carrier protein of different elements in the animal body (Tykva et al., 1993). Using hair mineral analysis as a means to assess the internal body burdens of environmental mineral pollutants, we investigated their bindings to soluble proteins isolated from the hair, skin, blood plasma, kidney, and liver, and the putative transport proteins. The protein extract from rat hair was subjected to discontinuous polyacrylamide gel electrophoresis (PAGE) in the presence of SDS (Votruba et al., 1985). The gels were then irradiated using synchrotron X-ray fluorescence (SXRF). The corresponding activity in the individual gel sections determined the putative bindings of the individual environmental pollutants (transformed previously to appropriate X-ray emitters) to proteins of the molecular weights corresponding to the active sections (Table 3.15). This method revealed a substantial distribution of lead in the region of more than 67 kDa previously determined by using ^{210}Pb. The results obtained suggest a good correlation between the distributions of lead and sulfur in hair proteins. Furthermore, e.g., copper was also found in the region of content of sulfur, etc. The combination of PAGE-SDS with activation analysis by SXRF thus makes it possible to search for bindings of different non-radioactive elements to extracted proteins. This could be of considerable relevance in human medicine, e.g., for blood samples obtained *in vivo.*

3.5 WHOLE BODY COUNTING

Whole-body monitors or counters began to be designed and built in the

TABLE 3.15. Distribution of Different Elements (%) in Several Regions of a Polyacrylamide Gel after Electrophoresis and Synchrotron X-Ray Fluorescence.

	Estimation in the Range of Mol. Weight (kDa)			
Element	>67	67–43	43–20	20–14
P	46	27	3	24
S	65	16	2	17
K	28	21	8	43
Fe	30	21	15	34
Cu	90	–	–	10
Zn	24	16	7	53
Pb	78	18	1	3

1950s when it was realized that nuclear weapons tests resulted in a worldwide spread of radionuclides in the biosphere. Some of these radionuclides have reached humans through contaminated air, water, and foodstuffs.

In addition to the assessment of internal radioactive contamination, the early whole-body monitors were also widely used for some medical and radiobiologic studies and examinations (Meneely and Linde, 1965). Medical applications included investigations directed at understanding the kinetic behavior and metabolism of trace elements in the human body, whether they occurred naturally as potassium ^{40}K or artificially as ^{137}Cs, ^{131}I, and ^{59}Fe.

Later, when nuclear medicine became well established and a number of specific methods were developed to meet the requirements of this new field or radionuclide applications, whole-body monitors came into use mainly for radiation protection purposes.

The development and construction of the first whole-body monitors fell just at the time when large sodium iodide crystals and multichannel analyzers began to be available, and they immediately found use in these instruments in which liquid and plastic scintillators were also employed. It may be interesting to note that in the beginning of the 1960s about 100 such monitors were at work in various laboratories and research centers as well as medical establishments throughout the world. Many of these machines are still in operation, although some of them may have been modified and now use more advanced electronics including computers.

The determination of radionuclides using the whole-body monitors permits a *simultaneous measurement in the whole organism.* These monitors can be adapted for measurements in the human body or, with smaller constructions, for laboratory animals. A detailed description of such a counting arrangement is to be found in the literature (Lössner, 1977). The application of these devices varies. From the viewpoint of low levels the counting in humans with radiation protection is important for the monitoring of internal contamination with radionuclides. It is also used in human medicine, mainly in diagnostics, using short-time radioindicators. However, in this case higher levels are usually involved. The same is true in counting in laboratory animals where whole body activity is usually followed after the application of radiotracers (Matsusaka et al., 1988).

A whole-body monitor consists typically of a shielded chamber, either fixed or mobile, with a bed or chair for positioning the person being measured. Several detectors, fixed or movable, are placed around the person or are moved on tracks. There are many different types of whole-body monitors depending on their actual applications and other special requirements. The most sensitive monitors are usually shielded by materials with the lowest possible contamination. For their construction, pre-1945 steel and lead are usually used.

Some whole-body monitors are of special design and they are used, for example, in the measurement of the lung deposition of actinides such as uranium, plutonium, and americium or in thyroid screening. There are also simple whole-body contamination monitors for routine checking of personnel working with open radioactive sources or in other facilities where persons may be contaminated. Some of these whole-body contamination monitors can distinguish radiation from radionuclides deposited on the surfaces of the clothes or skin.

The configuration and response of a conventional whole-body monitor are illustrated in Figure 3.1. The instrument typically uses four NaI scintillation detectors mounted on a ring around the body.

The modern whole-body monitors use both NaI(Tl) and Ge detectors to achieve maximum sensitivity and high resolution identification of radionuclides of complex mixtures. Some of them are of a scanning type generating count-vs-position information. Within a few minutes they can measure the whole-body activity with a sensitivity of 150–250 Bq in the case of ^{134}Cs and ^{137}Cs, the activity of ^{54}Mn, ^{58}Co, and ^{60}Co in the lungs with a sensitivity of 20–100 Bq, the activity of ^{131}I and ^{133}I located in the thyroid with a sensitivity of about 20 Bq, and the activity of ^{58}Co and ^{60}Co present in the GI tract with a sensitivity of better than 30 Bq (Canberra, 1992).

A simple whole-body counter for the determination of contamination of the human body (Berg et al., 1987) is provided with several relatively very large sodium iodide crystals (in Berg et al., 1990, with four 5-in. by 4-in. pieces). The measurement is usually performed using a bed-like geometry. The whole device is placed in a very large shielding box composed of lead, steel, and quartz sand. For measurement with a time interval of

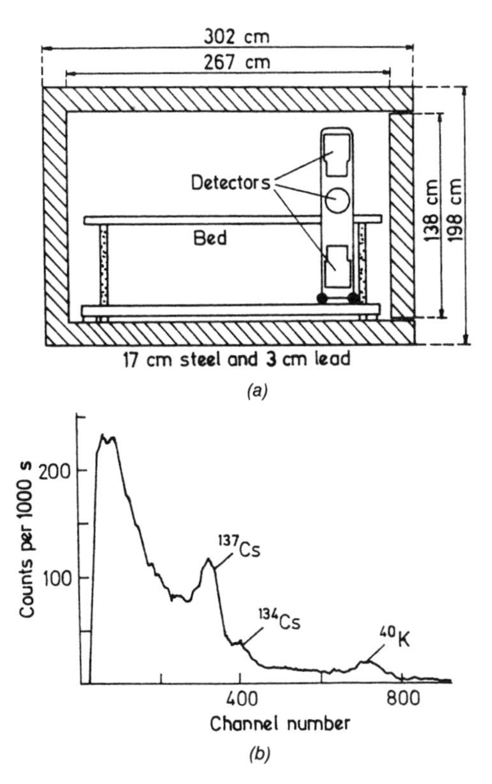

FIGURE 3.1. A conventional whole-body monitor: (a) a configuration, (b) a pulse height spectrum obtained from a person contaminated with ^{137}Cs and ^{134}Cs. From Hayball and Dendy (1991).

twenty-five minutes the detection limit of ^{137}Cs for an adult person was 70 Bq, applying the arrangement described. The error of determination of this level is also affected by poor reproducibility of the positioning of a measured person and it goes up to 10–20%.

The counter described was used in the area of Munich in Germany (Berg et al., 1987) for whole body measurement in children and adults after the Chernobyl accident (April 26, 1986). On May 6th the whole body retention of ^{131}I was 300 Bq decreasing with a half-life of eight to nine days. A steady increase of radiocesium was only detected four weeks after the accident. After 100 days a dynamic equilibrium was achieved at a level of 13 Bq per kg body weight for ^{137}Cs and approximately 6.5 Bq·kg^{-1} for ^{134}Cs. No significant influence of age or sex on radiocesium concentration could be observed. For adults a radiation dose to the thyroid of 0.8 mSv was estimated due to ^{131}I. From the whole body retention of radiocesium an effective dose equivalent of 0.07 mSv was estimated up to the end of 1986. On single occasions doses up to three times higher have been observed within the population examined.

The measured values of the body activity of ^{137}Cs fifteen months after the Chernobyl accident did not differ too much in a group of volunteers (Table 3.16). In the whole-body counter the activity of ^{40}K (30.9 $Bq \cdot g^{-1}$ of stable K) was also determined, which reflects the whole pool of stable K. The ratio of radiocesium to radiopotassium in the total body was four times higher than that in the urine which was measured using a low-level HPGe-detector.

In the study of biokinetics of Cs isotopes in humans a small group of volunteers was followed using whole body counting in combination with intake and excretion rate determinations (Henrichs et al., 1989). A highly contaminated sample of venison was examined. The experimental data are summarized in Table 3.17 where resolution of ^{134}Cs and ^{137}Cs is not carried out due to measurement by NaI(Tl) detectors (Section 2.3.4.2). For all subjects, a fast excretion via urine was found which reached the original value within three to five days after intake. This pattern is confirmed by the results of the whole body measurements showing a steep decrease during the first few days.

Large-scale routine examinations of many thousands of people were carried out in most countries affected by the Chernobyl accident. In these mea-

TABLE 3.16. Parameters of the Whole Body Retention of ^{131}I and the Dose of the Thyroid (D) to Fourteen Volunteers.

Volunteer No.	Activity (Bq)	Effective Half-Life (d)	D (mSv)
18	694	7.6	0.48
27	411	11.6	0.43
33	302	11.9	0.33
37	392	11.6	0.41
38	692	6.3	0.40
43	454	9.0	0.37
45	389	9.8	0.35
53	551	7.4	0.37
54	526	7.6	0.36
71	940	8.6	0.74
73	344	8.6	0.27
90	420	10.6	0.41
94	818	6.1	0.45
106	468	11.8	0.50
Mean ± S.D.	503 ± 190	9.2 ± 2.1	0.42 ± 0.11

Based on Berg et al. (1987).

TABLE 3.17. Personal Data of the Subjects Observed by Whole-Body Counting (Cs Activities as Sum of ^{134}Cs and ^{137}Cs) and Ingested Cs.

				Cs Activities before Meal (S.D.)		
Subject No.	Sex	Age (years)	Body Weight (kg)	Urinary Excretion (Bq d^{-1})	Total Body (Bq)	Cs Intake (Bq)
1	m	45	79	9 ± 1	1500 ± 300	1020
2	m	36	85	11 ± 1	2100 ± 400	1100
3	m	36	78	8.3 ± 0.7	1800 ± 300	1060
4	m	43	95	11 ± 2	1400 ± 300	1070
5	m	31	85	10 ± 2	1800 ± 400	1050
6	f	46	54	6.0 ± 0.6	730 ± 150	920
7	f	49	62	6.6 ± 0.6	1400 ± 300	1060
8	f	28	56	5.6 ± 0.8	1400 ± 300	1170
9	f	38	58	7.5 ± 0.5	880 ± 180	1030
10	f	35	61	6.8 ± 0.9	1500 ± 300	1030

Adapted from Henrichs et al. (1989).

surements the content of such radionuclides as ^{134}Cs and ^{137}Cs has been evaluated and studied for a long period of time. The results are used for assessment of the contribution of internal contamination to the total effective dose equivalent of exposed persons.

Another application of whole body counting involves determination of whole body content and turnover of cesium and potassium (Berg et al., 1990). The results obtained are given in Table 3.18.

A very good example of the use of the whole body counting in the radiation protection of people is the monitoring of ^{137}Cs transfer into mother's milk (Gall et al., 1991). Between May 1987 and December 1988 duplicates of daily food intakes, as well as corresponding mother's milk samples, were collected from twelve nursing mothers for two to four weeks in order to measure the ^{137}Cs activity. Once during the collection period the total body activity of each of the mothers involved was measured. Based on these results, ^{137}Cs is transferred into mother's milk not only directly from food intake but also from the accumulated body burden. Approximately 19% of the ^{137}Cs activity from the daily food intake and about 13% of the specific ^{137}Cs body activity of the mother are transferred into 1 L of mother's milk.

3.6 FIELD AND AREA MONITORS

Field and area monitors are commonly used for environmental *in situ* or working area measurements, including primarily the measurement of ex-

TABLE 3.18. Total Body Pools of Cesium and Potassium from Values Measured with the Whole-Body Counter.

Subject Sex/No.	Height (cm)	Weight (kg)	Age (years)	Body Cessium (μg)	Body Potassium (g)
M1	170	79	45	1690 ± 140	132 ± 2
M2	187	83	36	1680 ± 170	126 ± 2
M3	187	78	36	2060 ± 130	117 ± 2
M4	181	95	43	1590 ± 100	161 ± 2
M5	183	85	31	1830 ± 130	147 ± 2
F1	159	54	46	930 ± 60	77 ± 2
F2	170	62	49	1860 ± 40	100 ± 2
F3	158	56	28	1010 ± 100	100 ± 2
F4	165	58	38	780 ± 50	85 ± 2
F5	165	61	35	1700 ± 150	94 ± 3

Adapted from Berg et al. (1990).

ternal gamma radiation emitted by airborne radionuclides and radionuclides deposited on the ground.

There are also monitoring instruments which serve for the assessment of selected radionuclides present in the environment, especially in the air. While the first category of monitors usually includes portable types of instruments, the second category consists mostly of stationary or laboratory instruments, some of which may be transportable.

3.6.1 External Gamma Monitors

Monitors of gamma radiation are used for the measurement of the dose rates and doses from terrestrial radionuclides, radionuclides deposited on surfaces and radionuclides dispersed in the air. The response of these instruments usually also includes the contribution from cosmic rays. Their sensitivity must be below the natural gamma background so that any changes and anomalies in this background can be detected. To obtain a sufficient measuring range, starting from the background dose rate to the dose rates envisaged during emergency situations, sometimes two or more similar or different detectors are employed in one system.

In the event of any increase of radionuclides in the atmosphere and their deposition in the environment, measurement of the gamma dose rate can provide a rapid and straightforward indication of even very small changes in the radioactive contamination. Assessment of the need for further, more detailed, measurements or other actions depends many times on the results of external gamma radiation monitoring (Clark et al., 1993).

The gamma monitors usually measure air dose rate (air kerma rate) or exposure rate, and most of them can integrate their response so that results also in the form of total dose or exposure are available. These measurable quantities can be, in principle, converted into the main radiation protection quantities, namely the effective dose equivalent (ICRP, 1977) or effective dose (ICRP, 1991). This can be done, however, only when we know or assume at least some information as to the energy and direction of gamma radiation.

Essentially, two types of gamma monitors may be considered.

(1) Monitors are based either on *non-spectrometric detectors* or using spectrometric detectors, but not based on the energy spectra evaluation. These instruments register *"gross" gamma radiation* without relying on the information about its energy. Preferably their response should be energy independent; this is achieved simply by making use of suitable properties of a radiation sensor (e.g., ionization chambers), by appropriate modification of the detector response or a combination of two detection media (scintillation detectors), or using a special type and shape of detector shielding to modify its energy response (Geiger-Müller counters).

(2) Gamma monitors are based on *spectrometric detectors,* such as a scintillation detector (e.g., NaI or BGO) or a HPGe (portable version) where first a pulse-height spectrum is unfolded and then the final quantity is calculated. These instruments have a built-in microprocessor-controlled multichannel analyzer, or they use a small portable computer with other necessary accessories. So far, however, such sophisticated monitors are used more often in research (Minato and Kawano, 1970; Nagaoka, 1987; Okano et al., 1988) than in routine measurements, but in the near future we expect that they will constitute a very important part of radiation instrumentation.

At present, gamma monitors for environmental measurements are essentially based on the following three types of detectors:

(1) Pressurized ionization chambers
(2) Plastic and other types of scintillators
(3) Geiger-Müller counters

The monitors based on ionization chambers have an inherently satisfactory energy response and they are practically unaffected by temperature changes, but such chambers have to be very large to generate a measurable current at low radiation levels. Therefore, high-pressure argon- or nitrogen-filled chambers are normally used. Although some of these chambers are available in portable versions, most of them are installed at selected locations, usually around various nuclear facilities. In many cases they form a complex monitoring network where one control unit can collect data from several chambers which can work individually as well. Their sensitivity is typically about 10 nSv/h and their measuring range up to 5 mSv/h. The response is practically energy independent in the range from about 60 keV to several MeV.

Instruments using plastic scintillators usually operate in the current mode where the current is converted into a corresponding pulse rate. The pulses are then evaluated by a ratemeter or enumerated by a counter. These types of monitors have some advantages, e.g., a wide range of dose rates and a relatively small size, but also some disadvantages, mainly because their stability and reproducibility is affected by temperature changes.

A very good energy and directional response can be achieved by a special configuration of a main plastic scintillator with two inorganic scintillators ZnS(Ag) and NaI(Tl) (Viererbl et al., 1990). The composition and sizes of the scintillators as well as the capsule were selected so as to minimize the energy and directional dependence. The error due to energy dependence was found to be less than 5% in the energy range from 30 keV to 5 MeV, and the directional response was within $\pm$ 10% at an energy of 60 keV for angles between $-3/4\pi$ and $+3/4\pi$, which is better than can be obtained with scintillation detectors commonly used. The detector unit and its energy response are shown in Figure 3.2.

On the other hand, monitors using a compensated GM counter are inherently very stable against temperature changes. Their advantage also lies in their very simple electronics. An obvious disadvantage consists of the fact that they do not directly measure the quantity of interest, since their response depends neither on the ionization produced nor the energy deposited. It is well known that for the generation of an output pulse it is sufficient to produce in the sensitive volume of a GM counter just one pair electron-ion. The response of the GM counters, however, can be compensated in such a way that they can successfully compete with instruments based on ionization chambers or scintillation detectors. Typical sensitivity of a halogen quenched energy compensated GM counter is around 15–30 counts per second per μGy s^{-1}. Some area monitors may use two or more counters to increase their sensitivity and to cover a wider dose rate range.

In addition to portable gamma monitors, there are a number of various stand-alone continuous monitoring systems or measuring stations that are remotely controlled and programmed to deliver required information including alarm signals in the case of accident or emergency. These monitors are usually designed for outdoor applications and they are largely self-contained. They have batteries to be operable in case of power loss and are

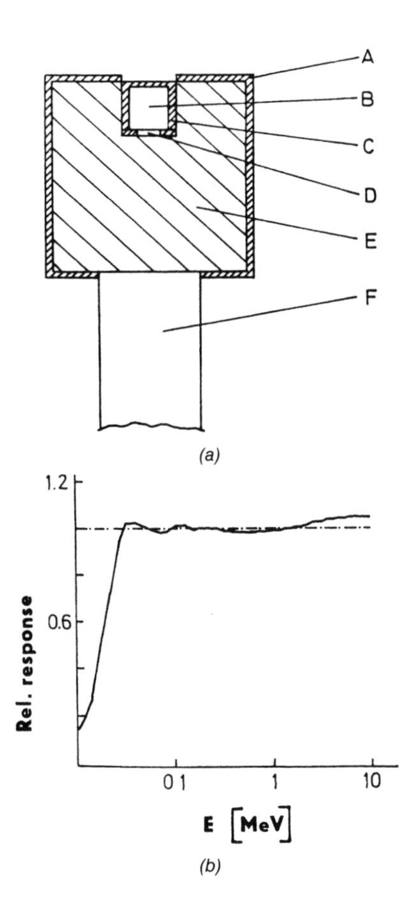

FIGURE 3.2. Scintillation detection unit and its energy response: (a) a combined scintillator [A—Zns(Ag), B—NaI(Tl), C—an iron absorption layer, D—a hole in the Fe absorption layer, E—a plastic scintillator, F—a photomultiplier], (b) an energy dependence of the combined scintillator detector.

equipped with a built-in modem for communication of data to a central computer. All components are housed in weatherproof enclosure. The minimum sensitivity is usually better than 10 nSv/h.

3.6.2 Radionuclide Monitors

Since radioactive material released into the atmosphere is transported very rapidly through the air, special attention should always be paid to monitoring this medium. For this purpose fully automatic monitoring of environmental radioactivity utilizing measuring networks covering large areas with stations which record all essential parameters on-line and automatically would be required. The instruments must be able to detect very low airborne radionuclide concentrations within a very short time, to analyze them and to transmit relevant data to the central station.

These monitors are usually permanently fixed and they continuously, or at certain selected time intervals, measure the concentration of some important radionuclides in the air, especially iodine, cesium, plutonium, and noble gases, but also other radionuclides, including tritium and radon/thoron and their decay products.

Special stationary universal units have been developed for the measurement of radioactive aerosols in the air, including alpha and beta emitters. Radionuclides are collected on one or more fiberglass or other special filters whose activity is then spectrometrically determined by two or more detectors. Usually, an internal air flow measurement capability is provided to correct the filter data for actual flow rates through the filter; other necessary corrections, some of which would be difficult to introduce in small portable instruments, are also made. These monitors are also characterized by a very high air-flow, which may be up to 1 m^3/h or even higher. To reduce the background due to radon when measuring other alpha radionuclides, sometimes a pseudocoincidence between the beta particle from the radon decay product ^{214}Bi and the subsequent alpha particle from ^{214}Po is used.

The requirements for the measurement of some alpha emitters, for example, ^{239}Pu, which should be detected on levels below 40 mBq/m^3 (this concentration corresponds to the value of the derived activity concentration for occupationally exposed persons), are quite strict. One has to realize that these low concentrations are supposed to be measured in the presence of background concentrations of radon decay products of several tens of Bq/m^3.

Other monitors are designed in order to measure gross alpha and gross beta particulate concentrations in air with the highest possible sensitivity required for the early detection of airborne radioactivity. These instruments can achieve gross alpha and beta detection limits ≥ 0.2 Bq/m^3 and ≥ 0.5 Bq/m^3, respectively.

The monitors for iodine are equipped with a disposable or a refillable charcoal cartridge to trap the radioiodine component. Aerosol pre-filtering, cartridge pre-heating (to remove moisture for more efficient capture of organic iodine compounds), and automatic gain stabilization ensure higher accuracy than in conventional monitors. It is possible to obtain detection limits better than 100 mBq/m^3 (EG&G, 1993).

For selective measurements of concentrations of some radionuclides, high-resolution gamma spectrometry is used. In a two-hour measurement cycle, detection limits as low as 100 mBq/m^3 for ^{131}I and ^{137}Cs in air can be achieved. For ^{60}Co the limit is around 30 mBq/m^3 obtained from one hour of measuring. Current automatic analyzers are capable of evaluating the spectra of up to 100 different radionuclides. On-line radionuclide identification is important for a reliable assessment of the actual situation in the case of an accident, which on the basis of such results can be better evaluated as to its origin and character.

Nuclear facilities release some amount of various radionuclides through their chimney stacks. These releases have to be measured in order to keep the effluent activities below the allowed emission values set by national or international standards. In most cases, the license for the operation of a nuclear power plant stipulates the amounts of radioactive materials that can be discharged into the environment. Some special monitors with various fixed as well as moving filters have been developed for this particular purpose.

The operation of reactors and critical facilities results in the production of fission products in the fuel and activation products in all materials and elements irradiated by high neutron fluences. Since water is a common coolant and moderator in most reactors, its exposure to neutrons results in the transformation of hydrogen and oxygen into tritium and ^{16}N. Neutron irradiation of boron and lithium also produces tritium. Air is often used as a coolant for reactors and critical facilities and is always trapped or dissolved in water in the reactor. The activation of oxygen in air also gives ^{16}N and the activation of noble gases leads to the generation of such radionuclides as ^{41}Ar, ^{82}Kr, ^{89}Kr, ^{135}Xe, and ^{137}Xe. These radioactive noble gases are essentially a source of external exposure. Their impact as internal contaminants is relatively unimportant. The most sensitive method for assaying the concentration of radioactive noble gases is the method based on an internal gas counting. Continuous monitoring of radioactive noble gas effluents released through stack ventilation systems is often required at nuclear power plants. Airborne radioactivity is, in this case, often measured continuously by a high-efficiency solid-state detector.

In some nuclear installations dedicated tritium monitors are required. Tritium air monitors are often based on ionization chambers, but much higher sensitivity (about 100 times) can be achieved by instruments using proportional flow-through counters with rise-time discrimination. This technique makes it possible to distinguish the responses of tritium from those of radioactive noble gases. In this way tritium can be measured simultaneously and separately with some other radioactive gases. The measuring range covers the interval from about 200 Bq/m^3 to more than 10 kBq/m^3. Tritium monitors are used in pharmaceutical radionuclide laboratories, accelerators, fusion research, and nuclear materials handling facilities.

3.7 ASSESSMENT OF RADON AND ITS DECAY PRODUCTS

The naturally radioactive noble gas radon is present in air and can accumulate in closed or poorly ventilated spaces, including buildings. As has already been mentioned (Section 1.1.2) the two significant isotopes of radon are ^{222}Rn, the immediate decay product of ^{226}Ra, deriving from the uranium series of natural radionuclides, and ^{220}Rn, the immediate decay product of ^{224}Ra, coming from the thorium series. These radionuclides are commonly known as radon and thoron. Thoron has a short half-life (55.6 seconds) and a low abundance relative to radon, whose half-life is 3.82 days. Because of their different characteristics and especially their different contribution to the total annual effective dose—radon and thoron and their decay products are responsible for 1.2 mSv and 0.07 mSv (world-wide average), respectively—radon and its decay products (progeny) are of primary interest. From the radiological point of view radon decay products are more important than radon itself. Actually, the effective dose from the inhalation of radon represents only about 5% of the total radon-related dose, i.e., 95% of this dose is delivered by the short-lived radon decay products ^{218}Po, ^{214}Pb, ^{214}Bi, and ^{214}Po. Consequently, for radiation protection purposes it is desirable to monitor preferably the presence and the concentration of radon decay products, while for the identification of the sources and origin of radon, the measurement of its concentrations in air or water, and sometimes its exhalation from soil and building materials, are more valuable.

3.7.1 Measurement of Radon in Air

There are many different methods of measuring radon in air; some are based on active, some on passive detectors which are in both cases specially modified. Most techniques rely on the detection and spectrometry of alpha particles emitted by radon or its progeny, but there are some methods based on the evaluation of gamma and also of beta radiations from decay products. Essentially, these methods can be divided into three categories:

(1) *Instantaneous (grab sample) method*—gives information about the radon concentration at a known reference time, which is usually the time when a certain volume of air was collected. The air sample is then measured and results are interpreted in terms of radon concentration related to the time of the sample collection.

(2) *Continuous method*—provides information about the radon concentration as a function of time.

(3) *Integrating (time-averaged) method*—is based on a measurement that takes from a few days to a year. The resulting value is the radon concentration averaged over the measuring time interval.

The precision and accuracy of the individual measuring techniques depend on several factors, such as the statistical character of radioactive decay, variations in detector response, interference with unmeasurable species, and unfulfilled assumptions concerning the atmosphere being sampled (Nazaroff, 1988).

The instruments and methods used for the measurement and estimation of radon and radon progeny concentrations in air have been discussed and summarized in several references (OECD, 1985; NCRP, 1988, Nevissi, 1987; Urban and Schmitz, 1991; Harley, 1992).

3.7.1.1 INSTANTANEOUS METHODS

These methods are based on the collection of a sample of air using a flexible foil bag or a container and its subsequent measurement by an ionization chamber, a scintillation detector or any other suitable detector or monitor.

In the first *ionization chambers* the current was measured by electrostatic electrometers which were later replaced by very sensitive electronic electrometers with capabilities as low as 10^{-15} A. Prior to entry into the chamber the air to be monitored is filtered in order to get rid of atmospheric aerosols including the radon decay products. The measured current or charge after its integration gives information about the radon concentration in air. If the measurement of the ionization current is measured immediately after the filling of the chamber, the current is observed to grow with time for about the first three hours. This is because the decay products' concentration in the chamber is growing to a state of equilibrium with its parent with a mean half-life of about 50 min. After that delay the current is directly proportional to the radon concentration in air. To avoid this delay, one can calculate the radon concentration from the measured current.

Various types of ionization chambers have been designed and constructed specially for random monitoring. They were usually in the form of a brass or steel cylinder with a central collecting anode. The minimum detectable concentration of current-mode chambers was about 4 Bq/m^{-3}; the actual parameters depended on the background and the volume of the chamber, which could be from less than 1 L to several liters. Today these types of ionization chambers are not used very often; they have been replaced by chambers working in pulse-mode.

Pulse ionization chambers have some problems with the slow mobilities of the negative O_2 ions and positive ions, which result in a very long rise-time of output pulses being in the range of several milliseconds. When electronegative impurities, principally oxygen, as well as water vapors are removed from the air to be measured, the rise time is shortened considerably and it is then possible to perform fast-pulse counting. For long measuring intervals – about fifteen to twenty hours – it is possible to monitor radon in air at concentrations below 1 Bq/m^3 (Harley, 1972). The ionization chambers for instantaneous radon monitoring are usually known as the *sealed chambers,* which are essentially laboratory instruments. Air samples are collected in the place of interest and then returned to the laboratory for evaluation.

The grab sample analysis can also be based on the use of an *electret ion chamber* which serves as an integrator of radon in a leak-tight container. This container is used as a sample collector. Using the data on initial and final voltage of the electret surface potential, and the period of integration, the radon concentration at the place of sample collection can be calculated (Rad Elec, 1992). Further details about the electret ion chamber will be given later in this chapter when integrating monitors are discussed.

Scintillation or *Lucas cells* have been very popular in radon monitoring since the mid-1950s (Van Dilla and Taysum, 1955; Lucas, 1957). These cells or flasks are lined on the inside surface with a powdered zinc sulfide silver-activated scintillator—ZnS(Ag)—which generates a flash of light when alpha particles emitted by radioactive nuclei present in the vessel strike it. These light pulses are converted in a photomultiplier, which is optically coupled to the cell window, in a measurable electrical signal.

The scintillation cell is actually a small metal, glass, or plastic chamber with a flat transparent bottom suitable for coupling the photomultiplier whose cathode is usually grounded. The original scintillation cell designed by Lucas (1957) is shown in Figure 3.3.

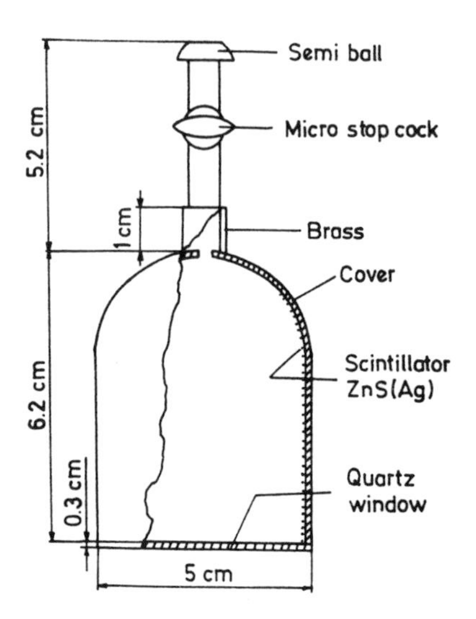

FIGURE 3.3. The original Lucas scintillation cell.

There are a variety of different shapes used: hemispherical, cylindrical, or conical. A filtered air sample is admitted into the cell using a pump, or the cell is first evacuated and then filled with measured air through a valve. A filter fixed at the entrance to the cell eliminates aerosols including radon decay products.

With a Lucas cell of volume 0.1 L radon concentrations on the level of about 10 Bq/m^3 with an uncertainty of about 30% can be measured. Many types of commercially available instruments are based on this principle. Some of them, in addition to the grab sampling mode, can work also as continuous radon monitors.

In order to increase the sensitivity, large volume cells (3 L) were developed (Cohen et al., 1983). The lower limit of these cells with a counting interval about four hours was 0.4 Bq/m^3.

Another method for the grab-sample measurement of radon in air is the so-called *two-filter method* where the radon concentration is deduced from pure radon decay products. In this case the sample air is pumped through a cylinder with a filter on each end at a known rate (typically about 10 L/min) for a relatively short period (about five minutes). The upstream filter removes all the decay products in the air sampled, while the second filter collects those decay products which have been produced by the decay of radon inside the cylinder. The alpha particles emitted by the deposited decay products (actually, practically only by ^{218}Po whose half-life is short enough) are then counted with an appropriate detection system, usually a scintillation detector based on ZnS(Ag). The resulting counts can finally be converted to radon concentration in air, taking into account the size of the cylinder, sampling rate and counting time. Using this method a sensitivity of about 4–10 Bq/m^3 can be achieved.

3.7.1.2 CONTINUOUS METHODS

Continuous radon monitors are commonly based on the following principles:

(1) Flow-through two-port scintillation cells
(2) Flow-through pulse multiwire ionization chambers
(3) Electrostatic collections of ^{218}Po
(4) Modified two-filter methods

In their simplest form, *continuous radon monitors based on a scintillation cell* consist of a cell itself having an inlet and an outlet, an optically coupled photomultiplier, counting electronics, and a pump. In operation, the air is pumped through a filter to remove radon decay products and particulates as they enter into and then leave the sensitive volume of the cell. As the sample air passes through the cell the radon decays into its products which in turn also decay and contribute to the final counts. It has been found that virtually all of the radon decay products produced in the cell are plated out on the cell walls. The presence of the decay products, however, causes a delay that affects the time correlation between the measured number of counts and the time to which the radon concentration is related. Therefore, it is not easy to calibrate these types of monitors.

A generalized set of equations to evaluate the mean radon concentration obtained from such continuous monitors has been described in Thomas and Countess (1979). The concentration of radon during a particular interval was derived on the basis of the measured net counts in that interval and the contribution from previous intervals. With a 1.46 L scintillation cell a continuous monitoring of radon concentrations from about 40 Bq/m^3 was possible.

Since the counts attributable to the radon decay products have a delayed time output, this results in a somewhat blurred time response with a delay of about thirty to forty-five minutes. The actual results can be taken only after about three hours following the beginning of the measurement. This time is necessary for the establishment of the equilibrium between radon and its decay products inside the cell. The delayed response may not be a very serious limitation in cases where radon concentration is supposed to change relatively slowly, but a substantial error may result under rapidly changing conditions.

The problem of *pulse ionization chambers* with slow pulses due to the long collection times of the ions has recently been solved using a combination of a specially designed electrode structure and pulse shaping electronics optimized for highest energy resolution and count rate (Baltzer, 1992). An energy resolution of 0.25 MeV FWHM has been achieved for the peaks associated with the alpha decay of radon and its decay products. This makes it possible to distinguish between ^{222}Rn and ^{220}Rn. Such a multi-wire ionization chamber shows a sensitivity corresponding to about 1 cpm for a radon concentration of about 50 Bq/m^3. This type of ionization chamber was built in two versions, one with a sensitive volume $10 \times 8 \times 8$ cm^3 and the other, more sensitive, with the dimensions $15 \times 18 \times 18$ cm^3.

A simplified block diagram of the pulse ionization chamber and associated devices including electronic blocks is shown in Figure 3.4. Air is pumped into the chamber by a membrane pump at a rate of about 1 L/min. Before entering the ionization chamber the air is cleaned while passing through two filters which remove radon decay products as well as any dust which could cause micro-discharges inside the chamber. It has been found that the chamber is sensitive to the humidity of the measured air. When the

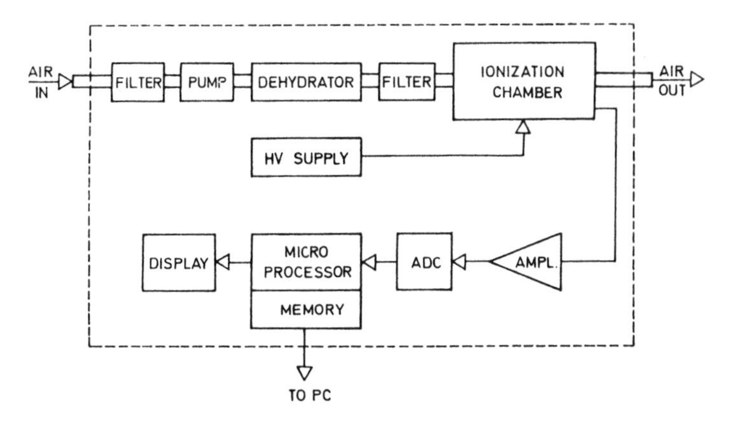

FIGURE 3.4. Principal arrangement of a pulse ionization chamber and associated components.

humidity of air at room temperature is about 60% or higher, microdischarges occur in the chamber which affect its normal operation. To prevent this, the air before reaching the chamber is therefore cooled to a temperature of +4°C using a temperature regulated Peltier cooler. The condensed water is conducted to the hot part of the Peltier element, where it is then evaporated. Having passed through the Peltier cooler, the air is again brought to room temperature by circulating it at the warm end of the Peltier element. The resulting relative humidity is then about 35% even if the entering room-temperature air has a humidity of 100%.

The count rate of the described ionization chamber system is about 1 per minute per 50 Bq/m^3. In order to measure a 150 Bq/m^3 concentration with a relative standard deviation of 10% a measuring time of about one hour is required.

A similar continuous measuring system based on a plane multiwire-electrode ionization chamber was also developed (Tanabe et al., 1986). This chamber has three rigid plane multiwire electrodes in a box of aluminum plates with holes to allow the air to flow freely through them. The slow rise-time pulses of alpha particles from the chamber are analyzed with a specially designed analog-to-digital converter which is controlled by a microcomputer. The energy spectra obtained have with a resolution of about 600 keV FWHM.

Other types of continuous radon monitors utilize the *electrostatic collection* of radon decay products by means of a detection surface maintained at a negative potential. The detection volume is usually defined primarily by a hemispherical polyurethane foam wall through which radon diffuses passively. Aluminized mylar film covers the detector, which is usually

ZnS(Ag) (Wrenn et al., 1975), or a silicon surface-barrier detector is used (Porstendorfer, 1980). The positively charged ^{218}Po formed in the detection volume, which is about 1 L, is collected on the outer layer of the film and then the alpha particles are detected. The sensitivity of these monitors is in the range of 3 to 7 Bq/m^3 for about a one hour counting period.

Later instruments based on electrostatic collection utilize a pump to ensure a defined air flow rate of about 1 L per minute and use solid state alpha spectrometry exclusively. In one such arrangement (Watnick et al., 1986), the air stream is first dried and filtered before entering a 2 L detection volume containing a Teflon-housed silicon surface-barrier detector of a 300 mm^2 active area. A voltage of 3.5 kV is applied between the chamber walls and the detector. The background counting rate of this system is less than 0.1 cph. The instrument can measure radon concentrations at the level of 40 Bq/m^3 with a relative standard deviation of about 20%.

A number of different continuous monitors for the assessment of radon in air based on a *two-filter method* were also developed. In principle, these instruments consist of an inlet filter removing radon decay products, a chamber or vessel (usually a cylinder with a filter on either end), an exit filter for the collection of radon decay products produced within the chamber, and a pump to force air through the system. The activity measured on the second filter is practically independent of the degree of equilibrium between radon and its decay products in air before it is admitted into the chamber. The number of counts from the exit filter measured by a ZnS(Ag) scintillation or silicon detector is a measure of radon concentration in air. The results, however, have to be corrected for growth and decay of decay products and for loss of decay products deposited on the wall of the chamber outside the detector. Some of the two-filter monitors using advanced electronics are reported to have a sensitivity of better than 0.4 Bq/m^3 for ^{222}Rn with no interference from ^{220}Rn (Schery et al., 1980).

Actually, when interference due to the presence of ^{220}Rn is expected, it can be eliminated by placing a decay chamber of sufficient volume in front of the monitor to allow thoron to decay before it can reach the first filter.

The principal arrangement of a two-filter continuous radon monitor is illustrated in Figure 3.5.

3.7.1.3 INTEGRATING METHODS

These methods are widely used in indoor and outdoor radon monitoring in order to obtain a mean value of the concentration averaged during a relatively long period of time, which may range from a few days to one year.

The monitors used as integrating devices can be divided into the following categories:

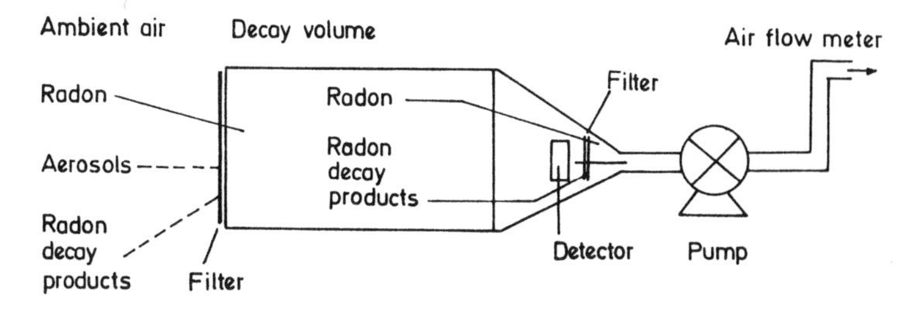

FIGURE 3.5. Radon monitor based on a two-filter method.

(1) Monitors based on *solid-state nuclear track detectors*
(2) Monitors with *thermoluminescence detectors*
(3) Monitors using *activated charcoal canisters*
(4) Passive *electret ion chamber* radon monitors
(5) Monitors integrating the response of *active radon detectors*

The most widely used integrating radon monitors use *solid-state nuclear track detectors* (SSNTD), known also as etched-track detectors, as sensors for the registration of alpha particles emitted by radon or its decay products. When such an alpha particle strikes a certain type of plastic, a radiation damage track is produced. After appropriate chemical or electrochemical etching these tracks can be made detectable. The tracks are magnified and counted either manually (visually) or automatically using, for example, *spark counters* or sophisticated *image analysis techniques.*

In principle, two configurations of radon chambers using SSNTDs are possible:

(1) *Diffusion radon chamber*—a closed chamber into which only radon (but not radon decay products) can diffuse
(2) *Radon chamber without a filter*—essentially the same as a diffusion chamber but in this case both radon and its decay products can reach the sensitive volume of the chamber

In a *diffusion chamber* a detector foil is usually placed on its bottom. The chamber is closed by a hydrophobic fiberglass or other suitable filter. A special cover is used to prevent contamination of the filter by large dust particles.

The cross section of a typical diffusion chamber is presented in Figure 3.6 (Urban et al., 1985). Figure 3.7 shows the areas from which alpha particles emitted by ^{222}Rn, ^{218}Po, and ^{214}Po can reach the detector with energies in the range 0.5 to 2 MeV. Alpha particles with such energies have the best chance of being registered by the SSNTD.

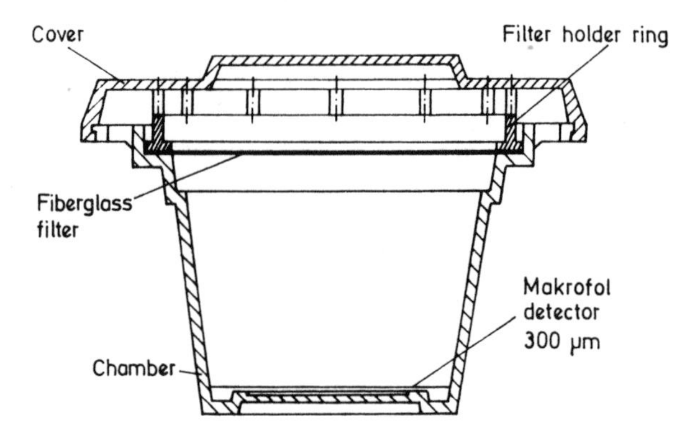

FIGURE 3.6. Cross section of the KfK diffusion radon chamber. From Urban et al. (1985).

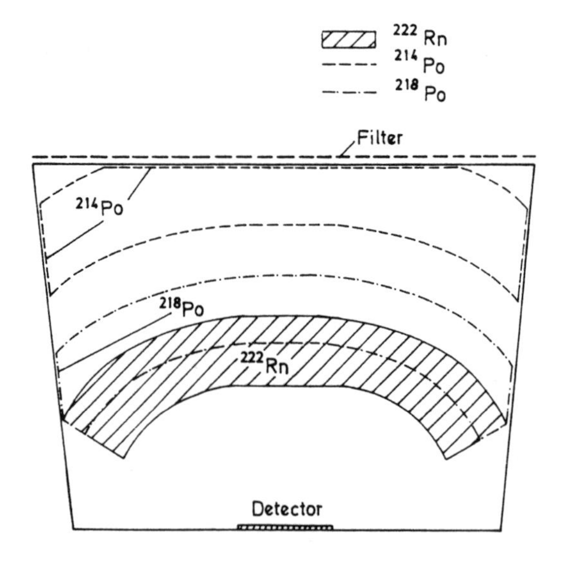

FIGURE 3.7. Cross section of a diffusion chamber with the range of alpha particles of ^{222}Rn, ^{218}Po and ^{214}Po. From Urban et al. (1985).

Although at present some 150 different track detector materials are known, only a few of them are suitable for radon chambers. The most common in research and routine applications are LR 115 (cellulose nitrate), CR 39 (allyldiglycol polycarbonate) and MAKROFOL (polycarbonate). All these materials are insensitive to beta or gamma radiation.

When an alpha particle strikes an unconducting detector foil it creates along its pathway a tiny track of damaged material. The diameter of this track is in the range of 5–10 nm. The damaged part of the material reacts faster than the rest of the material when certain chemicals are applied. The chemical treatment of detector materials aimed at the enlargement of the tracks into conical or cylindrical holes with diameters from 1 to 30 μm is called *chemical etching.* When during this etching (conventional etching) an AC electrical field is applied, it is possible to further enlarge the track size so that they can be seen by the eye. Such a process is called *electrochemical etching.* The etching processes are affected by many parameters, especially by the type of solutions, temperature, detector material and, in the case of electrochemical etching, also by the electric field and applied frequency. The characteristics of the most often used detector materials with respect to the etching conditions and track diameters, including a required magnification for their counting, are shown in Table 3.19.

The results of the measurement give the number of tracks which can be converted into the mean radon concentration $\overline{C}_{Rn}$ using the following procedures. The basic relationship between the mean value $\overline{C}_{Rn}$ in Bq m^{-3} and the radon exposure X_{Rn} in Bq m^{-3} d and the irradiation (exposure) time T in days can be written in the form

$$C_{Rn} = \frac{X_{Rn}}{T} = \frac{\int_0^T C_{Rn}(t)\mathrm{d}t}{T} \quad (3.1)$$

Assuming the linear relationship between the radon exposure and the number of corresponding tracks, the sensitivity ϵ of the diffusion radon chamber may be expressed as

$$\epsilon = \frac{\dfrac{N_1}{A_1} - \dfrac{N_2}{A_0}}{X_{Rn}} \quad (3.2)$$

where ϵ is given in (number of tracks/cm^2)/(kBq m^{-3} d); N_1 and N_0 are the total number of tracks counted in the field A_1 and the number of background tracks counted in the field A_0, respectively. Both the areas A_1 and A_0 are in

TABLE 3.19. Etching Procedures and Track Characteristics Related to the Three Most Common Detector Materials.

Detector Material	Etching Techniques		Track Di. (μm)	Magni-fication
	Conventional	Electrochemical		
LR 115	10% NaOH 60°C, 2 hours	–	8	400×
CR 39	90% 6N KOH 10% C_2H_5OH 20°C, 30 min	–	10	400×
CR 39	90% 6N KOH 20°C, 30 min	10% C_2H_5OH 20°C, 3 hours 600 V_{eff}, 5 kHz	50	40×
MAKROFOL	80% 6N KOH 20°C, 1 hour	20% C_2H_5OH 20°C, 3 hours	120	20×

From Urban et al. (1985).

cm^2. The sensitivity of the KfK diffusion radon chamber is as follows (Urban et al., 1985):

$$\epsilon = 16.2 \frac{\text{tracks/cm}^2}{\text{kBq m}^{-3}\text{ d}} = 5.9 \frac{\text{tracks/cm}^2}{\text{Bq m}^{-3}\text{ y}} \tag{3.3}$$

The radon chambers without a filter or simple bare foils detect both the radon and the radon decay products present in the ambient air. Their calibration and their interpretation are more complicated. Once the equilibrium factor is known at least approximately the results of monitoring can be used for the assessment of both radon and radon decay product concentrations. These types of monitors are not suitable for reliable low-level radon measurements.

Integrating radon monitors can also be designed and built using *thermoluminescence detectors* (TLD), which are sensitive to the energy deposited by radiation. The TLD chips are used in both diffusion chambers and in two-filter chambers where they detect radiation emitted by radon decay products collected on the surface of the second filter.

In one system, the so-called passive environmental radon monitor (PERM), radon diffuses through a 2 kg silica gel bed and a filter paper into an inverted 1.5 L volume metal funnel. The neck of the funnel is sealed with a rubber stopper through which a brass rod is inserted to function as a cathode (George and Breslin, 1977; NCRP, 1988). A molded plastic pedestal attached to the end of the rod holds a TLD chip. About a 900 V negative collecting potential is applied to the rod using dry cell batteries. The positively charged ^{218}Po and ^{214}Po ions produced by radon decay in this sensitive volume are attracted by the TLD chip. After an exposure period, the chip is removed and evaluated by a TLD reader. The result is corrected for gamma background which is measured by another TLD chip shielded from alpha particles. The usual exposure period of this radon monitor is about one month. The lower detection limit is about 1 Bq·m^{-3} of ^{222}Rn for a one-week exposure time.

Other systems based on a similar principle were developed later, for example (Nyberg and Bernhardt, 1983; Maiello and Harley, 1987), but they have not found mass applications in routine radon monitoring.

Another type of integrating radon monitor is based on the adsorption of radon by activated charcoal. The principle was known long ago (Rutherford, 1906) but not used for many years until the 1950s (Hursh, 1954) and especially the 1980s when this method was revived and introduced in routine radon monitoring (Cohen and Cohen, 1983; George, 1984; Prichard and Mariën, 1985; Cohen and Nason, 1986; Jenkins, 1991).

The radon monitors using activated charcoal have a similar configuration as radon diffusion chambers with SSNTDs. During the exposure radon diffuses through the filter, preventing the entry of radon decay products into the canister containing charcoal. After the exposure period the canister is collected and measured by means of gamma spectrometry. From the amount of gamma radiation emitted by radon decay products produced inside the canister (originating from radon that diffused into the charcoal and was adsorbed there), the radon concentration in air can be assessed. The counting of gamma radiation is usually carried out by a NaI(Tl) spectrometry system, which is set to register gamma radiation from ^{214}Pb (energies 295 keV and 352 keV) and ^{214}Bi (energy 609 keV).

This method is suitable for large-scale monitoring of radon concentrations from about 10–20 Bq/m^3 with an integration period of a few days to one week. The exposure time cannot usefully be longer because of the half-life of radon (3.8 days). The results may be affected partly by humidity and temperature as well as by the fact that the adsorption of radon is not linear during the exposure time.

During the last few years electret ion chambers were developed and are now widely used as radon monitors suitable for measuring of radon concentrations in air and in water as well as carrying out some other radon related and environmental evaluations (Kotrappa et al., 1988, 1990, 1992).

The ion electret system includes the following three essential components:

(1) An electrostatically charged *electret,* which creates an electric field and collects the charge produced in the chamber

(2) An *ion chamber* made of conductive plastic into which an electret is loaded

(3) A *reader* to measure the surface potential (voltage) of the electret before and after exposure

The electret disk is mounted in an electrically conductive plastic holder and normally is covered by a "keeper" cup when not used in the chamber and stored. The surface voltage of the electret changes during the exposure and the drop in its reading can be related to the radon concentration in air.

Different types of electrets and ion chambers are now commercially available (Rad Elec, 1992) for short- and long-term monitoring. Short-term electrets—with a 1.542 mm thickness—have fading (unexposed voltage loss during storage or in a closed chamber) in the range of 2 to 4 V per month and are especially recommended for use with large-volume chambers for short-term measurements. Long-term electrets having a 0.127 mm thickness, with a stability of about 1 V per month, are intended for long-term measurement, from a few months to one year.

The electret ion chamber operates as a true integrator wherein the electret serves both as a source of electrostatic field and as a sensor. The charge produced inside the chamber, into which only radon that has passed through a filter that has removed atmospheric aerosols is allowed, causes a reduction in the electret surface potential. This reduction is proportional to the ionization integrated over the period of exposure. The detection sensitivity and the dynamic range of the monitor depend on the type of electret and the volume of the chamber. Figure 3.8 shows schematic views of three different commercial chambers (Rad Elec, 1992). The S-type chamber has an on/off mechanism that can be used to close and open the electret from outside.

The disadvantage of electret ion chambers used for radon measurements is their sensitivity to ambient gamma fields. This response under certain circumstances (Kotrappa et al., 1992) can also be used for environmental gamma exposure monitoring. In the case of radon monitoring the interference of gamma radiation has to be taken into consideration and final results corrected. The correction depends on the exposure rate at the place where the chamber is positioned for radon measurement. It is therefore necessary to use a suitable gamma survey meter to take a reading of the actual exposure rate. The response of electret ion chamber is not linear and depends on the actual state of the electret surface potential. The applied correction factors should be based on accurate measurements of the chamber response with electret potential covering the whole working range (Sabol et al., 1993).

The lower limit of detection using electret ion chambers depends on the chamber size, electret type and measuring time interval. These limits,

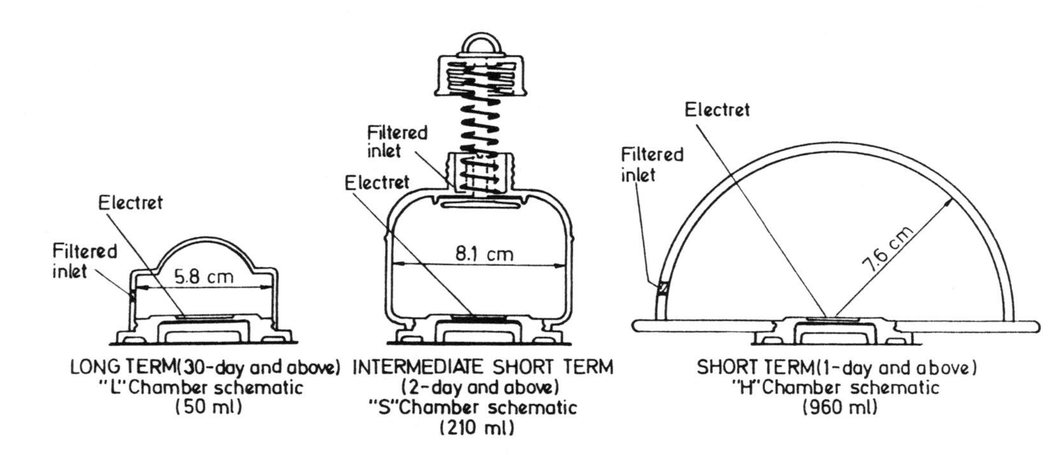

FIGURE 3.8. Different commercially available electret ion chambers.

TABLE 3.20. Minimum Measurable Radon Concentrations at Stated Relative Standard Deviation δ Using Different Electret Ion Chambers, Electrets, and Measuring Periods (EIC—Electret Ion Chamber, ST and LT—Short-Term and Long-Term Electret, Respectively).

EIC	ET	Exposure Time (days)	Minimum Measurable C_{Rn} (Bq/m³) 50%	25%	10%	δ (%) at 150 Bq/m³
S	ST	2	9.3	25.9	111	8.6
		7	5.9	14.8	40.7	6.4
S	LT	30	6.7	18.5	77.7	8.1
		90	6.7	14.8	37.7	6.7
L	LT	90	11.1	29.6	140.6	9.8
		365	7.0	14.8	40.7	6.8
H	ST	2	7.0	14.8	40.7	6.5

given in terms of a minimum measurable concentration at stated error for different arrangements are summarized in Table 3.20 (Rad Elec, 1992). There are three principal sources which contribute to the total error associated with radon measurements made by means of an EIC: (1) error due to system component imperfections, (2) error in the voltage reading, and (3) error caused by gamma background uncertainty.

In addition to integrating radon monitors based on passive detectors, such as SSNTDs, TLD or EIC, all continuous radon electronic systems can also in most cases measure average radon concentrations over a selected period of time.

3.7.2 Radon in Water

There is now growing concern over the health hazard associated with the presence of dissolved radon in public and private water supplies (Kotrappa and Jester, 1993). In the USA, for example, a new regulation has been proposed by the U.S. Environmental Protection Agency restricting the concentration of radon in public water supplies to 11 Bq/L. High water-borne radon concentrations may directly elevate radon concentrations in air inside residential and public buildings and therefore it is important to measure the content of radon in water in order to check whether its quality complies with the regulatory requirements.

Several different methods for the measurement of radon in water have been developed and used. The following techniques are the ones most often applied:

(1) Liquid scintillation counting
(2) Gamma-ray spectrometry
(3) Gas extraction method
(4) Method based on the use of SSNTDs
(5) Electret ion chamber monitors

The first method makes use of the fact that radon is readily soluble in suitable organic solvents. Using this procedure, a small amount of water (10 mL) is pipetted into a PTFE-coated polyethylene LSC vial containing 10 mL of a water-immiscible cocktail based on a mineral oil (Prichard and Gesell, 1983; Schönhofer, 1992). Following capping, the vial is shaken in order to extract radon into the organic phase. The vial is then cooled and after three hours an equilibrium between radon and its decay products is established. This method was recently used in Austria in a large-scale monitoring program (Schönhofer, 1992). The lowest value in this survey was about 0.4 Bq/L, while the highest concentration of radon in water was around 760 Bq/L.

The second method for the measurement of radon concentrations in water usually uses a large NaI(Tl) scintillation detector for the spectrometric evaluation of gamma photons emitted by radon short-time decay products (Countess, 1978). This method requires relatively large samples of water (8 to 15 L). The calibration of the system, in terms of photopeak efficiencies at the fixed geometry, is carried out using standard ^{134}Cs and ^{152}Eu reference solutions with certified concentrations (Farai and Sanni, 1992). In this method gamma radiation from ^{214}Pb can also be detected. Besides scintillation spectrometers Ge(Li) or HPGe detectors may be used. The presence of the interfering ^{226}Ra can be evaluated by repeating the measurement after about thirty days, when the radon originally present will have decayed and the only remaining radon will be that which is in secular equilibrium with ^{226}Ra. The direct gamma-ray counting method can be used for measuring radon concentrations in water above 1–2 Bq/L.

Low-level radon concentrations in water can be measured by a gas extraction method (Mathieu et al., 1988) based on the extraction of radon as a gas from a water sample using helium which is bubbled through the water, stripping the radon. The mixture of gases is then trapped on cold charcoal from which, after heating, radon is released and collected into an evacuated scintillation cell for counting.

A fourth and a fifth method are in principle similar: both methods are based on measuring radon in the air above a water sample in a closed container. One of these methods uses a SSNTD for measuring the radon released from water, while the other takes advantage of newly developed electret ion chambers (Kotrappa and Jester, 1993).

3.7.3 Radon Decay Products

Unlike radon, which is more or less homogeneously distributed in space, its decay products are attached to dust particles or deposited on the surfaces of surrounding materials. Most methods of monitoring radon decay products are based on their collection on a filter which is then measured using a detector for simple counting; more preferably, a spectrometry system is used for the evaluation of alpha particles emitted by ^{218}Po and ^{214}Po. In this arrangement, the sampling air with a known rate is drawn through a suitable filter which can collect radon products with high efficiency. In addition to the filter, the decay products can also be collected electrostatically and then detected. Besides active detectors for the registration of particles from decay products, some passive detectors can be used as well.

Methods for measuring radon decay products can be divided into three principal categories:

(1) *Grab-sampling* or *semi-instantaneous method* is based on the measurement of the limited volume of air sampled during a relatively short time period.
(2) *Semi-continuous method* usually relies on a modified grab-sample technique with successive repetition of sampling. In principle, a continuously moving filter that collects decay products while simultaneously measuring them is another possibility. In any case, however, the response of a semi-continuous monitor will always be delayed.
(3) *Integrating method* consists of the measurement of a time-averaged concentration of radon decay products. The integration may be carried out either on a filter using an integrating detector, or the results from semi-continuous monitors may be evaluated to obtain the average concentration of radon decay products.

Most of these methods can be used for measurements which can be interpreted in terms of the potential alpha-energy concentration (PAEC) or the equilibrium-equivalent decay-product concentration (EEDC). In order to determine these concentrations it is necessary to find the concentrations of individual decay products. Since the half-life of ^{214}Po is too short (164 μs), its contribution to both PAEC and EEDC is negligible and it may not be considered. Ideally, it would be highly desirable to know the individual concentrations of ^{218}Po, ^{214}Pb, and ^{214}Bi with half-lives of 3.11 minutes, 26.8 minutes, and 19.9 minutes, respectively. In practice, however, a direct measurement of all concentrations is very complicated and difficult. Instead of measuring simultaneously both alpha and beta radiation (or also gamma), practically all monitors of decay products rely only on the detection of alpha particles because they can be measured selectively. Older monitors usually measured the total gross alpha particles using, for example, a ZnS(Ag) scintillation detector, but nowadays most instruments are based on alpha spectrometry so that they can distinguish between alpha particles emitted by ^{218}Po and ^{214}Po. Their emission rate from ^{214}Po actually represents the activity of ^{214}Bi.

Further, some monitors using different approaches to measure radon decay products following the above mentioned principles will be described briefly.

One of the first applications of the *grab-sampling method* was based on the counting of alpha particles from the deposited decay products in three consecutive time intervals (five, fifteen and thirty minutes) following air sampling (Tsivoglou et al., 1953). The air sampling is taken typically for ten minutes at a rate of 5 L/min. The concentrations of individual decay products ^{218}Po, ^{214}Pb, and ^{214}Bi are then calculated by solving three simultaneous equations.

Many different modifications of this method have been developed and used for measurements of radon decay products (Thomas, 1970; Thomas, 1972; Cliff, 1978; Busigin and Phillips, 1980; Nazaroff, 1983). The various methods differ in the duration of the sampling as well as in the selection of counting intervals. Using this technique an EEDC of 1 Bq/m^3 can be measured with a 20% relative standard deviation using sequential sampling and counting intervals (Nazaroff, 1983, 1988).

Unlike gross alpha counting, alpha spectrometry using silicon surface-barrier detectors offers further potential for radon products monitoring (Duggen et al., 1968; Martz et al., 1969). The spectrometry method can easily distinguish between ^{218}Po and ^{214}Po, and also ^{212}Po so that thoron decay products can be evaluated as well.

In one arrangement (Martz et al., 1969) a semiconductor detector and multichannel analyzer are used to measure counts due to ^{218}Po and ^{214}Po collected on a membrane filter sample at two counting periods, five and thirty minutes following the sampling. In this case only two simultaneous equations are needed (in the Tsivoglou method three equations are required) to determine the relative concentrations of radon decay products. Compared to the gross alpha method the spectrometry technique gives more reliable results, e.g., the relative standard deviation for the ^{218}Po concentration is about 8%, while the error in the Tsivoglou method is about 30%.

Now several types of universal computer-based instruments are commercially available for both radon and thoron decay product measurements, for example Alpha Smart (Alpha Nuclear, 1992). Recently, the accuracy of grab-sampling methods based on alpha spectrometry for the evaluation of radon decay products has been analyzed and reviewed (Weng et al., 1992). Alpha spectrometry systems have been used successfully in indoor radon and thoron progeny measurements in homes and public buildings (Tu et al., 1992).

Instruments for the *continuous,* or actually rather *semi-continuous,* monitoring of radon (and in some cases also thoron) decay products use grab-sample methods with such modifications as to permit automatic sampling and counting, including analysis and evaluation of results in appropriate units. A system such as Alpha Smart, mentioned before, can be used for continuous programmed measurements with the possibility of selecting suitable algorithms with regard to counting intervals, gross alpha or beta counting, or alpha spectrometry with selected regions of interest. Another type of instrument based on both alpha spectrometry and beta counting for the simultaneous measurement of radon and thoron decay products has recently been developed for use in field conditions (Rolle, 1992).

Integrating monitors of radon decay products as an integrating detector ususally use a TLD, a SSNTD or an electret ion chamber. In all these cases a pump draws air at a low rate (typically 0.1 to 1 L/min) through a filter collecting radon decay products. The monitors of such a configuration are known as the *radon progeny integrating sampling units* (RPISU).

As a thermoluminescence dosimeter, usually a thin crystal of LiF, fixed in a holder facing the upstream side of the filter, is used (Schiager, 1974). A similar arrangement can also be used with a SSNTD (Johnson, 1970).

In addition to the techniques mentioned above, a method based entirely on SSNTDs without using a pump has also been proposed (Fleischer, 1984). The system was designed for long-term, integral measurements that are appropriate for radon product levels experienced in homes over periods of months to a year. It is also suitable for short-term measurements in mines where higher radon concentrations are present. This approach is based on an array of four detectors covered by absorbers of varying thicknesses and mounted in a diffusion box.

Another approach in radon progeny integrating sampling units uses a monitor with an electret located in an electret ion chamber serving as a sensor for measuring the charge produced by alpha particles from radon decay products collected on a one-inch diameter filter (Kotrappa et al., 1990). A conventional low-flow rate air sampling pump is used to collect the radon products on the filter mounted on the side of the electret ion chamber such that the deposited products ionize the air inside the chamber. The negative ions are collected by a positively charged electret causing the electret voltage drop. This change in the electret voltage which occurs during the sampling period is proportional to the time integrated radon product concentration. The corresponding calibration factor has ranged from 1.2 to 1.6 V per mWL·d when sampled at the rate 1 L/min.

3.8 REFERENCES

Alpha Nuclear. 1993. *Model 770–Radon and Thoron Progeny Monitoring System* (Operating Manual). Alpha Nuclear Co., 1125 Derry Road East, Mississauga, Ontario, Canada.

Baltzer, P., K. G. Görsten and A. Bäcklin. 1992. "A Pulse-Counting Ionization Chamber for Measuring the Radon Concentration in Air," *Nucl. Instrum. Meths. Phys. Res.*, A317:357-364.

Belot, Y. 1986. "Transfer of Long-Lived Radionuclides through Marine Food Chains: A Review of Transfer Data," *J. Environ. Radioactivity*, 4:83–90.

Berg, D. et al. 1987. "Radioactive Iodine and Cesium in Bavarian Citizens after the Nuclear Reactor Accident in Chernobyl," *Trace Substances in Environmental Health–XXI*, D. D. Hemphill, ed., Columbia: University of Missouri, pp. 219-225.

Berg, D. et al. 1990. "Whole Body Content and Turnover of Cs and K," in *Biological Trace Element Research*, G. N. Schrauzer, ed., Totowa, NJ: The Humana Press Inc., pp. 249-256.

Bunzel, K., W. Kracke and G. Vorwohl. 1988. "Transfer of Chernobyl-Derived ^{134}Cs, ^{137}Cs, ^{131}I and ^{103}Ru from Flowers to Honey and Pollen," *J. Environ. Radioactivity*, 6:261–269.

Busigin, A. and C. R. Phillips. 1980. "Uncertainties in the Measurement of Airborne Radon Concentrations," *Health Phys.*, 39:943.

Canbera Catalog. 1992. Canberra Industries, Inc., One State Street, Meriden, CT 06450.

Clark, M. J., P. H. Burgess and D. R. McClure. 1993. "Dose Quantities and Instrumentation for Measuring Environmental Gamma Radiation during Emergencies," *Health Phys.*, 64(5):491-501.

Cliff, K. D. 1978. "The Measurement of Low Concentrations of Radon-222 Daughters in Air, with Emphasis on RaA Assessment," *Phys. Med. Biol.*, 23:55–65.

Cohen, B. L. and E. S. Cohen. 1983. "Theory and Practice of Radon Monitoring with Charcoal Adsorption," *Health Phys.*, 45:501–508.

Cohen, B. L. and R. Nason. 1986. "A Diffusion Barrier Charcoal Adsorption Collector for Measuring Rn Concentrations in Indoor Air," *Health Phys.*, 50(4):457–463.

Cohen, B. L., E. M. Granayni and E. S. Cohen. 1983. "Large Scintillation Cells for High Sensitivity Radon Concentration Measurement," *Nucl. Instrum. Meth.*, 212:403.

Countess, R. J. 1978. "Measurement of Rn in Water," *Health Phys.*, 34:390-391.

Desmet, G. and C. Myttenaere. 1988. "Considerations on the Role of Natural Ecosystems in the Eventual Contamination of Man and His Environment," *J. Environ. Radioactivity*, 6:197–202.

Dreicer, M. and C. S. Klusek. 1988. "Transport of ^{131}I through the Grass-Cow Milk Pathway at a Northeast US Dairy Following the Chernobyl Accident," *J. Environ. Radioactivity*, 7:201–207.

Duggan, M. J. and D. M. Howell. 1968. "A Method for Measuring the Concentrations of Short-Lived Daughter Products of ^{222}Rn in the Atmosphere," *Int. J. Appl. Radiat. Isot.*, 19:865–873.

EG&G. 1993. *Nuclear and Radiation Protection Systems.* EG&G Instruments General Catalog, Nuclear Products Group, 100 Midland Road, Oak Ridge, TN 37830.

Eisenbud, M. 1963. *Environmental Radioactivity.* New York: McGraw-Hill, pp. 69–131.

Farai, I. P. and A. O. Sanni. 1992. "^{222}Rn in Groundwater in Nigeria: A Survey," *Health Phys.*, 62(1):96–98.

Fleischer, R. L. 1984. "Theory of Passive Measurement of Radon Daughters and Working Levels by the Nuclear Track Technique," *Health Phys.*, 47(2):263–270.

Gall, M., S. Mahler and E. Wirth. 1991. "Transfer of ^{137}Cs into Mother's Milk," *J. Environ. Radioactivity,* 14:331–339.

George, A. C. 1984. "Passive, Integrated Measurement of Indoor Radon Using Activated Carbon," *Health Phys.*, 46(4):867–872.

George, A. C. and A. J. Breslin. 1977. "Measurement of Environmental Radon with Integrating Instruments," in *Workshop on Methods for Measuring in and around Uranium Mills,* E. D. Harward, ed., Washington, DC: Atomic Industrial Forum, p. 177.

Halford, D. H. 1987. "Effect of Cooking on Radionuclide Concentrations in Waterfowl Tissues," *J. Environ. Radioactivity,* 5:229–233.

Harley, J. H. 1972. *HASL Procedures Manual* (USDOE Report HASL-300). New York, NY: Health and Safety Laboratory.

Harley, J. H. 1992. "Measurement of ^{222}Rn: A Brief History," *Radiat. Prot. Dosim.*, 54(1/4):13–18.

Hauschild, J. and D. C. Aumann. 1989. "Iodine-129 in the Environment of a Nuclear Fuel Reprocessing Plant: V. The Transfer of ^{129}I and ^{127}I in the Soil-Pasture-Cow-Milk/Meat Pathway, as Obtained by Field Measurements," *J. Environ. Radioactivity,* 9:145–162.

Hayball, M. P. and P. P. Dendy. 1991. "Baseline Levels of Total-Body Radioactivity by Whole-Body Monitoring," *Radiat. Prot. Dosim.*, 36(2/4):93–96.

Held, J., S. Schubeck and W. Rauert. 1992. "A Simplified Method of ^{85}Kr Measurement for Dating Young Groundwaters," *Appl. Radiat. Isot.*, 43(7):939–942.

Henrichs, K. et al. 1989. "Measurements of Cs Absorption and Retention in Man," *Health Physics,* 57:571–578.

Holm, E. and R. B. R. Persson. 1977. "Radiochemical Studies of ^{241}Pu in Swedish Reindeer Lichens," *Health Phys.*, 33:471–473.

Hötzl, H. and R. Winkler, 1987. "Activity Concentrations of ^{226}Ra, ^{228}Ra, ^{210}Pb, ^{40}K and ^{7}B and Their Temporal Variations in Surface Air," *J. Environ. Radioactivity,* 5:445–458.

Hursh, J. B. 1954. "Measurement of Breath Radon by Charcoal Absorption," *Nucleaonics,* 12(1):62.

ICRP. 1977. *Recommendations of the International Commission on Radiological Protection* (ICRP Publication 26). Oxford: Pergamon Press.

ICRP. 1991. *1990 Recommendations of the International Commission on Radiological Protection* (ICRP Publication 60). Oxford: Pergamon Press.

Itawi, R. K. and Z. R. Turel. 1987. "Determination of Thallium in Biological Samples by Thermal Neutron Activation Analysis and Low-Level Beta-Counting," *J. Radioanal. Nucl. Chem. Articles,* 115:141–147.

Jackson, D., P. J. Coughtrey and D. F. Crabtree. 1987. "Predicted Concentrations ^{137}Cs, ^{131}I, ^{129}I, ^{241}Pu and ^{241}Am in Various Foodstuffs Following Deposition to Ground," *J. Environ. Radioactivity,* 5:143–158.

Jenkins, P. H. 1991. "Equations for Calculating Radon Concentration Using Charcoal Canisters," *Health Phys.*, 61(1):131–136.

Johnson, D. R., R. H. Boyett and K. Becker. 1970. "Sensitive Automatic Counting of Alpha Particle Tracks in Polymer and Its Applications in Dosimetry," *Health Phys.*, 18:424.

Katagiri, H. et al. 1990. "Low Level Measurements of ^{129}I in Environmental Samples," *J. Radioanal. Nucl. Chem. Articles,* 138:187–192.

Kathren, R. L. 1991. *Radioactivity in the Environment–Third Edition.* Chur, Switzerland: Harwood Academic Publishers, pp. 193–242.

Kotrappa, P. and W. A. Jester. 1993. "Electret Ion Chamber Radon Monitors Measure Dissolved ^{222}Rn in Water," *Health Phys.*, 64(4):397–405.

Kotrappa, P., T. Brubaker, J. C. Dempsey and L. R. Stieff. 1992. "Electret Ion Chamber System for Measurement of Environmental Radon and Environmental Gamma Radiation," *Radiat. Prot. Dosim.*, 45(1/4):107–110.

Kotrappa, P., J. C. Dempsey, J. R. Hickey and L. R. Stieff. 1988. "An Electret Passive Environmental ^{222}Rn Monitor Based on Ionization Measurement," *Health Phys.*, 54:47–56.

Kotrappa, P., J. C. Dempsey, R. W. Ramsey and L. R. Stieff. 1990. "A Practical E-PERM (Electret Passive Environmental Radon Monitor) System for Indoor ^{222}Rn Measurement," *Health Phys.*, 58(4):461–467.

Kotrappa, P., J. C. Dempsey, L. R. Stieff and R. W. Ramsey. 1990. "An E-RPISU (Electret Radon Progeny Integrating Sampling Unit)—A New Instrument for Measurement of Radon Progeny Concentration in Air," in *The 1990 International Symposium on Radon and Radon Reduction Technology* (Atlanta, GA, Feb. 19–23, 1990), USEPA.

Libby, W. F. 1955. *Radiocarbon Dating, Second Edition.* Chicago: The University of Chicago Press.

Lössner, V. 1977. "Detection of Low Energy Photon Emitters by a Whole Body Counter in SAAS" (in German), *Radiobiol. Radiother.*, 4:397–426.

Lucas, H. F. 1957. "Improved Low-Level Alpha-Scintillation Counter for Radon," *Rev. Scient. Instrum.*, 28:680.

Maiello, M. L. and N. H. Harley. 1987. "EGARD: An Environmental Gamma-Ray and Radon Detector," *Health Phys.*, 53:301.

Manjón, G., A. Martínez-Aquirre and M. García-León. 1992. "Low-Level Radioactivity Studies in the Marine Environment of the South of Spain," *Nucl. Instrum. Methods Phys. Res.*, A312:231–235.

Martin, C. J., B. Heaton and J. D. Robb. 1988. "Studies of ^{131}I, ^{137}Cs and ^{103}Ru in Milk, Meat and Vegetables in North East Scotland Following the Chernobyl Accident," *J. Environ. Radioactivity,* 6:247–259.

Martz, D. E., D. F. Holleman, D. E. McCurdy and K. J. Schiager. 1969. "Analysis of Atmospheric Concentrations of RaA, RaB, and RaC by Alpha Spectroscopy," *Health Phys.*, 17:131.

Mathieu, G. G., R. A. Biscaye, R. A. Lupton and D. E. Hammond. 1988. "System for Measurement of Rn at Low Levels in Natural Waters," *Health Phys.*, 55:989–992.

Matsusaka, N. et al. 1988. "Influence of Zinc Deficiency on the Whole-Body Retention of ^{65}Zn in Young and Adult Mice," *Jpn. J. Vet. Sci.*, 50(4):966–967.

Meneely, G. R. and S. M. Linde, eds. 1965. *Radioactivity in Man.* Springfield, IL: C. C. Thomas.

Mills, C. F. 1985. "Changing Perspectives in Studies of the Trace Elements and Animal Health," in *Proc. Fifth Int. Symp. on Trace Elements in Man and Animals,* C. F. Mills, I. Bremner and J. K. Chesters, eds., Slough, United Kingdom: Commonwealth Agricultural Bureaux, pp. 1–10.

Minato, S. and M. Kawano. 1970. "Evaluation of Exposure due to Terrestrial Gamma-Radiation by Response Matrix Method," *J. Nucl. Sci. Technol.*, 7:401–406.

Mondspiegel, K., J. Sabol and R. Tykva. 1990. "Contribution to the Evaluation of Radiation Hazard in the Surrounding of the Uranium Ore Mill MAPE Mydlovary" (in Czech), *Report of the Institute of Landscape Ecology.* České Budějovice, Czech Republic: Academy of Sciences, p. 15.

Moser, H. and W. Rauert. 1980. *Isotopenmethoden in der Hydrologie.* Berlin: Gebrüder Borntraeger.

Nagaoka, T. 1987. "Intercomparison between EML Method and JAERI Method for the Measurement of Environmental Gamma-Ray Exposure Rates," *Radiat. Prot. Dosim.*, 18(2):81–88.

Nakayama, S. and D. M. Nelson. 1988. "Comparison of Distribution Coefficients for Americium and Curium: Effects of pH and Naturally Occurring Colloids," *J. Environ. Radioactivity,* 8:173–181.

Nazaroff, W. W. 1983. "Optimizing the Total-Alpha Three-Count Technique for Measuring Concentrations of Radon Progeny in Residences," *Health Phys.*, 46:395.

Nazaroff, W. W. 1988. "Measurement Techniques," in *Radon and Its Decay Products in Indoor Air,* W. W. Nazaroff and A. V. Nero, Jr., eds., New York, NY: John Wiley and Sons, pp. 491–504.

NCRP. 1988. *Measurement of Radon and Radon Daughters in Air* (NCRP Report No. 97). Bethesda, MD: National Council for Radiation Protection and Measurements.

Nevissi, A. E. 1987. "Methods for Detection of Radon and Radon Daughters," in *Indoor Radon and Its Hazards,* D. Bodansky, M. A. Robkin and D. R. Stadler, eds., Seattle: University of Washington Press, pp. 30–41.

Newman, M. C. and I. L. Brisbin, Jr. 1990. "Variation of ^{137}Cs Levels between Sexes, Body Sizes and Collection Localities of Mosquitofish, *Gambusia holbrooki* (Girard 1859), Inhabiting a Reactor Cooling Reservoir," *J. Environ. Radioactivity,* 12:131–141.

Niese, S. 1985. "Application of Multi-Sample Beta-Gamma Coincidence Spectroscopy for High Sensitivity Activation Analysis," *J. Radioanal. Nucl. Chem. Articles,* 88(1):7–12.

Noordijk, H. et al. 1992. "Impact of Ageing and Weather Conditions on Soil-to-Plant Transfer of Radiocesium and Radiostrontium," *J. Environ. Radioactivity,* 15:277–286.

Nyberg, P. C. and D. E. Bernhardt. 1983. "Measurement of Time-Integrated Radon Concentration in Residences," *Health Phys.*, 45:539.

OECD. 1985. *Metrology and Monitoring of Radon, Thoron and Their Daughter Products.* Paris: Organization for Economic Cooperation and Development.

Okano, M., K. Izumo, T. Katon and M. Wada. 1988. "Measurement of Environmental Radiations with a Scintillation Spectrometer Equipped with Spherical Scintillators of Several Kinds," *Radiat. Prot. Dosim.*, 24(1/4):301–305.

Petit, D., M. Thomas and L. Lamberts. 1987. "Origin of Heavy Metal Fluxes to the Meuse River in the Southern Belgium Using ^{210}Pb-Dated Water-Meadow Sediments," *J. Environ. Radioactivity,* 5:303–316.

Petra, M., G. Swift and S. Landsberger. 1990. "Design of a Ge-NaI(Tl) Compton Suppression Spectrometer and Its Use in Neutron Activation Analysis," *Nucl. Instrum. Methods Phys. Res.*, A299:85–87.

Poletilo, C. 1990. "Methodology for Quality Assurance in Measurement the Radioactivity in the Marine Environment," *Radiat. Protect. Australia,* 8(4):97–103.

Porstendorfer, J., A. Wicke and A. Schraub. 1980. "Methods for a Continuous Registration of Radon, Thoron, and Their Decay Products In- and Outdoors," in *Proceedings of the Symposium on the Natural Radiation Environmental III,* T. F. Gesell and W. M. Lowder, eds., Technical Information Center, CONF-780422.

Prichard, H. M. and K. Mariën. 1985. "A Passive Diffusion ^{222}Rn Sampler Based on Activated Carbon Adsorption," *Health Phys.*, 48:797.

Rad Elec. 1992. *E-Perm System Manual.* Rad Elec, Inc., 270 Technology Park, Frederick, MD.

Rangarajan, C., R. Madhavan, Smt. S. Gopalakrishnan. 1986. "Spatial and Temporal Distribution of Lead-210 in the Surface Layer of Atmosphere," *J. Environ. Radioactivity,* 3:23–33.

Robens, E. and D. C. Aumann. 1988. "Iodine-129 in the Environment of a Nuclear Fuel Reprocessing Plant: I. ^{129}I and ^{127}I Contents of Soils, Food Crops and Animal Products," *J. Environ. Radioactivity,* 7:159–175.

Rolle, R. 1992. "Efficient Measurement of Radon Daughters," *Radiat. Prot. Dosim.*, 45(1/4):57–60.

Rösner, G. et al. 1984. "Low Level Measurements of Natural Radionuclides in Soil Samples around a Coal-Fired Power Plant," *Nucl. Instrum. Methods Phys. Res.*, 233:585–589.

Rutherford, E. 1906. "Absorption of Radio-Active Emanations by Charcoal," *Nature,* 74:634.

Sabol, J., C. H. Mao and P. S. Weng. 1993. "Response of Electret Ion Chambers and Electrets to Gamma Rays," *Nucl. Sci. J.*, 30(3):197–205.

Schiager, K. L. 1974. "Integrating Radon Progeny Air Sampler," *Am. Ind. Hyg. Assoc. J.*, 35:165.

Schönhofer, F. 1992. "Measurement of ^{226}Ra in Water and ^{222}Rn in Water and Air by Liquid Scintillation Counting," *Health Phys.*, 45(1/4):123–125.

Sheppard, S. C. and W. G. Evenden. 1988. "The Assumption of Linearity in Soil and Plant Concentration Ratios: An Experimental Evaluation," *J. Environ. Radioactivity,* 7:221–247.

Šilar, J. and R. Tykva. 1991. "Charles University, Prague: Radiocarbon Measurements I," *Radiocarbon,* 33(1):69–78.

Simon, S. L. and S. A. Ibrahim. 1987. "The Plant/Soil Concentration Ratio for Calcium, Radium, Lead and Polonium: Evidence for Non-Linearity with Reference to Substrate Concentration," *J. Environ. Radioactivity,* 5:123–142.

Tadmor, J. 1986. "Radioactivity from Coal-Fired Power Plants: A Review," *J. Environ. Radioactivity,* 4:177–204.

Thomas, J. W. 1970. "Modification of the Tsivoglou Method for Radon Daughters in Air," *Health Phys.*, 19:691–696.

Thomas, J. W. 1972. "Measurement of Radon Daughters in Air," *Health Phys.*, 23:783–789.

Tsivoglou, E. C., H. E. Ayer and D. A. Holaday. 1953. "Occurrence of Nonequilibrium Atmospheric Mixtures of Radon and Its Daughters," *Nucleonics,* 11(9):40–45.

Tu, K. W., A. C. George, W. M. Lowder and C. V. Gogolak. 1992. "Indoor Thoron and Radon Progeny Measurements," *Radiat. Prot. Dosim.*, 45(1/4):557–560.

Tuaková, S. and R. Tykva. 1994. "Effect of Cigarette Smoke on the Measured Equivalent Volume Activity of ^{222}Rn in Air," *J. Radioanal. Nucl. Chem., Letters,* 187(2):131–135.

Tykva, R. et al. 1990. "Special Silicon Detectors for the Actinide Assay in Snow Fields in High Tatras," in *20 Journées des Actinides, Abstracts.* Prague: Charles University, pp. 76–77.

Tykva, R. et al. 1993. "Deposition of Heavy Metals in Rat Hair and Organs Using Radiotracers," in The Significance of Hair Mineral Analysis as a Means for Assessing Internal Body Burdens of Environmental Pollutants (Report on an IAEA Co-Ordinated Research Programme, NAHRES-18). Vienna: International Atomic Energy Agency, pp. 101–116.

Urban, M. and J. Schmitz. 1991. "Radon and Radon Daughters Metrology: Basic Aspects," *Fifth Interat. Symp. on the Natural Radiation Environment,* Salzburg, Austria, Sept. 22–28, 1991.

Urban, M., D. A. C. Bins and J. J. Estrada. 1985. *Radon Measurements in Mines and Dwellings.* KfK 3866, Kernforschungs-zentrum Karlsruhe, Karlsruhe (Germany).

van Bergeijk, K. E. et al. 1992. "Influence of pH, Soil Type and Soil Organic Matter Content on Soil-to-Plant Transfer of Radiocesium and Strontium as Analyzed by a Nonparametric Method," *J. Environ. Radioactivity,* 15:265–276.

van Dilla, M. A. and D. H. Taysum. 1955. "Scintillation Counter for Assay of Radon Gas," *Nucleonics,* 13(2):68.

Viererbl, L., O. Nováková and L. Jursová. 1990. "Combined Scintillation Detector for Gamma Dose Rate Measurement," *Radiat. Prot. Dosim.*, 32(4):273–277.

Votruba, I., J. Veselý and R. Tykva. 1985. "The Putative Lead-Binding Carrier Protein in Rat Epidermis Using ^{210}Pb," *J. Radioanal. Nucl. Chem. Letters,* 96(5):557–566.

Watnick, S., N. Latner and R. T. Graveson. 1986. "A ^{222}Rn Monitor Using Alpha Spectrometry," *Health Phys.*, 50:645.

Weng, P. S., C. J. Chen, T. C. Chu and Y. M. Lin. 1992. "On the Accuracy of Grab-Sampling Methods for Radon Daughters," *Radiat. Prot. Dosim.*, 45(1/4):523–526.

Wirth, E. and A. Kaul. 1989. "Kontamination Fierisher Produkte mit Radionukliden," in *Rückstände in von Tieren stammenden Lebensmitteln,* D. Grossklaus, ed., Berlin: Verlag Paul Parey, pp. 145–177.

Wrenn, M. W., H. Spitz and N. Cohen. 1975. "Design of a Continuous Digital-Output Environmental Radon Monitor," *IEEE Trans. Nucl. Sci.*, NS-22:645.

Xilei, L., B. Rietz and K. Heydorn. 1991. "Limit of Detection for the Determination of Pt in Biological Material by RNAA Using Electrolytical Separation of Gold," *Int. Topical Conf. on Methods and Applications of Radioanalytical Chemistry II. Abstracts.* Kona: American Nuclear Society, p. 32.

Yu-Fu, Yu., B. Salbu and H. E. Bjornstad. 1991. "Recent Advances in the Determination of Low Level Plutonium in Environmental and Biological Materials," *J. Radioanal. Nucl. Chem. Articles,* 148:163–174.

INDEX